SHERIDAN

A History of
the American Light Tank

Volume 2

by
R.P. Hunnicutt

Line Drawings
by
D.P. Dyer

Color Drawing
by
Uwe Feist

Foreword
by
Colonel Dana B. Dillon, U.S. Army (Retired)

ECHO POINT BOOKS & MEDIA, LLC

Published by Echo Point Books & Media
Brattleboro, Vermont
www.EchoPointBooks.com

Copyright © 1995, 2015 R. P. Hunnicutt
ISBN: 978-1-62654-154-2

Cover image by Uwe Feist

Cover design by Adrienne Núñez,
Echo Point Books & Media

Editorial and proofreading assistance by Christine Schultz,
Echo Point Books & Media

Printed and bound in the United States of America

CONTENTS

ACKNOWLEDGEMENTS

Among the many people who contributed material for this book, I particularly would like to thank Colonel Dana Dillon, USA Retired, who also agreed to write the Foreword. No officer has greater experience in the employment of airborne armor.

For the story of the Sheridan during Operation Desert Shield/Desert Storm, I am indebted to Lieutenant Colonel Charles A. Donnell and Command Sergeant Major Dennis C. Wilson who led the 3/73rd Armor during this period. Captain Scott Womack contributed photographs of the Sheridan taken during his service in the 3/73rd which included Operation Just Cause in Panama.

Much of the development history of the early light tanks came from the United States Army Tank Automotive Command. Major General Oscar Decker was particularly helpful during his period of command. Clifford Bradley and Leon Burg, both now deceased, made major contributions to this part of the story. Dr. Richard McClelland provided valuable information on the engines of the various vehicles.

John Purdy, Charles Lemons, and Katie Talbot at the Patton Museum, Fort Knox, Kentucky, made the resources of their library available and obtained measurements of some vehicles in the museum collection. XM551 pilot number 4 was included among these. Jon Clemens of Armor Magazine provided some excellent photographs of the Sheridan and the XM8.

At Aberdeen Proving Ground, Major General Andrew Anderson was a great help and Dr. J. Britt McCarley, Historian of the Test and Evaluation Command, located some rare photographs of the experimental vehicles. Dr. William F. Atwater, Director of the Ordnance Museum, also supplied background information and photographs of the light tanks during their test program at the Proving Ground.

Details of the 76mm gun tank T92 came from AAI Corporation through the courtesy of Charles Lehner. Information on the AAI proposal for the AR/AAV resulted from the efforts of John Hebert and Nelson Conner, Jr. at AAI. William Jones at TRADOC located sources of information on several of the vehicles.

Photographs taken by James Loop in Vietnam and Russell Vaughan in Europe showed the M41 and the Sheridan on active service. Other photographs were contributed by Fred Crismon, Michael Green, Robert Lessels, James Mesko, and Greg Stewart. Don Loughlin located much of the data on the various cannon and their ammunition.

Background information on the various AGS candidates was supplied by Richard G. DuVall at Teledyne Continental Motors, Tom Jambriska and Dick Hughes at Cadillac Gage, and by representatives of Hägglund USA. At United Defense (formerly FMC Corporation) Dick Brewster, Bruce Heron, William Highlander, and Anthony Lee provided data and photographs of the XM8. Also at United Defense (formerly BMY) Judy McIlvaine obtained material on the self-propelled artillery and the artillery support vehicles. Craig Vanbebber of Loral Vought Systems supplied photographs and data regarding the installation of the LOSAT turret on the AGS chassis.

Phil Dyer prepared the four view drawings despite a limited amount of information on some of the experimental vehicles. As usual, Uwe Feist produced the color drawing of the Sheridan as it appeared during Operation Desert Storm.

Also, I am indebted to Allan Millar and Shirley Wong for help in the preparation of the text and photo captions.

FOREWORD

by
Colonel Dana B. Dillon, U.S. Army (Retired)
Honorary Colonel, 73rd (Airborne) Armor Regiment

Dick Hunnicutt's first volume, "A History of the American Light Tank", documents the development of light armored vehicles from their early beginning in World War I to the end of the Second World War. This volume 2 traces not only the development of the light tank, but also its transition from a lightly armored vehicle supporting infantry to its current in extremis role for the Army's early entry forces.

The centerpiece of volume 2 is the M551, General Sheridan, armored reconnaissance/airborne assault vehicle (AR/AAV). The Sheridan was a revolutionary vehicle; a tank that could swim, fire missile and caseless conventional ammunition from the same stabilized gun-launcher system, and be delivered to the combat area by parachute. All of these capabilities had to be developed and integrated into a vehicle weighing roughly sixteen tons. The Sheridan was the most complex armored vehicle ever developed by the Army at that time. It was to be the most controversial vehicle ever developed.

Mr. Hunnicutt does an outstanding job documenting the difficulties encountered in the development and fielding of the Sheridan. He also accurately records the divergent opinions held by the two main users of the vehicle, the cavalryman and the paratrooper. For the cavalryman, posted within kilometers of his General Defense Position, the Low Velocity Air Drop (LVAD) capability was unreasonable and of no use. For the paratrooper, with no idea of where he might deploy next, the LVAD capability was his only guarantee of armor support in the airhead. Their discussions were somewhat like the ubiquitous TV commercial featuring two beer connoisseurs yelling at each other "———TASTES GREAT!! LESS FILLING!!———".

I joined the Army's only Sheridan equipped tank battalion, the 4th Battalion (Light) (Airborne) 68th Armor in the summer of 1978. It was later redesignated as the 3rd Battalion (Airborne) 73rd Armor and I would serve in that unit off and on over the next ten years as the Operations Officer, Executive Officer, and Battalion Commander.

Despite all of the stories I had heard to the contrary, I found the Sheridan to be an incredibly robust tank. It can withstand the G forces of Low Altitude Parachute Extraction (LAPE) and Low Velocity Air Drop (LVAD). Once on the ground, it can operate for extended periods with only the crew to provide maintenance. No armored vehicle in the U.S. Army must operate under such harsh and trying conditions. It is efficient, traveling over 300 miles on 175 gallons of fuel. Placed in the hands of a well trained crew, the Sheridan is a worthy opponent for most Eastern Block armored vehicles. To quote a paratrooper, "The Sheridan and four paratroopers is the most lethal weapon on the battlefield."

To my knowledge, the horse is the only mount the Army campaigned longer than the General Sheridan. The Sheridan stood guard on the East-West German border for most of the Cold War, guarded the DMZ in Korea for over ten years, fought in the Republic of Vietnam almost four years, and in 1970, spearheaded the invasion into Cambodia. It stood ready to deploy to the Suez Canal in 1973, Zaire in 1978, and to Iran in 1980. It deployed to Honduras in 1989 to demonstrate U.S. resolve to Nicaragua, was parachuted into Panama during Operation Just Cause, was the first American armor to deploy to South West Asia during Operation Desert Shield/Desert Storm, and elements of the Airborne Armor Battalion were deployed on Operation Urgent Fury. Regardless of it's shortcomings and reputation, no one can deny that the Sheridan has served as the vanguard of the Armor Corps for almost three decades.

The XM8 armored gun system (AGS) is due to replace the Sheridan sometime toward the end of this century. Like the Sheridan, it is a revolutionary vehicle; a three man crew, a U.S. produced 105mm cannon, an autoloading gun, and a state of the art fire control system. The AGS will be capable of LVAD from both strategic and tactical airlift. As this is written, the AGS is undergoing exhaustive testing, with much better results than the Sheridan enjoyed.

INTRODUCTION

At the end of World War II, the War Department Equipment Review Board included a new light tank among the requirements for the postwar army. This new vehicle was intended to be a replacement for the last wartime light tank, the M24. It was to serve in the same role, primarily as a highly mobile, powerfully armed, reconnaissance vehicle. The postwar development program produced the M41 series of 76mm gun tanks to meet this requirement. Although popular with the troops, the M41 was considered too heavy at 25 tons and too limited in cruising range. With the introduction of the Soviet T54 tank, its armament also was considered to be inadequate. Several proposals were presented to increase the firepower of the M41 series, but none were adopted. The search for a lighter weight tank produced the T71 and T92 designs, both of which retained the 76mm gun. Two of the T92 vehicles were constructed for test purposes and they weighed about 19 tons. However, information on the new Soviet PT76 amphibious tank resulted in the additional requirement that the new light tank have swimming capability. Since the T92 had been designed to be as compact as possible in order to maximize protection at the lowest possible weight, there was no practical way of increasing the volume to provide sufficient displacement for flotation. As a result, the T92 program was terminated.

A new program was initiated to provide a light armored vehicle which would be amphibious and could serve both in the reconnaissance role and as a lightweight assault vehicle for airborne operations. Although this was precisely the role of a light tank, it was considered that if the term tank was used in any form, it would be confused with the main battle tank. Thus, the new vehicle was officially named the armored reconnaissance/airborne assault vehicle (AR/AAV). However, the troops frequently referred to it as a light tank.

The development program for the new AR/AAV, designated as the XM551 General Sheridan, was extremely complex. The design of the vehicle itself was relatively straightforward, although a number of problems were encountered. However, the new weapon system with which it was to be armed was another story. The Shillelagh system consisted of a 152mm gun-launcher which fired either conventional ammunition or served as a launcher for a guided missile. Problems with the missile delayed the program and additional development was required to increase the range.

However, the most serious difficulty was with the conventional ammunition. To eliminate the problem of the empty cartridge case in the turret after firing, this ammunition used a combustible cartridge case. Unfortunately, these combustible cases were fragile and extremely sensitive to humidity. When exposed to moisture, they would leave a smoldering residue in the chamber after firing. Such a residue could easily ignite a subsequent round before the breech was closed with disastrous results. Before these problems were completely solved, the Sheridan was ordered into production. This resulted in a long and expensive development and retrofit program before the Sheridan was considered to be satisfactory for troop use.

Because of its complexity, the Sheridan required a high standard of maintenance. Thus troops that did not require its wide range of capability preferred a simpler, more easily maintained, vehicle. As a result, it was phased out of service with one exception in 1978. The exception was the airborne armor battalion in the 82nd Airborne Division. Here no other vehicle existed that could fill the requirement and the Sheridan remains on active duty today after successful combat in Panama and in the Persian Gulf. It will continue in this role until the new XM8 armored gun system is in production later in this decade. Once again, the army prefers not to use the term light tank to describe the new vehicle.

Other light tracked combat vehicles included in this volume are the self-propelled artillery, antitank, and antiaircraft vehicles. Lightweight recovery, engineer, and artillery support vehicles also are described.

In Vietnam, armament was installed in the M113 personnel carrier and it was employed in a combat role as the armored cavalry assault vehicle (ACAV). In many of these operations it served as a light tank. However, it has been excluded from this volume as it would be more appropriate for another volume devoted to tracked carriers and reconnaissance vehicles as well as infantry and cavalry fighting vehicles. The tracked amphibian landing vehicles should also be covered in the same volume.

Since this is primarily a development history, only a brief description is included covering the combat record of the Sheridan and the other light vehicles. The few actions described are used to illustrate the advantages and disadvantages of the various vehicles and are not intended to be a complete combat history.

R. P. Hunnicutt
Belmont, California
April 1995

PART I

APPLYING THE LESSONS OF WORLD WAR II

At the end of World War II, all tank production in the United States was terminated. The only exceptions were a few experimental vehicles. The production run of the light tank M24 had ended in August 1945 after the completion of 4731 vehicles. They were destined to serve in the United States Army until after the Korean War. However, prior to the end of World War II, consideration was given to the future requirements of the Army. On 2 January 1945, the Army Ground Forces Equipment Review Board met in Washington, D.C. Later, in November of the same year, the War Department Equipment Review Board was convened under the leadership of General Joseph W. Stilwell. Both of these boards included a light tank in their recommendations for new development.

The proposed light tank was to be about 25 tons in weight with increased firepower and mobility compared to the light tank M24. Since the latter was considered to be deficient in firepower, the new light tank was to be armed with a gun approximately three inches in caliber capable of penetrating five inches of homogeneous steel armor at 30 degrees obliquity and a range of 1000 yards.

The design of a light tank to meet the requirements of the Stilwell Board was initiated by the Research and Development Division at Detroit Arsenal during July 1946. Meetings at the Arsenal between personnel from the Ordnance Department and the Army Ground Forces on the 8th and 9th of July defined additional requirements for the new

tank. On 27 September 1946, Item 31059 of the Ordnance Committee Minutes (OCM) designated the new vehicle as the light tank T37 and recommended the manufacture of three pilot tanks. This was later reduced to two by OCM 31554 dated 1 May 1947.

By early 1949, the design for the T37 was complete, a wooden mock-up constructed, and the drawings released for production of the two pilot tanks. The new vehicle was manned by a crew of four with the commander, gunner, and loader in the turret. The driver was located in the left front hull with four T17 periscopes in the hull roof providing vision through an arc of about 200 degrees. The right front hull was occupied by a stowage rack for main armament ammunition.

The T37 was powered by the AOS-895-1 gasoline engine manufactured by Continental Motors as part of a new family of engines for military vehicles. As its designation indicated, this six cylinder power plant was air-cooled with an opposed configuration and supercharged. With a displacement of approximately 895 cubic inches, it developed 500 gross horsepower at 2800 revolutions per minute (rpm). This engine had many components in common with the more powerful members of the same family such as the 12 cylinder AV-1790 intended for the medium tank. The AOS-895-1 drove the T37 through the CD-500 cross drive transmission manufactured by General Motors Corporation. The power pack was installed with the final drives and the sprockets

The mock-up of the light tank T37 appears below and the dimensions of the proposed vehicle are shown in the sketch at the right.

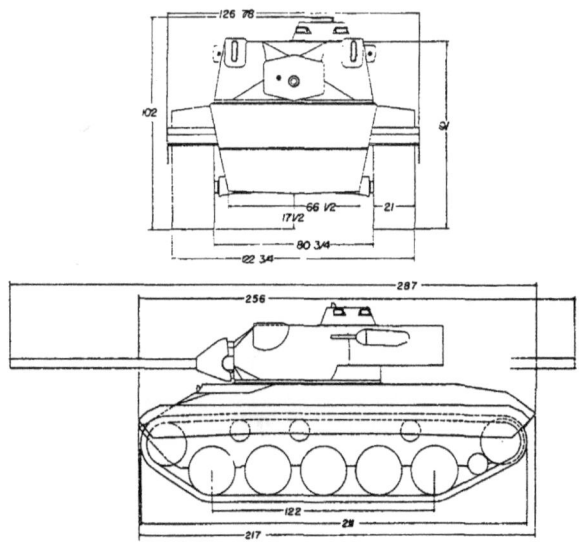

The drawing at the left illustrates the space required for the new power pack consisting of the air-cooled engine and the CD-500 transmission (upper) compared to that for the power pack from the M26 medium tank (lower). Both engines developed 500 gross horsepower.

$\frac{3}{8}$ inches on the bottom rear. A torsion bar suspension with five dual road wheels per side supported the vehicle which was fitted with 21 inch wide, forged steel, T91, center guide, single pin tracks. A small track tension idler was installed between the rear road wheel and the sprocket on each side. The upper track run was supported by three rollers per side.

As originally outlined, the turret development program for the light tank T37 was divided into three phases. The Phase I design utilized the 76mm gun T94 in the T137 combination gun mount. This weapon had a muzzle velocity of 2600 feet per second with armor piercing (AP) ammunition. The primary aiming device for the cannon was the stereoscopic optical range finder T37 installed in the top front of the turret. It was operated by the gunner located on the left side of the cannon in front of the tank commander. The loader's station was on the right side of the turret. A .50 caliber coaxial machine gun was installed on the right side of the combination gun mount. A .30 caliber machine gun in a blister mount was located on each side wall of the turret. These guns followed the main armament through its elevation range of +20 to −9 degrees and could be further

at the rear of the vehicle. The combination of the AOS-895-1 engine and the CD-500 transmission provided a very compact package compared to earlier designs of equivalent power. The rear drive in the new tank also eliminated the drive shaft under the fighting compartment, reducing the height of the vehicle. It also resulted in more available space in the front of the tank. The hull of the T37 was assembled by welding rolled homogeneous steel armor plate which ranged in thickness from 1¼ inches on the upper front to

Below, the pilot light tank T37 (USA 30163666) is at Aberdeen Proving Ground on 3 June 1949. The T94 76mm gun is fitted with a single baffle muzzle brake.

These views of the T37 at Aberdeen show the vehicle with all of its machine gun armament. Note the long barrel of the .50 caliber coaxial weapon. The stereoscopic range finder is installed in the turret housing.

elevated to +45 degrees. The gunner's periscopic sight was used to aim these weapons. Another .50 caliber machine gun was pedestal mounted on the turret roof.

The Phase II turret was to be armed with the higher velocity 76mm gun T91 and equipped with the Vickers fire control system featuring azimuth and elevation stabilization. The vehicle with this turret was subsequently redesignated as the light tank T41.

The Phase III turret was to incorporate either the T91 gun with an automatic loader or a stubby case 76mm gun proposed for development. The fighting compartment on the Phase III vehicle was to be completely integrated with an IBM stabilization system and power loading. This fire control system was to include a stereoscopic range finder, an automatic lead computer, and an automatic ballistic corrector.

The first pilot light tank T37 was shipped to Aberdeen Proving Ground for tests beginning on 30 May 1949. The test program continued until August 1950 when the T37 was returned to Detroit Arsenal for a production study after the outbreak of war in Korea.

The large bustle on the T37 turret is obvious in these photographs also dated 3 June 1949. Note the track tension idler between the sprocket and the rear road wheel.

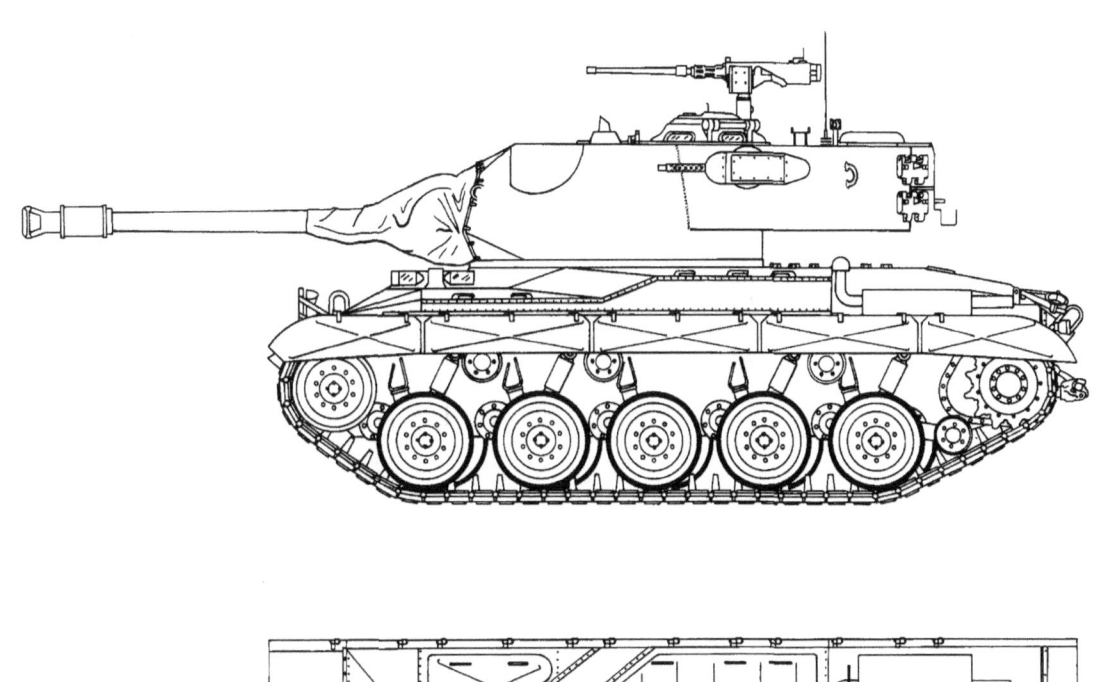

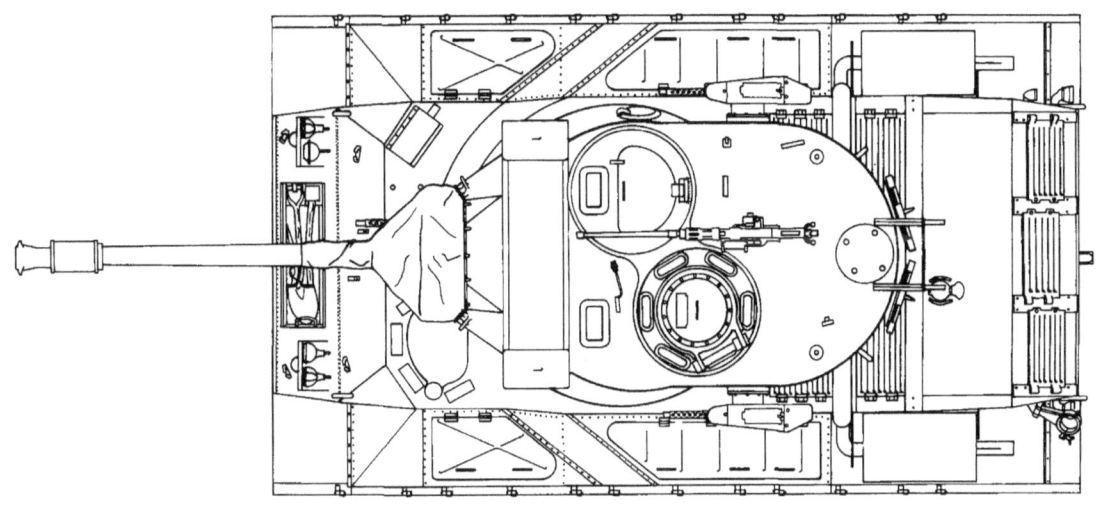

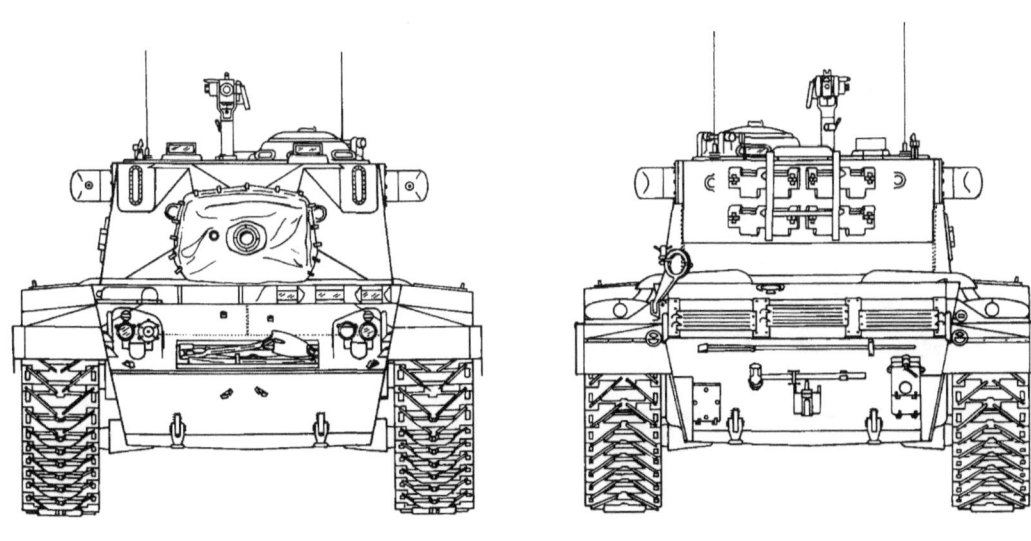

Scale 1:48

©D.P. Dyer

Light Tank T37

The location of the gunner and tank commander on the left side of the T37 turret is revealed by the position of the cupola in these views. Note the flat turret roof without any reinforcement.

These photographs show the T37 without the stereoscopic range finder installed.

15

The first pilot light tank T41 (USA 30162852) appears above and at the right during tests at Aberdeen Proving Ground. At the bottom of the page, the T41 is being demonstrated for President Harry S. Truman during his visit to the Proving Ground on 17 February 1951.

Three T41 light tanks were completed and evaluated at Aberdeen and Fort Knox along with the T37. Armed with the high velocity 76mm gun T91, the original turret on the first T41 pilot was similar to that on the T37, but the flat roof was reinforced by two external stiffeners installed at an angle of 60 degrees with the turret center line. A later version of the T41 turret had a roof which peaked at the center and sloped toward each side eliminating the need for the stiffeners. The Vickers fire control system stabilized the turret in azimuth and the gun in elevation. It included a dual color, coincidence type, range finder and an automatic lead computer. The track tension idler was eliminated on the suspension of the T41 pilots.

16

The diagonal reinforcements on the flat roof of the original turret on light tank T41, pilot number 1, are visible in both of these photographs. The turret bustle is even longer than that on the light tank T37 and the track tension idler has been eliminated from the suspension system.

Light tank T41 pilot number 2 (USA 30162853) with the later peaked roof turret appears above. The internal arrangement of this vehicle is illustrated in the sectional drawings below.

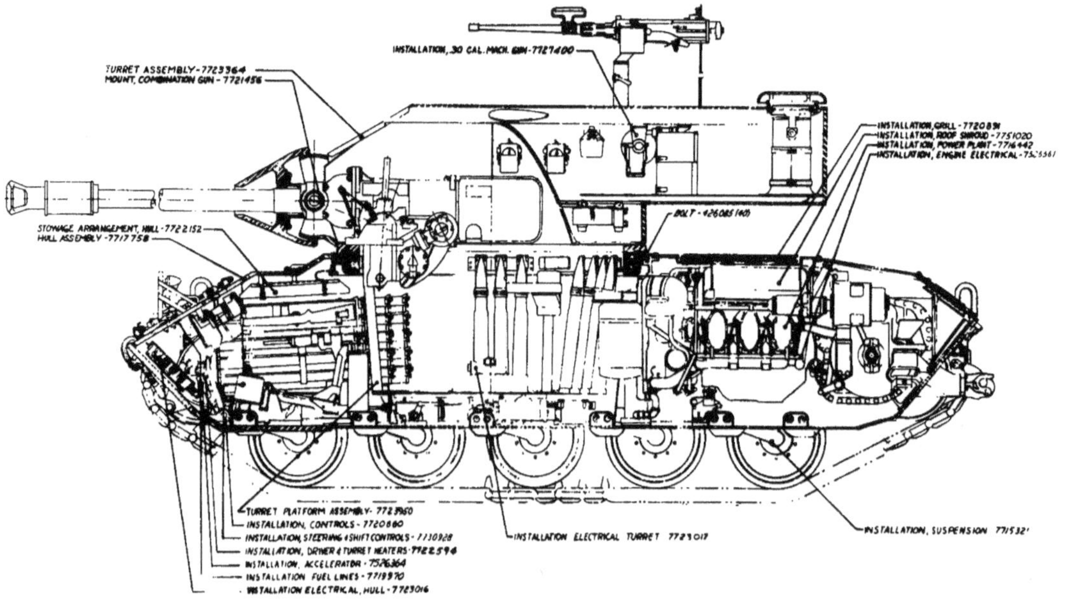

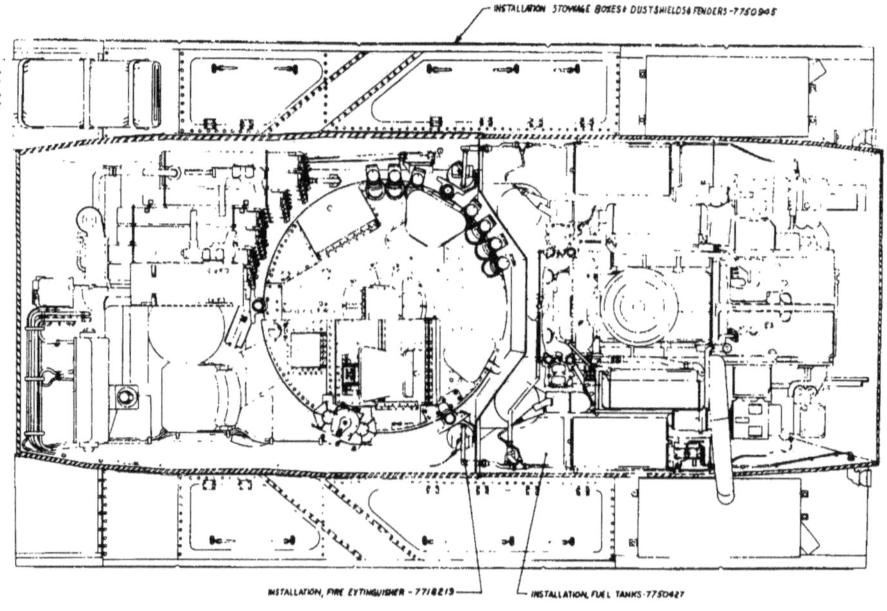

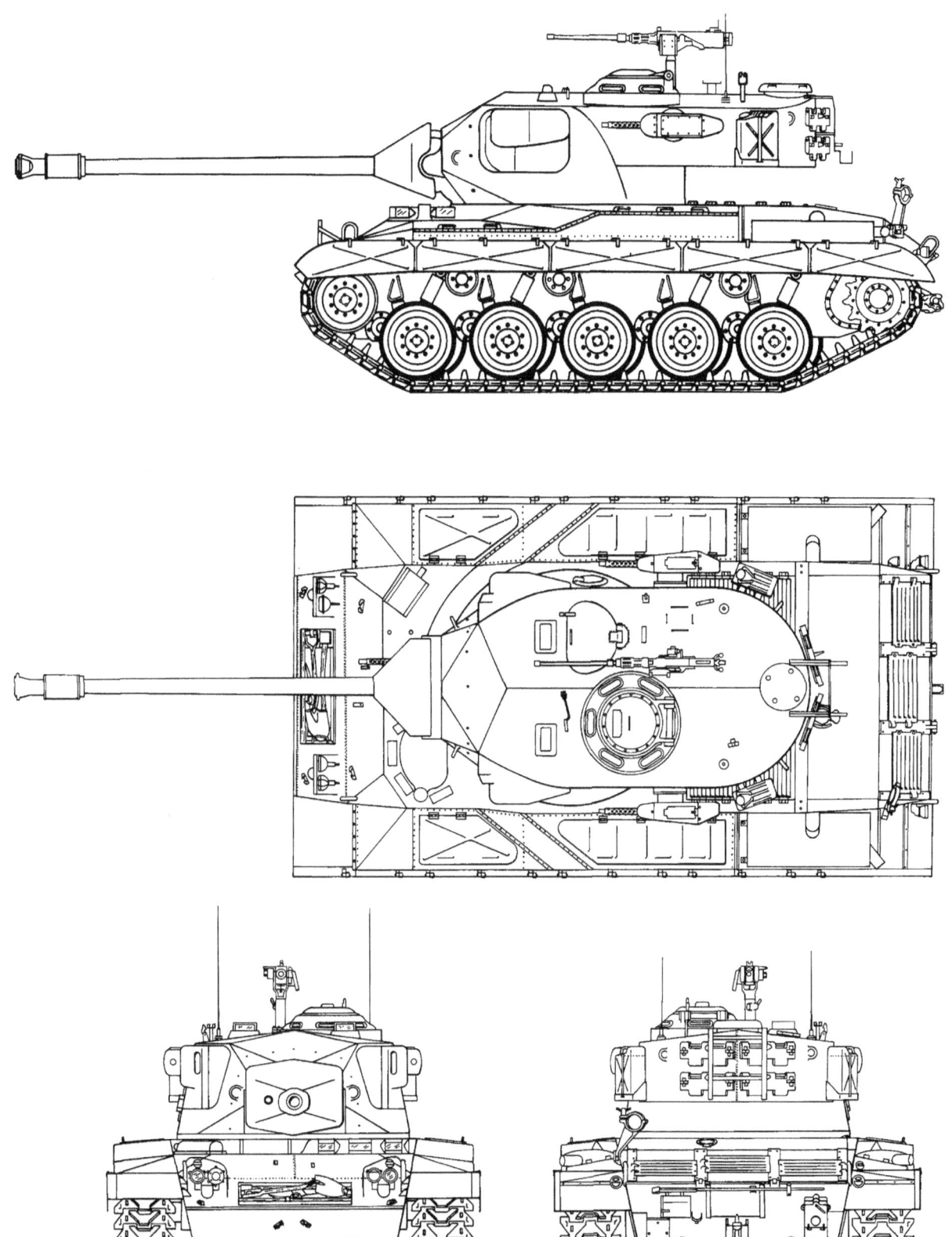

Scale 1:48

©**D.P. Dyer**

Light Tank T41

19

Above, light tank T41 pilot number 1 is on display by the Ordnance Museum at Aberdeen Proving Ground. Note that it is now fitted with the later peaked roof turret.

The test program indicated that the range finder was a definite advantage and essential in obtaining first round hits at long range. However, the Vickers fire control system still required considerable development before it was ready for production. As a result, the requirement for the stabilization system was dropped, but the range finder was retained. Unfortunately, the dual color, coincidence type, range finder on the T41 was an integral part of the Vickers fire control system and could not be used separately. Thus it was replaced by a modified version of the stereoscopic range finder from the light tank T37.

Numerous failures during the test program revealed the need for various modifications to improve the reliability of the power train. It also was recommended that an auxiliary generator be installed to provide power when the main engine was shut down. In the turrets of the light tanks T37 and T41, the gunner and the tank commander were located on the left side of the cannon with the loader on the right. This followed the practice of the earlier light tanks. As a result of the test program, it was recommended that the positions be reversed placing the gunner and the tank commander on the right side of the turret with the loader on the left. This provided better utilization of the space within the turret and coincided with the crew arrangement in the medium tanks. The blister mounted .30 caliber machine guns also were eliminated from the turret side walls and the inside diameter of the turret ring was increased from 69 inches to 73 inches. After a review of the various recommendations, Detroit Arsenal redesigned the T41 to incorporate the new features and the vehicle was redesignated as the light tank T41E1.

In 1948, the Cadillac Motor Car Division of General Motors Corporation had been selected to be the primary producer of light tanks in the event of a future emergency and they had completed a survey of the facilities, tools, and labor required for the manufacture of the T41. In January 1950, Cadillac received a contract to build approximately 100 T41E1 light tanks in order to work out the various production techniques required. This contract specified that two pilot tanks were to be built and tested before the balance of the production run was completed in order to incorporate any necessary modifications. Unfortunately, this orderly development program was disrupted world events. War broke out in Korea during June 1950 before either of the pre-production pilot tanks was completed.

Although the actual fighting was in Korea, the greatest danger was perceived to be in Europe with the possibility of all-out war with the Soviet Union. Under these conditions, the situation paralleled that of ten years earlier when tank production was initiated on a crash basis prior to the entry of the United States into World War II. Like then, it was considered essential to start production immediately and make any corrective modifications later. In 1940, this proved to be a wise decision. In 1950, all-out war did not occur and the result was a long expensive retrofit program before the new tanks were accepted by the Army Field Forces. However, at the time, the possibility of another world war did not permit any other course of action and Cadillac's educational contract for about 100 T41E1s was replaced by a new order for approximately 1000 light tanks. This was to be only the beginning. To simplify production, Ordnance dropped the requirement for the range finder and Cadillac redesigned the turret lowering its silhouette and reducing the weight. On 9 November 1950, a change in nomenclature occurred. Tanks were no longer classified by weight, but by the caliber of their main armament. Thus under OCM 33476, the light tank T41E1 became the 76mm gun tank T41E1.

Above are the wooden hull and turret mock-ups of the 76mm gun tank T41E1. These detailed wooden mock-ups were used to determine the proper location and fit of the various components prior to the construction of the actual tank.

PRODUCTION AND MODIFICATION

The Cadillac Motor Car Division began production of the T41E1 in mid 1951 at the Cleveland Tank Plant. This continued with various modifications of the original design until 1954 with a total run of 3729 tanks. The early production tanks were immediately submitted for engineering tests and user evaluation. Any necessary changes resulting from the test program were introduced onto the production line as rapidly as possible. However, the tests were not completed until March 1952 and, by that time, more than 900 tanks had been completed. These vehicles had to be reworked prior to acceptance by the Army Field Forces. A major problem concerned the turret and gun control system. The original specification required only that there be complete power control of the turret in azimuth and the gun in elevation. These conditions were met by the T41E1's pulsing relay control system. However, in November 1951, the Army Field Forces specified that the control system should be able to place the main gun on a target within five seconds after

Below, the turret is being installed on an early production 76mm gun tank T41E1 at the left and the new tanks are being loaded on railway cars for shipment at the right.

Above, an early production 76mm gun tank T41E1 is at Aberdeen Proving Ground for evaluation. Note the early muzzle brake and turret design.

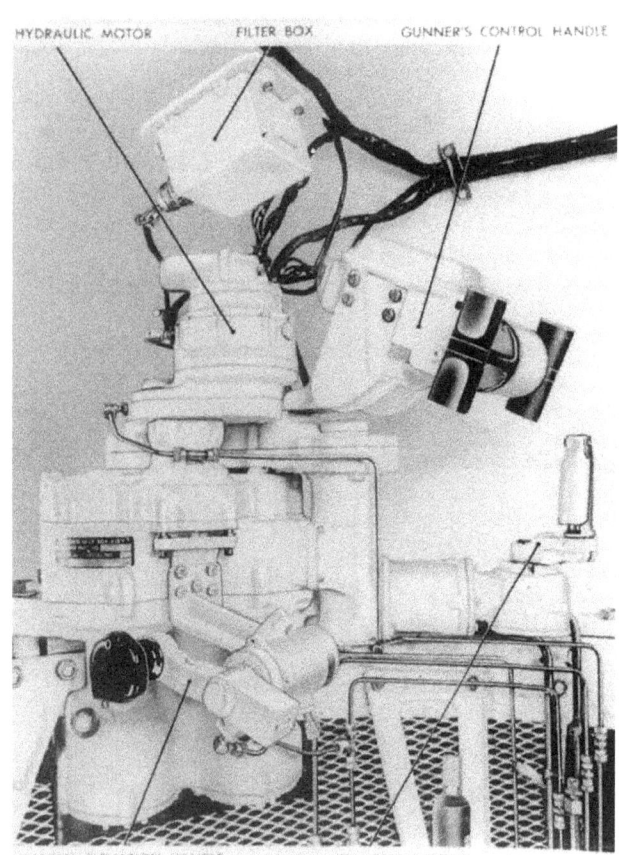

the decision was made to engage. At that time, the pulsing relay control system was unable to meet this requirement. To obtain the necessary performance, Cadillac designed a new turret and gun control system. This consisted of an oil gear turret traverse mechanism which could be operated by either the gunner or the tank commander. The tank commander could override the gunner's control in an emergency. The gunner had only a manual elevation control, but the tank commander could elevate the cannon using the electric slewing motor. In the original pulsing relay system, the gunner had power and manual controls for both elevation and traverse. The tank commander could override the gunner's controls in both elevation and traverse. After modification, the late version of the pulsing relay control system provided only manually operated hydraulic elevation for the gunner as well as power and manual traverse. The tank commander could operate the power traverse with his override control, but he could not elevate the cannon. The new Cadillac control system was less complex, occupied less space, and could meet the five second requirement for engaging a

The view at the left shows the early version of the gunner's pulsing relay gun control system on the T41E1. Compare this photograph with the later design on the opposite page.

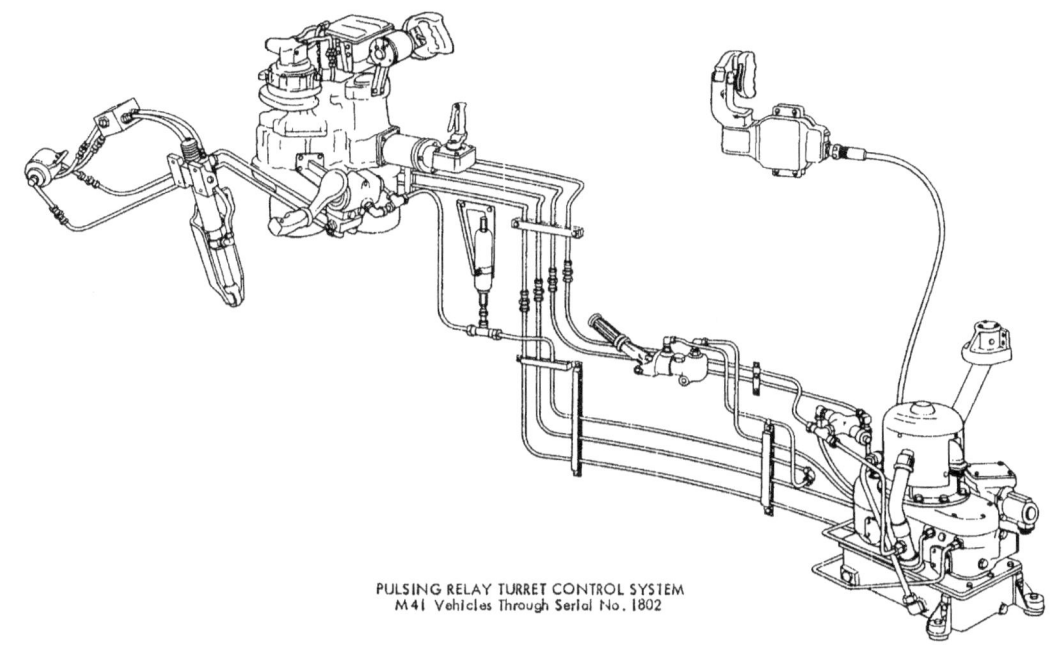

PULSING RELAY TURRET CONTROL SYSTEM
M 41 Vehicles Through Serial No. 1802

The modified pulsing relay gun control system is illustrated in the sketch above and in the photograph at the bottom right. At the bottom left is the Cadillac designed gun control system installed in the T41E2.

target. This system was first introduced onto the production line on tank number 1419 and the vehicle was redesignated as the 76mm gun tank T41E2. Eleven more T41E2s (serial numbers 1466, 1474, 1482, 1528, 1532, 1533, 1535, 1537, 1548, 1550, and 1553) were mixed in with the T41E1 production before the complete changeover to the T41E2 beginning with tank number 1803. At that time, it was planned to convert all of the T41E1s to the new control system. However, after 416 of the earlier T41E1s had been reworked to

the T41E2 standard, modification of the pulsing relay system combined with improved training showed that it could meet the five second time limit for target engagement. As a result, no further T41E1s were converted to T41E2s. On 29 May 1953, it was directed that the T41E1 and the T41E2 be standardized as the 76mm gun tanks M41 and M41A1 respectively. Originally referred to as the Little Bulldog, the new tank was named the Walker Bulldog after the death in Korea of Lieutenant General Walton H. Walker.

A — BALLISTIC DRIVE M 4 (T23)
B — GUN FIRING CONTROL BOX
C — TRAVERSING HAND DRIVE HANDLE
D — AZIMUTH INDICATOR M31 (T24)
E — TURRET TRAVERSING LOCK
F — ELEVATING HAND PUMP
G — TELESCOPE COVER CONTROL HANDLE
H — TELESCOPE M97 (T156)
J — PERISCOPE M20 (T35) or M20A1 (GUNNER'S)
K — GUNNER'S POWER TRAVERSING CONTROL HANDLE
L — HYDRAULIC DUMP VALVE TOGGLE SWITCH

The 76mm gun tanks M41A1 in the photographs above and at the right are fitted with the later design turret. Compare the weld joint pattern on these turrets with that on the tank in the top photograph on page 22. The tank above retains the early single baffle muzzle brake and the vehicle at the right is equipped with the later fabricated design. The early fenders on the tank above can be compared to the later type on the vehicle at the right. Also, note the location of the auxiliary generator engine muffler on the right front fender.

In addition to the new turret and gun control system, numerous other changes were introduced during the production run. The turret was a welded assembly of castings and rolled homogeneous steel armor preformed to the desired shape. The components of this assembly were somewhat different on the early and late production tanks. The weld line between the side plate and the turret ring casting was much higher on the early turrets and the bottom of the side plate extended back in a straight line to the bottom of the turret bustle. On the later production turrets, this side weld was lower and the bottom of the side plate angled upward to join the turret bustle.

The diagram at the right illustrates the welding sequence used to assemble the later design turret.

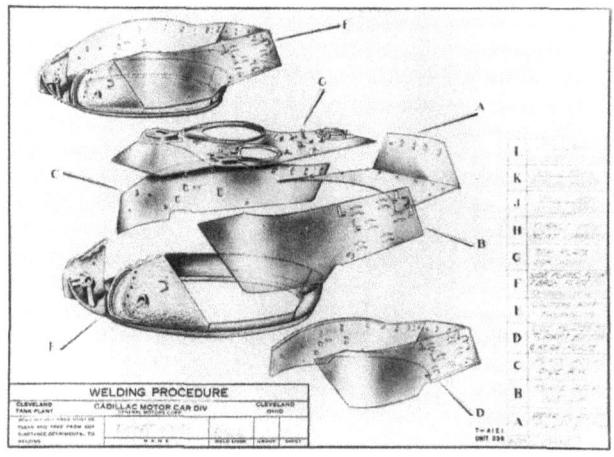

WELDING PROCEDURE

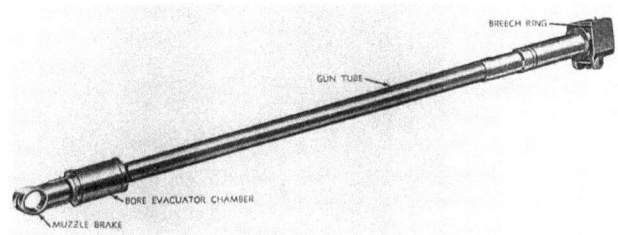

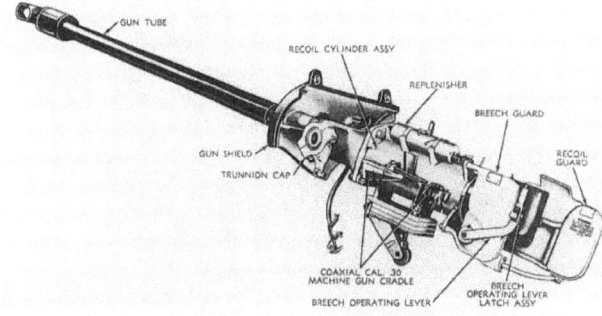

Above is the 76mm gun M32 with the later fabricated, single baffle, muzzle brake. The two upper views at the right show the weapon installed in the M76 mount. The difference on the right side of the M76A1 mount can be seen at the lower right.

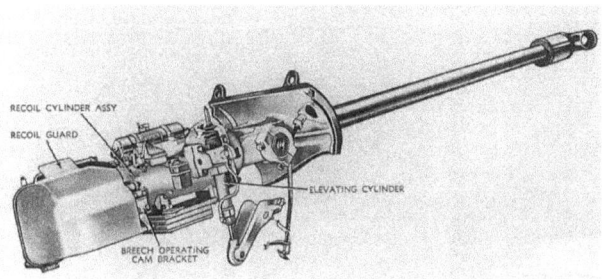

The main armament of the production tanks was the 76mm gun M32 (T91E3). This weapon had a muzzle velocity of 3200 feet per second with a 15 pound AP round and much higher velocities with hypervelocity armor piercing (HVAP) ammunition. Originally a round, cast, muzzle brake was installed just in front of the bore evacuator. Later, it was replaced by a fabricated, T shape, design. Both were single baffle muzzle brakes. The 76mm gun was installed in the combination gun mounts M76 (T138E1) and M76A1 (T138E2) in the M41 and M41A1 tanks respectively. The difference in the mounts was to accommodate the gun elevation mechanisms in the two control systems. Originally, the production tanks were armed with a .50 caliber coaxial machine gun on the left side of the cannon. However, this weapon was replaced later by a .30 caliber machine gun. A .50 caliber antiaircraft machine gun was pedestal mounted on the turret roof.

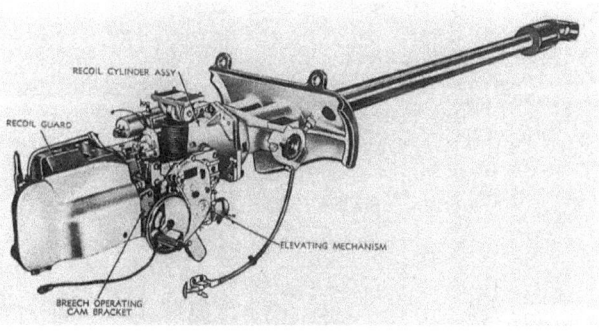

The ammunition available for use in the 76mm gun M32 is illustrated below. The 76mm rounds in the left photograph, from top to bottom, are the AP-T M339, the HVAP-T M319, and the HVAP-DS-T M331A2. The right photograph shows the HE M352, the WP M361 (smoke), and the canister M363 in the same sequence.

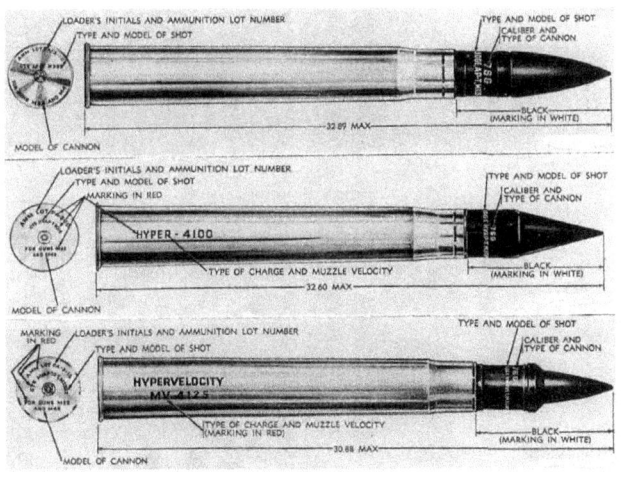

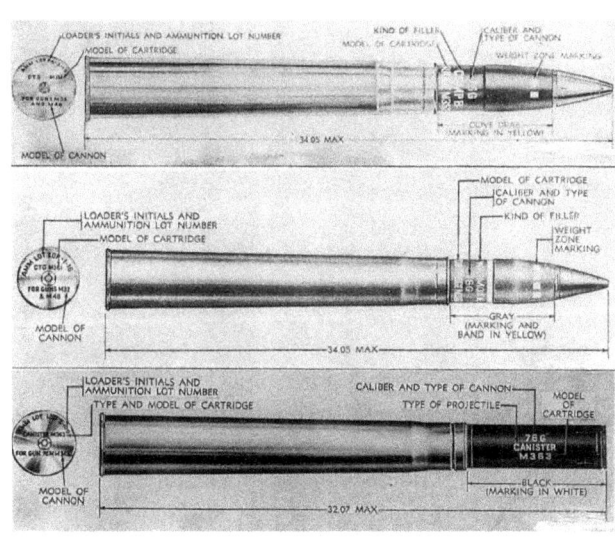

The compact design of the Cadillac turret and gun control system provided additional space in the M41A1 turret. New turret ammunition racks took advantage of this space to stow 21 76mm rounds compared to 13 in the M41. Combined with the 11 round ready rack and the 33 round hull rack, the total 76mm ammunition stowage was 57 rounds for the M41 and 65 rounds for the M41A1.

Above is another late production M41A1 and below are views of the 76mm ammunition turret racks. The ready rack is at the left. Note the difference between the M41 and M41A1 vehicles in the right photograph.

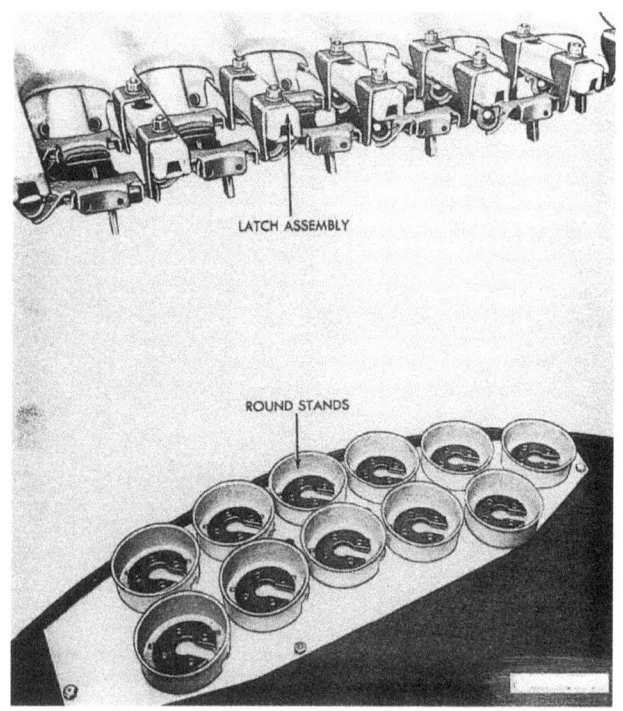

LATCH ASSEMBLY

ROUND STANDS

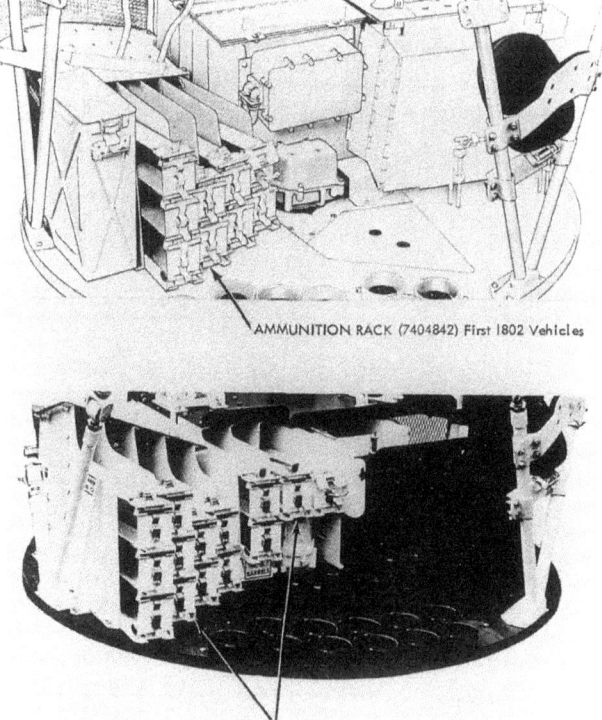

AMMUNITION RACK (7404842) First 1802 Vehicles

AMMUNITION RACKS (7982226 and 7982273) Vehicle 1803 and up

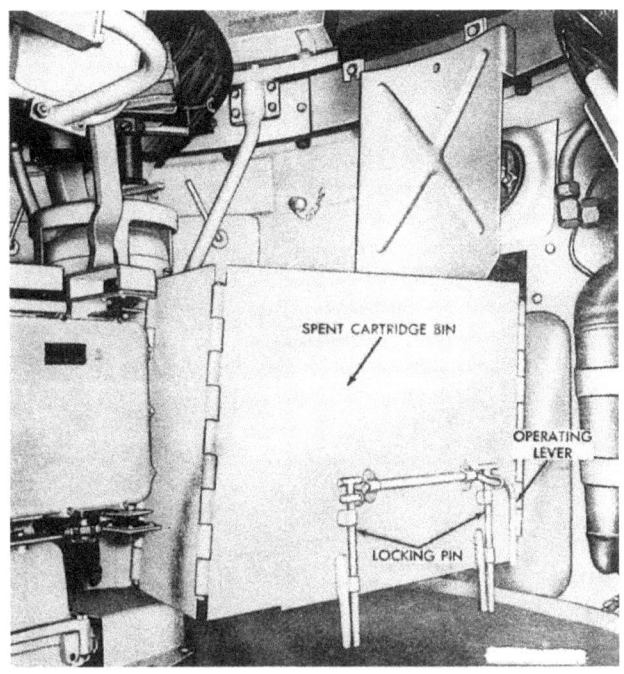

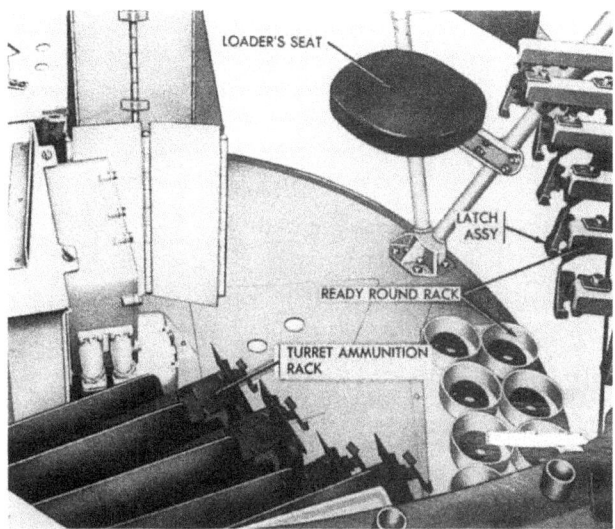

The spent cartridge case bin in the M41 (above and top right) can be compared with that in the M41A1 (middle right and below).

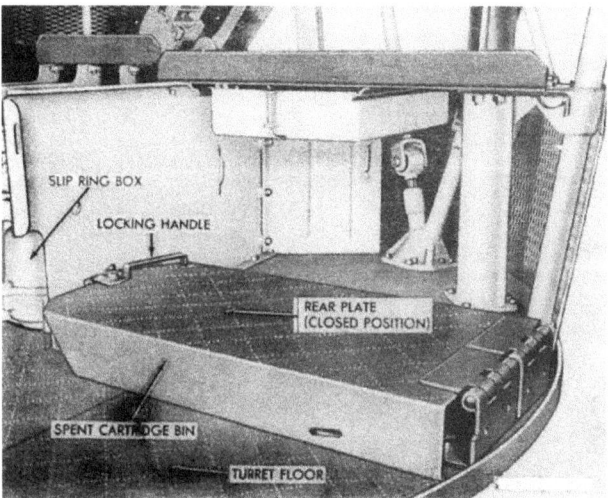

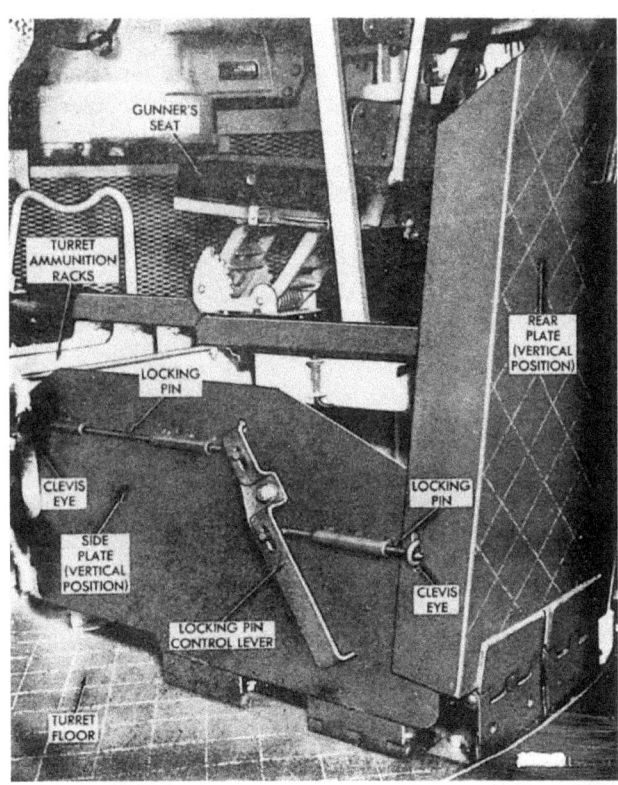

At the right is a view of the 76mm ammunition rack in the right front hull.

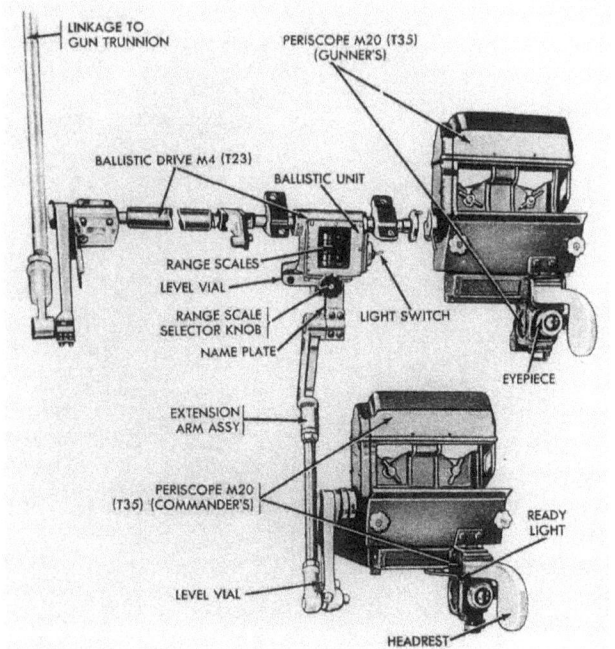

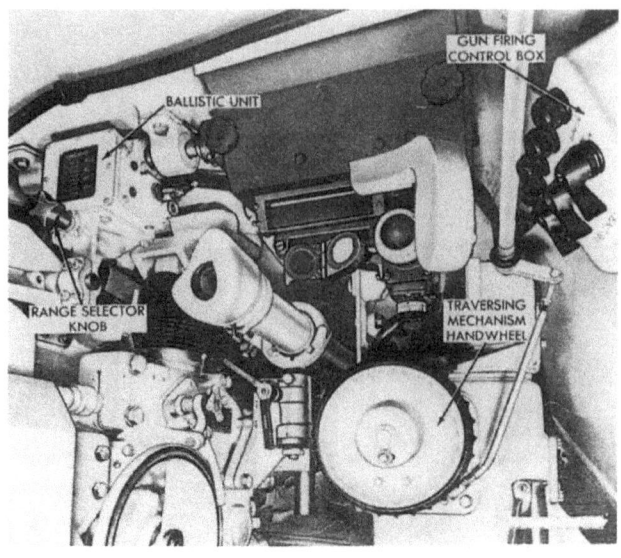

The ballistic drive and the two M20 periscopes appear at the top left and the gunner's telescope in the M41A1 is above. Below is a view of the loader's hatch in the turret roof.

The primary fire control system for the main armament consisted of two M20 (T35) periscopic sights, one for the gunner and one for the tank commander. They were linked to the gun mount through the ballistic drive M4 (T23). An M97 (T156) telescopic sight was installed on the right side of the combination gun mount for secondary use in direct fire. The commander's cupola was fitted with five direct vision blocks which, with the M20 periscope, provided 360 degree vision. An M13 or M13B1 periscope was mounted in the turret roof for use by the loader. The AN/GRC-3, -4, -5, -6, -7, or -8 radio was located in the turret bustle.

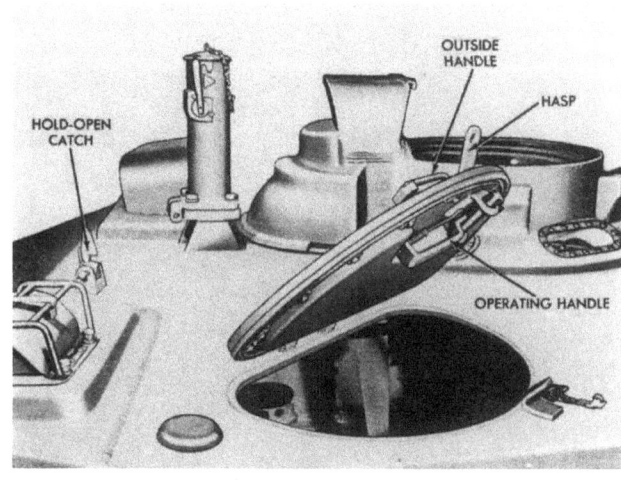

Below, the radio is installed in the turret bustle and at the bottom right is the interphone box on the outside of the rear hull.

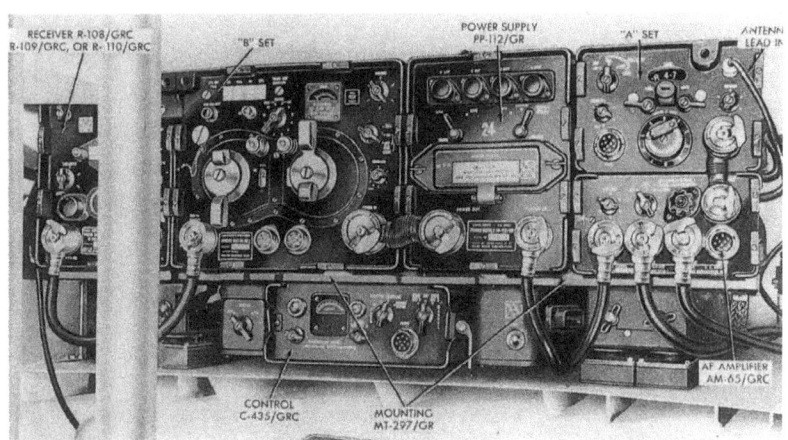

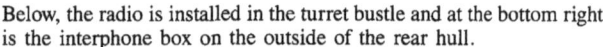

28

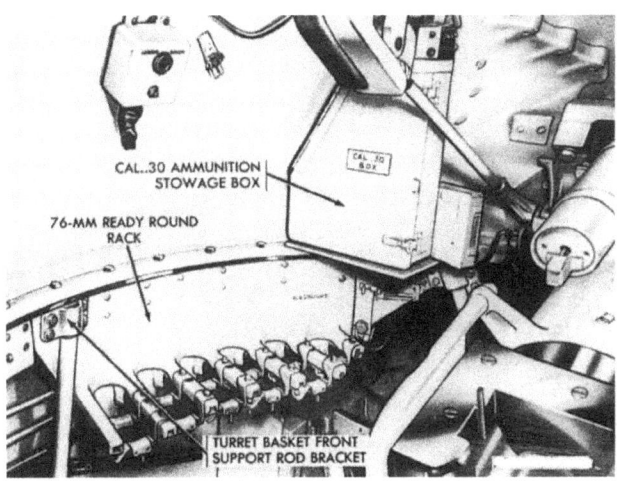

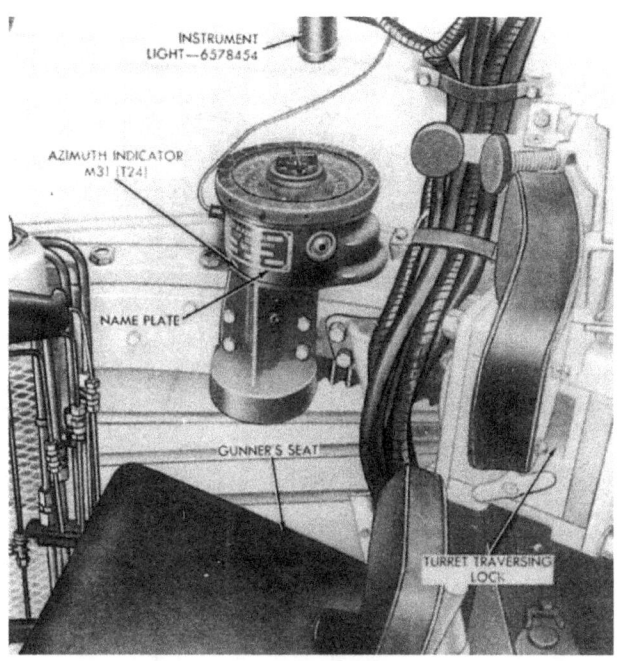

In the left front of the turret above, the ammunition stowage for the coaxial machine gun can be seen above the 76mm ready round rack. At the right is the gunner's seat and the azimuth indicator on the right side of the turret.

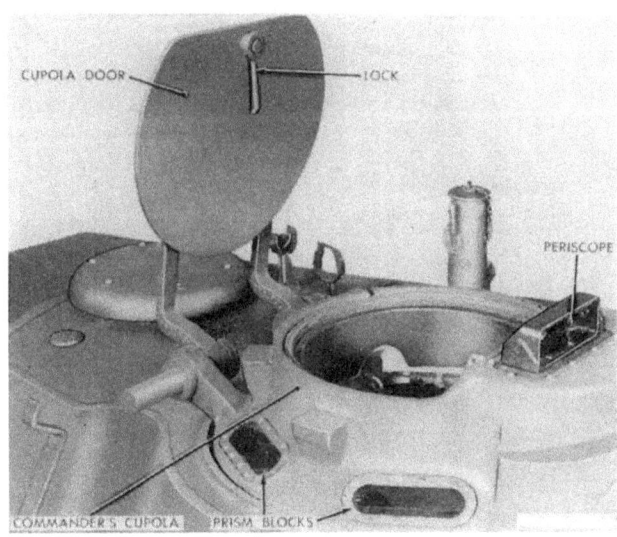

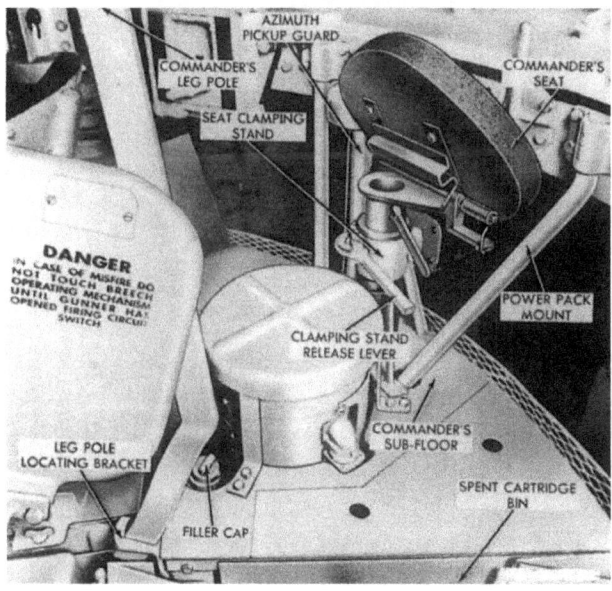

Above, the commander's cupola appears at the left and his seat is at the right in the folded position. The arrangement of the components in the M41 turret can be seen in the drawing at the right.

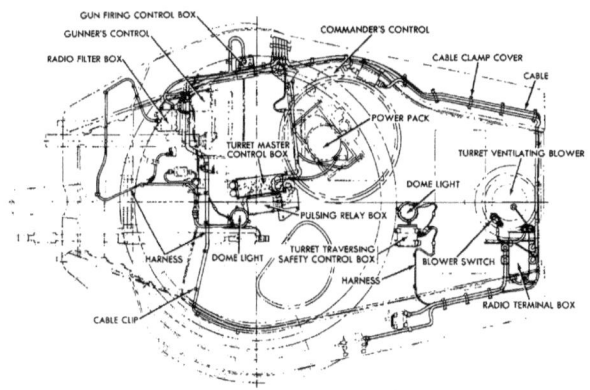

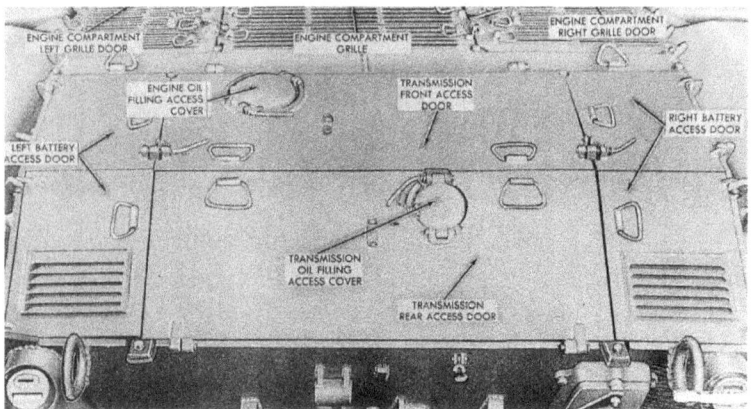
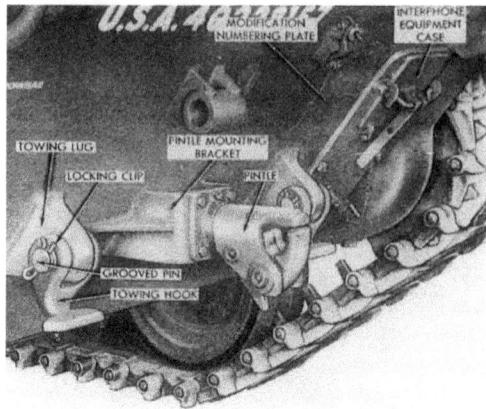

Above, the top rear of the hull and the towing pintle are at the left and right respectively. Below, the auxiliary generator and engine are at the left with the controls at the right.

A —CARBURETOR GUARD
B —EXHAUST PIPE
C —CARBURETOR AIR INTAKE TUBE
D —MULTIPLE CABLE RECEPTACLE
E —NAME PLATE
F —GROUND CABLE TERMINAL POST
G —BATTERY POSITIVE CABLE TERMINAL POST
H —REAR SHROUD PANEL
J —FUEL SUCTION LINE
K —GENERATOR SHROUD
L —CRANKCASE DRAIN PLUG
M —MOUNTING CHANNEL
N —MAGNETO TIMING HOLE COVER
P —GOVERNOR TIMING HOLE COVER
Q —LEFT SIDE UPPER PANEL
R —TRUNNION
S —THROTTLE AND CHOKE CONTROL LEVER

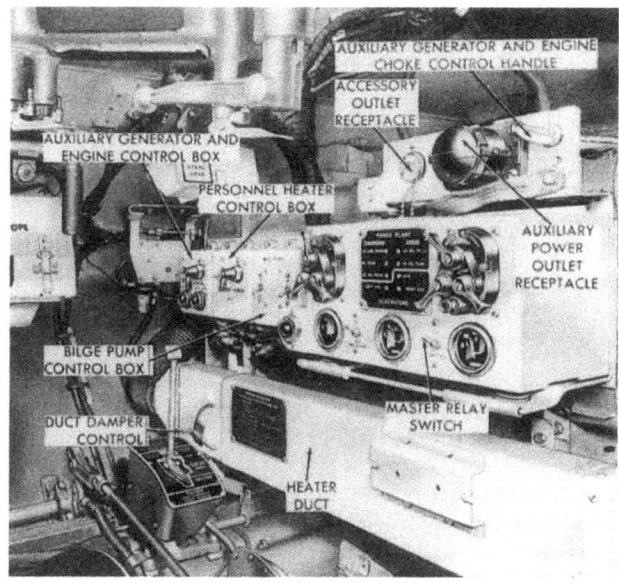

Below is a sketch of the auxiliary generator exhaust system and cooling duct assembly. The M19 infrared periscope is pictured at the right.

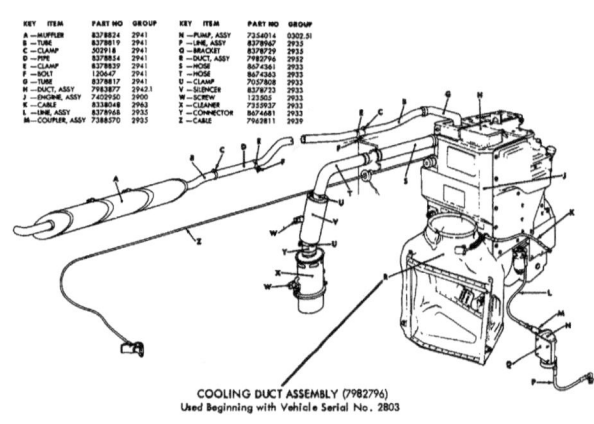

KEY	ITEM	PART NO	GROUP	KEY	ITEM	PART NO	GROUP
A	—MUFFLER	8378824	2941	N	—PUMP, ASSY	7354014	0302.51
B	—TUBE	8378819	2941	P	—LINE, ASSY	8378967	2933
C	—CLAMP	302918	2941	Q	—BRACKET	8378729	2935
D	—PIPE	8378854	2941	R	—DUCT, ASSY	7982796	2952
E	—CLAMP	8378839	2941	S	—HOSE	8674361	2933
F	—BOLT	120647	2941	T	—HOSE	8674363	2933
G	—TUBE	8378817	2941	U	—CLAMP	7057808	2933
H	—DUCT, ASSY	7982877	29421	V	—SILENCER	8378733	2933
J	—ENGINE, ASSY	7402950	3900	W	—SCREW	123503	2933
K	—CABLE	8338048	2963	X	—CLEANER	7355937	2933
L	—LINE, ASSY	8378968	2935	Y	—CONNECTOR	8674681	2933
M	—COUPLER, ASSY	7388570	2935	Z	—CABLE	7963811	2939

COOLING DUCT ASSEMBLY (7982796)
Used Beginning with Vehicle Serial No. 2803

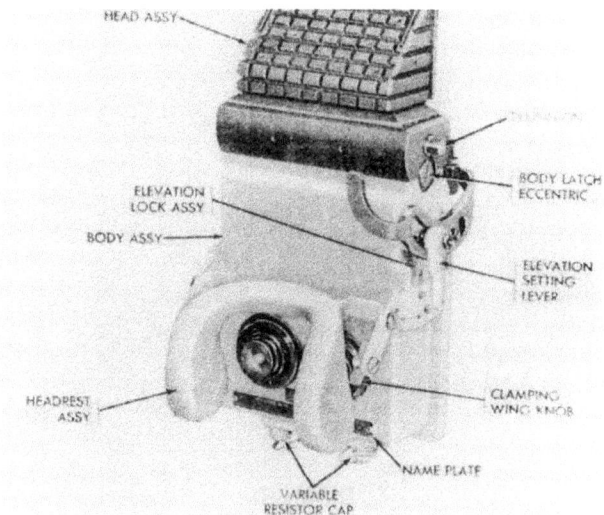

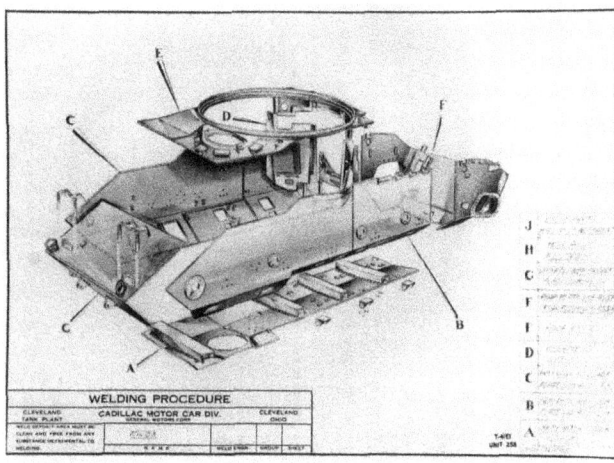

The diagram above illustrates the weld assembly procedure for the T41E1 hull.

Like the light tanks T37 and T41, the hull on the M41 and M41A1 was a welded assembly of rolled and formed homogeneous steel armor plate. The driver was seated in the left front hull with four M17 periscopes around his overhead hatch providing vision for approximately 270 degrees. A solid flat hatch cover was installed on M41s 1 through 260. On tanks 261 through 3333, the driver's flat hatch cover was fitted with a mount for the M19 infrared periscope. Later tanks were provided with a formed steel hatch cover which also could mount the M19 periscope. The flat hatch on vehicles 1 through 3333 was opened and closed by a cam actuated lever. Tank 3334 and later vehicles used a gear operated mechanism.

The formed steel driver's hatch cover which was installed on the later tanks is illustrated below.

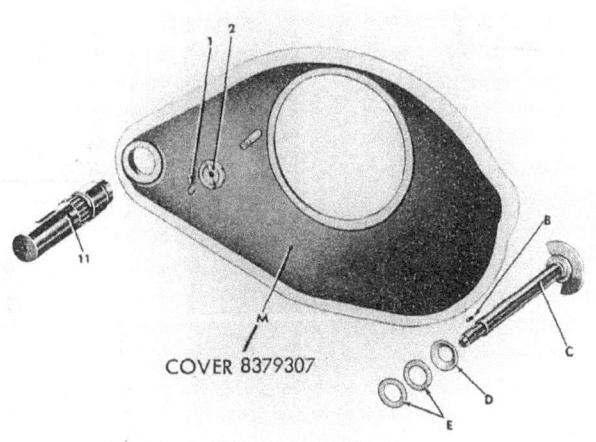

COVER 8379307

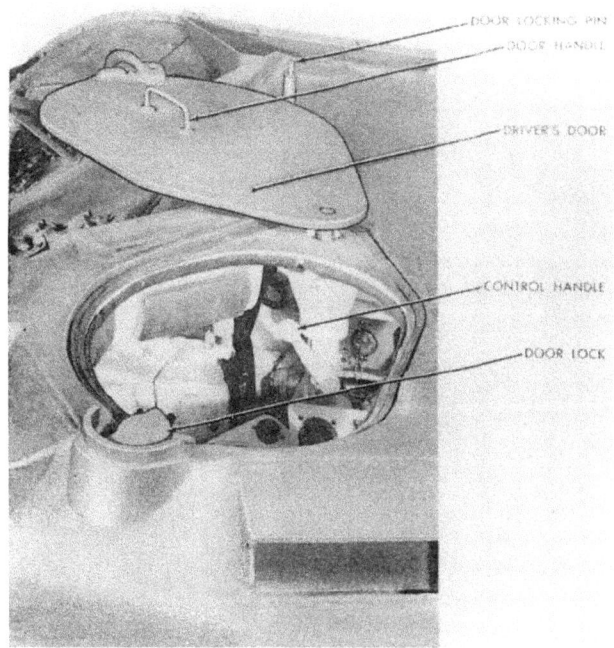

The early driver's hatch without the mount for the M19 periscope is shown above. The controls for the driver's hatch cover on the first 3333 tanks can be seen below.

Below are the controls for the driver's hatch which were introduced on tank number 3334.

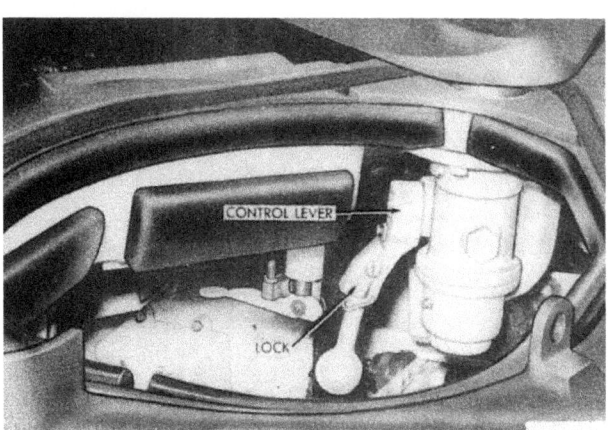

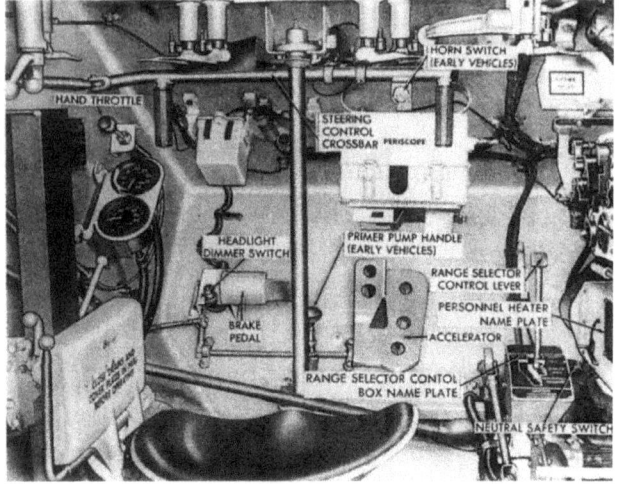
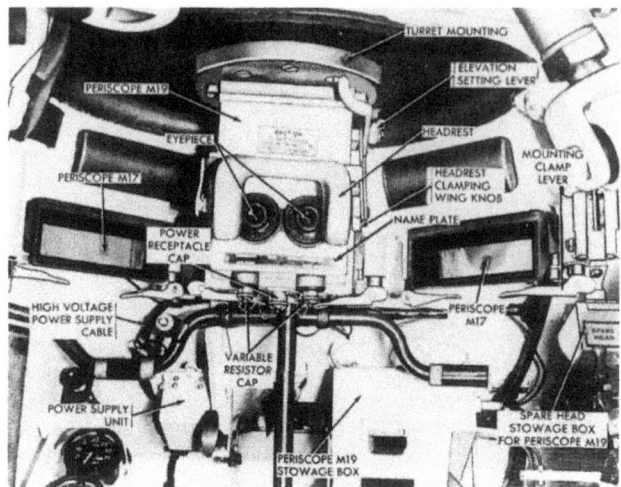

The driver's controls on the early and late production tanks can be seen above at the left and right respectively. Note the changes in the steering control crossbar. Below, the driver's seat is at the left with the seat back removed.

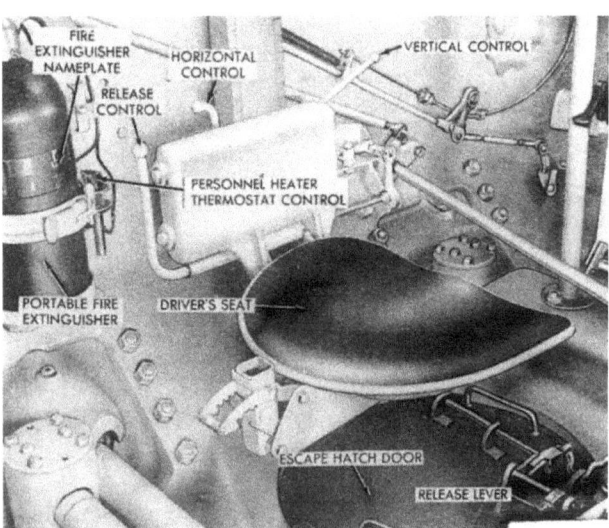

The driver's steering control crossbar on M41s 1 through 601 had vertical handgrips. However, the crossbar frequently fractured in service near the horn button. This area was reinforced on tanks 602 through 2452, but the vertical handgrips were retained. Later vehicles used a new control crossbar with horizontal handgrips which reduced driver fatigue, particularly when driving with the head out of the hatch. The driver's seat on the first 1490 vehicles had a four notch adjuster on the back rest. Unfortunately, the fourth notch allowed the back rest to contact the turret guard resulting in damage to the cushion or injury to personnel. To correct this, the fourth notch was eliminated on the later vehicles and it was recommended that a steel bar be tack welded to the adjuster to prevent use of the fourth notch on the earlier tanks.

The fighting compartment occupied the center of the tank hull with the turret mounted on a 73 inch inside diameter ring. In the engine compartment at the rear, the Continental AOS-895-3 air-cooled engine drove the vehicle through the Allison CD-500-3 cross drive transmission. This power pack was removable from the vehicle as a unit. The auxiliary generator and its engine were located in the right front corner of the engine compartment with the exhaust pipe protruding through the top of the hull. No mufflers were installed on the auxiliary engine exhaust on vehicles 1 through 1367. On tanks 1368 through 2289, a small muffler was mounted on top of the right main engine muffler. This installation was unsatisfactory with the auxiliary engine muffler often being damaged by heat from the main engine muffler. Beginning with tank 2290, a larger muffler mounted on the right front fender was provided for the auxiliary engine.

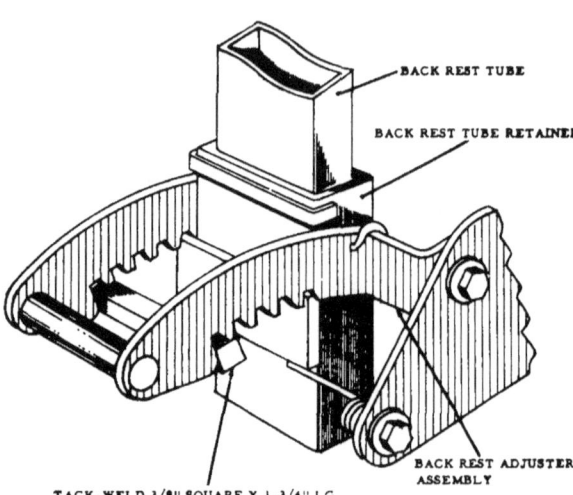

The modification to the driver's seat back adjuster is shown in the sketch at the left.

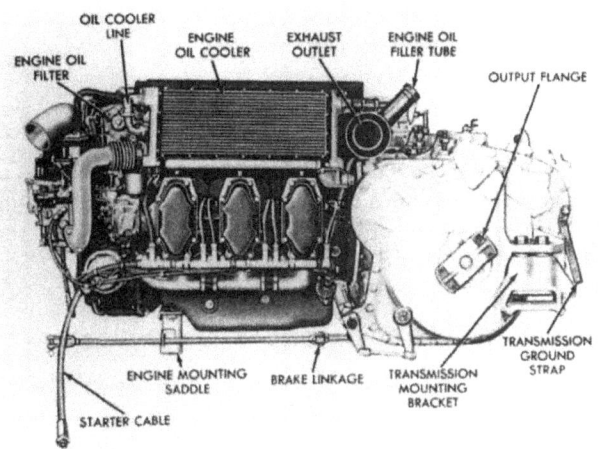

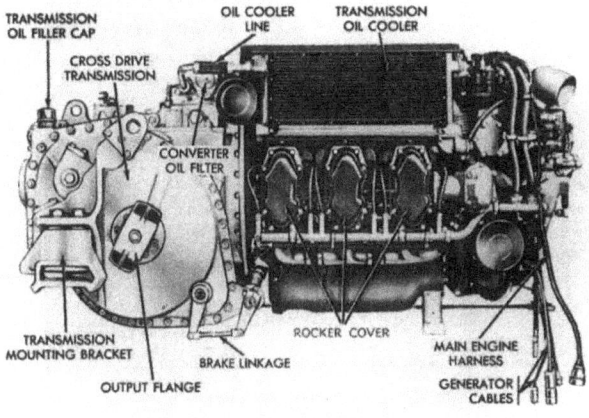

Both sides of the power pack can be seen above. At the left below is a view of the early auxiliary generator engine muffler installation on top of the main muffler shield. The sketch below at the right shows the general arrangement of the power train components.

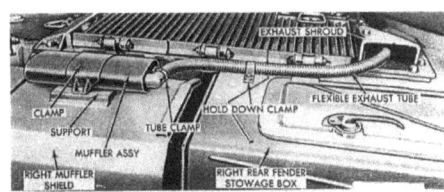

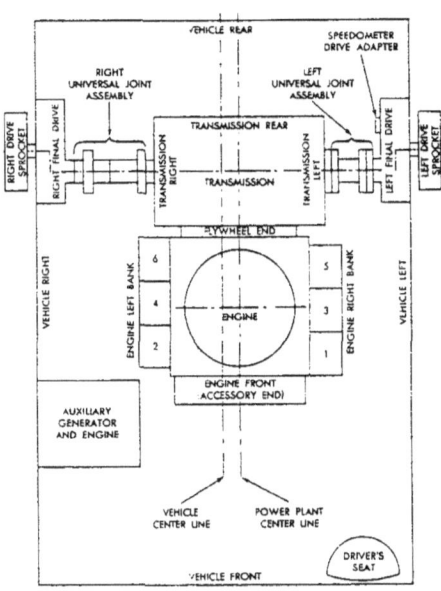

The late production auxiliary generator engine muffler installation is at the left. Below are views of the engine compartment empty and with the power pack installed.

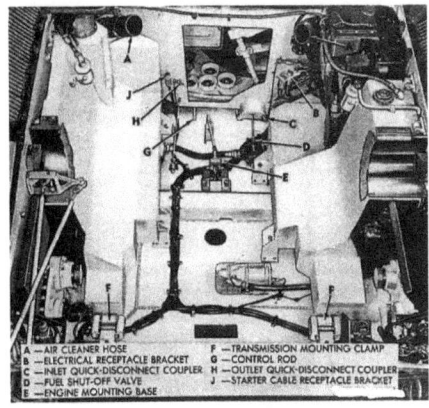

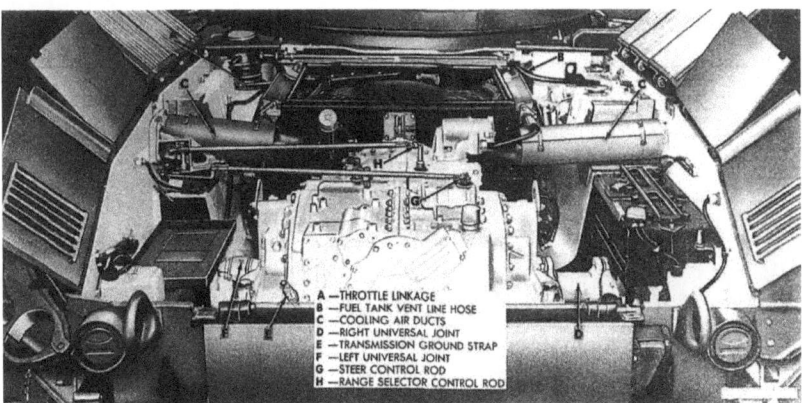

33

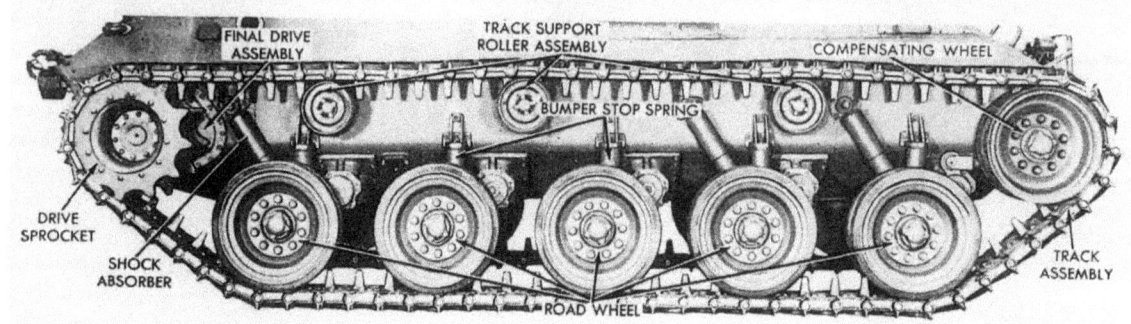

Details of the light tank suspension can be seen in the photograph above. Below, the T91E3 track shoe is shown assembled and disassembled.

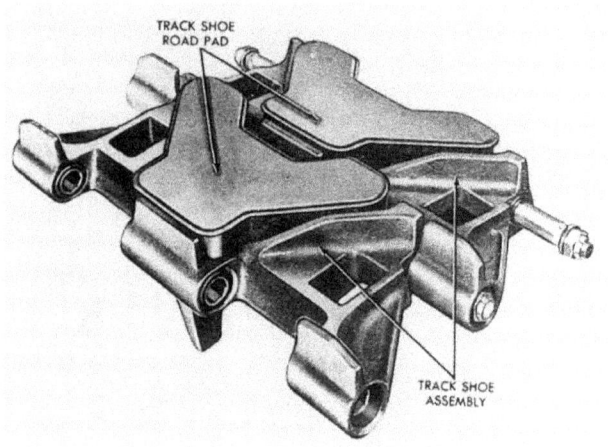

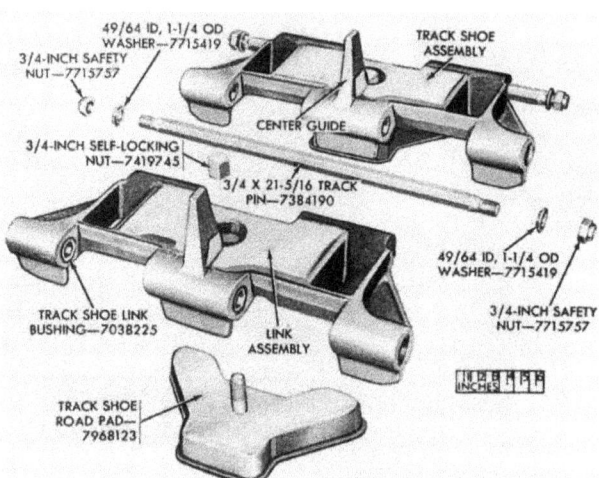

The torsion bar suspension supported the tank on ten dual road wheels, five per track. Shock absorbers were installed on the first two and last road wheel arms on each side. The upper track run was carried on three dual support rollers per track. On tanks 1 through 3375, these rollers consisted of a flat disc and rubber tire. Beginning with tank number 3376, a contoured disc and tire was installed. The new design reduced hull vibration and increased tire life. A dual compensating idler wheel was mounted at the front of each track. These idler wheels were steel without tires on tanks 1 through 2611. Vehicles 2612 through 3475 used rubber tired idlers identical with the road wheels. Later tanks were fitted with idlers having contoured rubber tires to reduce hull vibration and stone damage. The T91 single pin

tracks on the light tanks T37 and T41 were of forged steel. This same construction applied to the T91E1 and T91E2 tracks which appeared on the early T41E1 tanks. They were then replaced by the cast steel T91E3, single pin, tracks which were adopted for all future production. All of these center guide tracks were 21 inches wide with a six inch pitch and they featured a detachable rubber pad. The original production tanks were fitted with 75 track shoes per side. However, this was soon reduced to 74. With the introduction of the rubber tired compensating idler wheel on tank number 2612, the number of track shoes was again increased to 75 to compensate for the larger diameter of the rubber tired idler.

Below are the early and late configurations of the compensating idler (left) and the track support rollers (right).

VEHICLES BELOW SERIAL NO. 2593 WITH SQUARE FENDERS AND SKIRTS

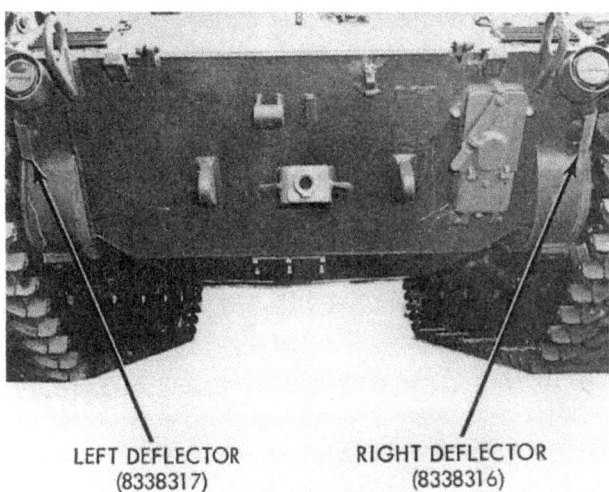

LEFT DEFLECTOR
(8338317)

RIGHT DEFLECTOR
(8338316)

Above, the early and late fender designs are compared. At the right, the installation of the track deflectors (upper) and the track skids (lower) can be seen.

SKID (8654516, left)
(8674521, right)

It was noted that whenever a track was completely thrown, it climbed the sprocket, hooked onto the corner of the hull, and damaged the track, hull, and fenders. To prevent this, track deflectors were welded to the left and right sides of the rear hull on the production tanks starting with serial number 1791. Like many of the other modifications, it was recommended that the track deflectors be retrofitted to the earlier production tanks.

Fenders with square corners and track skirts were standard on the first 2592 tanks. Starting with tank number 2593, the front and rear of the fenders were cut at an angle and the track skirts were eliminated. Beginning with M41A1 number 2908, track skids were installed underneath the fenders. They eliminated much of the damage from track slap and that caused by foreign objects.

Tanks with the early (right) and late (below) fenders appear in these photographs.

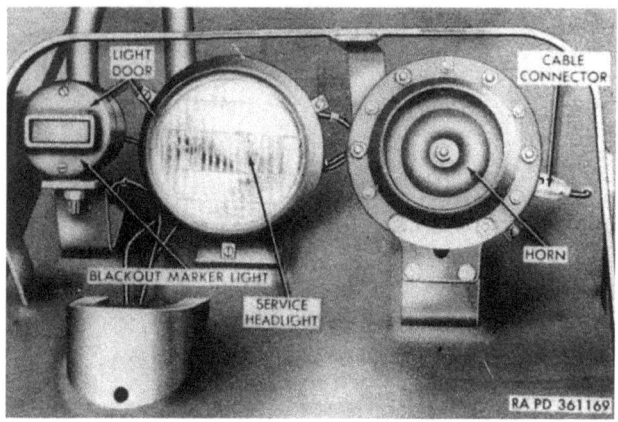 

The early headlight arrangement (above) is compared here with the later installation (below). Note the addition of the infrared blackout headlight on the later vehicle.

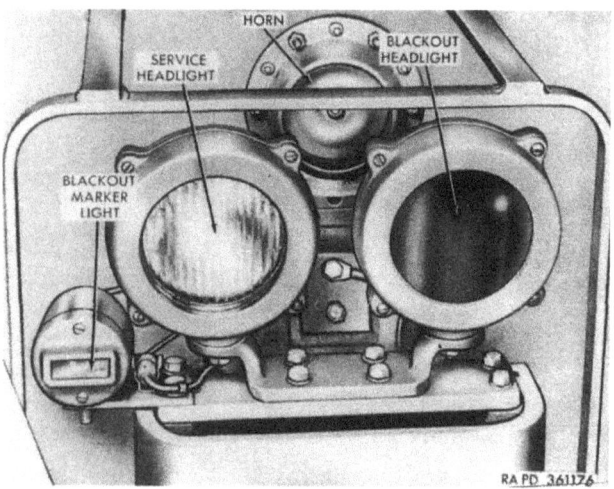

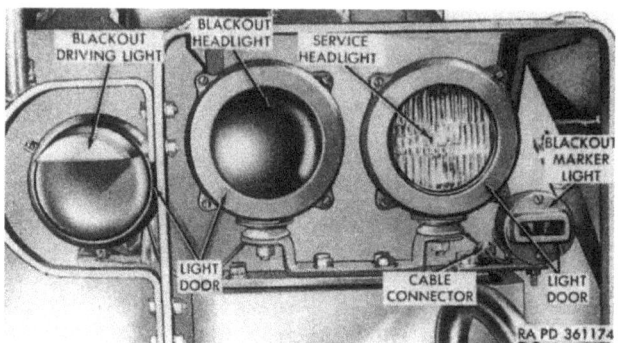

At the right is the late taillight arrangement. Below is a sketch of the new fire control linkage designed to eliminate thermal expansion errors.

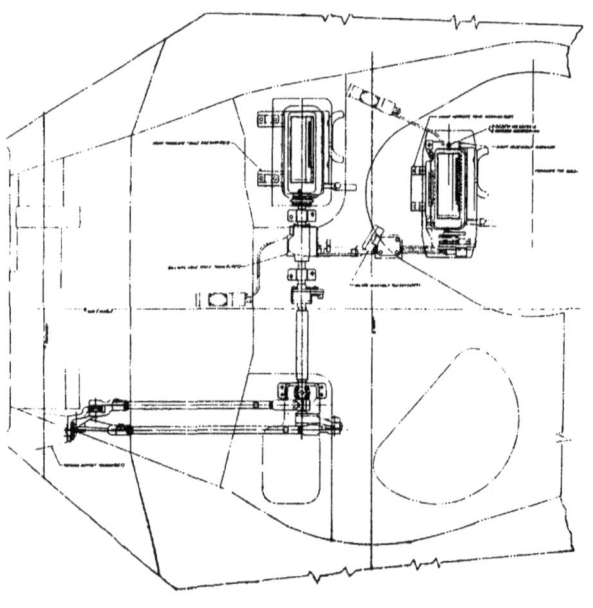

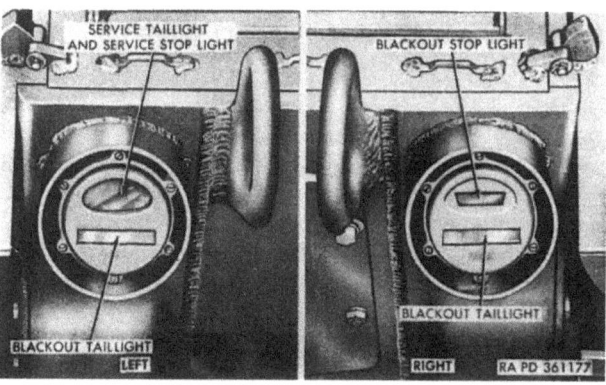

The front of the early M41s was fitted with service headlights as well as blackout marker lights and a blackout driving light. The latter were intended for use under dim out conditions. The later tanks were provided with infrared blackout headlights for use with the driver's M19 infrared periscope under complete blackout conditions.

After production of the M41A1 had ended, the Cadillac Motor Car Division, in cooperation with Detroit Arsenal, was engaged in a long term product improvement program. This program covered a variety of projects, but two were of particular importance. The first of these involved the periscopic sights for gunner and tank commander. Since the linkage for these sights was attached to the turret roof, thermal expansion of the top armor introduced errors into the

36

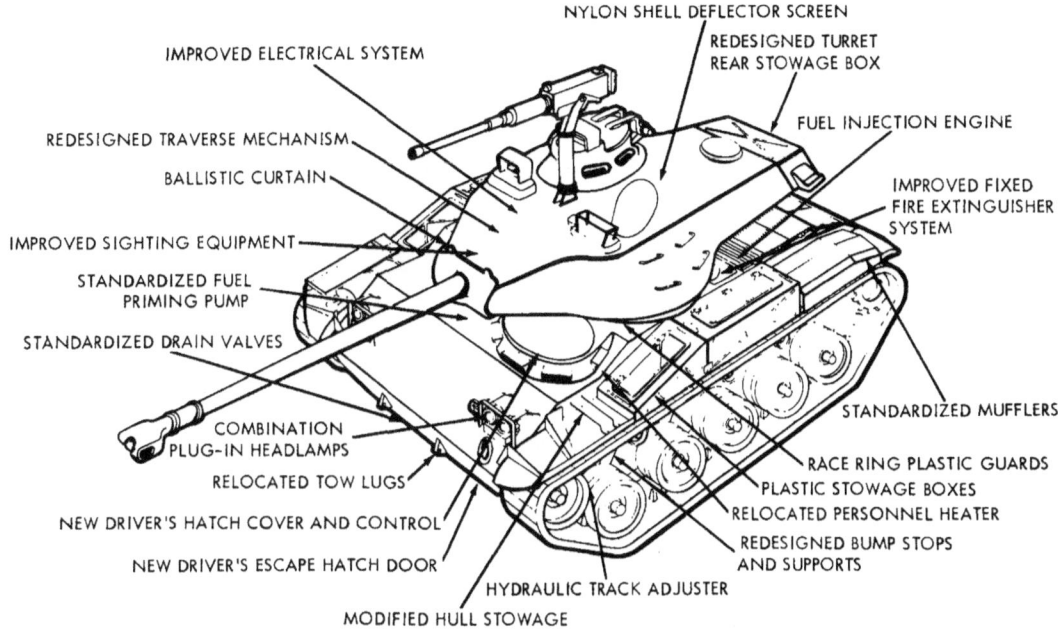

IMPROVED ELECTRICAL SYSTEM

NYLON SHELL DEFLECTOR SCREEN

REDESIGNED TURRET REAR STOWAGE BOX

FUEL INJECTION ENGINE

REDESIGNED TRAVERSE MECHANISM

BALLISTIC CURTAIN

IMPROVED FIXED FIRE EXTINGUISHER SYSTEM

IMPROVED SIGHTING EQUIPMENT

STANDARDIZED FUEL PRIMING PUMP

STANDARDIZED DRAIN VALVES

COMBINATION PLUG-IN HEADLAMPS

RELOCATED TOW LUGS

NEW DRIVER'S HATCH COVER AND CONTROL

NEW DRIVER'S ESCAPE HATCH DOOR

MODIFIED HULL STOWAGE

HYDRAULIC TRACK ADJUSTER

REDESIGNED BUMP STOPS AND SUPPORTS

RELOCATED PERSONNEL HEATER

PLASTIC STOWAGE BOXES

RACE RING PLASTIC GUARDS

STANDARDIZED MUFFLERS

The modifications proposed for future production of M41 series tanks are illustrated above.

line of sight. Cadillac designed a new linkage arrangement which was separated from the turret roof eliminating the effects of temperature changes.

The second major problem concerned the limited cruising range of the M41 and M41A1. Further development of the Continental engine resulted in the replacement of the carburetors on the AOS-895-3 by a fuel injection system. Tests of the new engine, designated as the AOSI-895-5, during the Spring of 1955 at Fort Knox showed a decrease of approximately 20 per cent in fuel consumption. As a result of these tests, it was directed that the existing AOS-895-3 engines be converted to the fuel injection type by the installation of a kit under a Field Service Modification Work Order. When powered by the AOSI-895-5 engine, the M41 and M41A1 were redesignated as the 76mm gun tanks M41A2 and M41A3 respectively.

Other features considered under the product improvement program included a redesigned Cadillac turret and gun control system, the Cadillac Gage (no relation to the Cadillac Motor Car Division of General Motors) constant pressure hydraulic turret and gun control system, fiber glass reinforced plastic stowage boxes, a new driver's hatch, an hydraulic track adjuster, and many others. These modifications were included in the design of a future production version of the tank if it should be required. However, no further M41 series tanks were produced. Although they did not fully satisfy the requirements of the Army Field Forces, they remained in service as an interim vehicle pending the development of an improved light tank.

At the right is a drawing of the Cadillac Gage fire control system proposed for installation in the M41 series tanks.

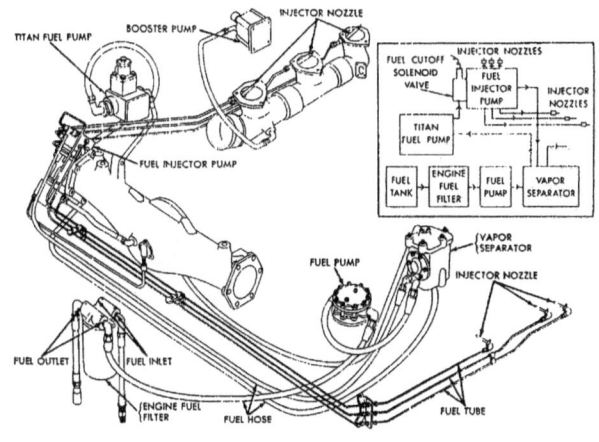

The sketch above shows the fuel injection system installed on the AOSI-895-5 engine.

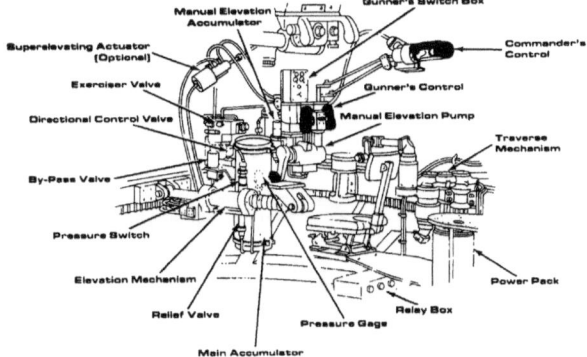

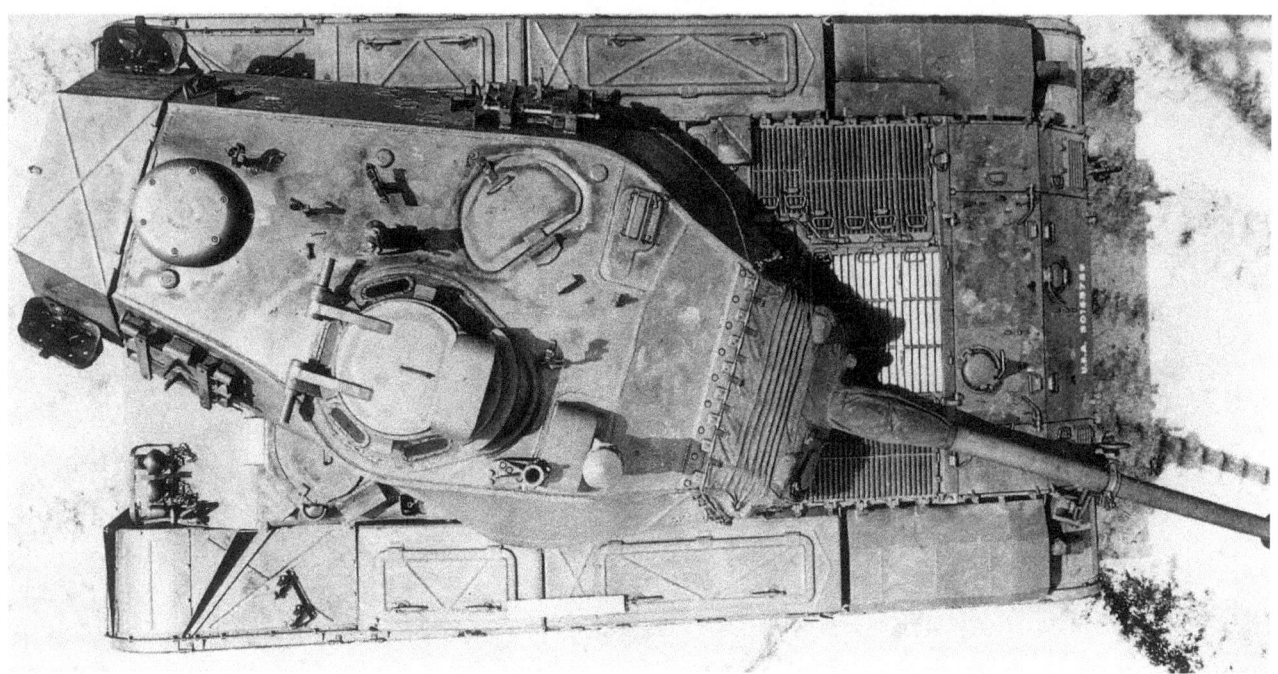

Details of the early production 76mm gun tank M41 can be seen in the top view above. Below, both tanks are fitted with the early muzzle brake, but the tank at the left has the later turret design while the one at the right retains the original configuration. A sectional drawing of the early production M41 appears at the bottom of the page.

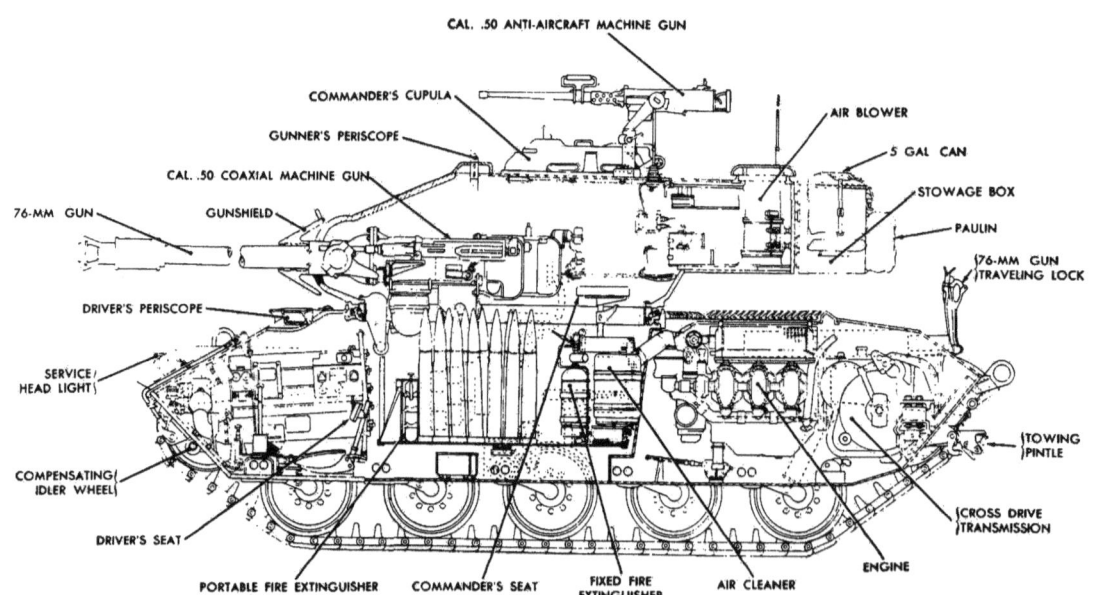

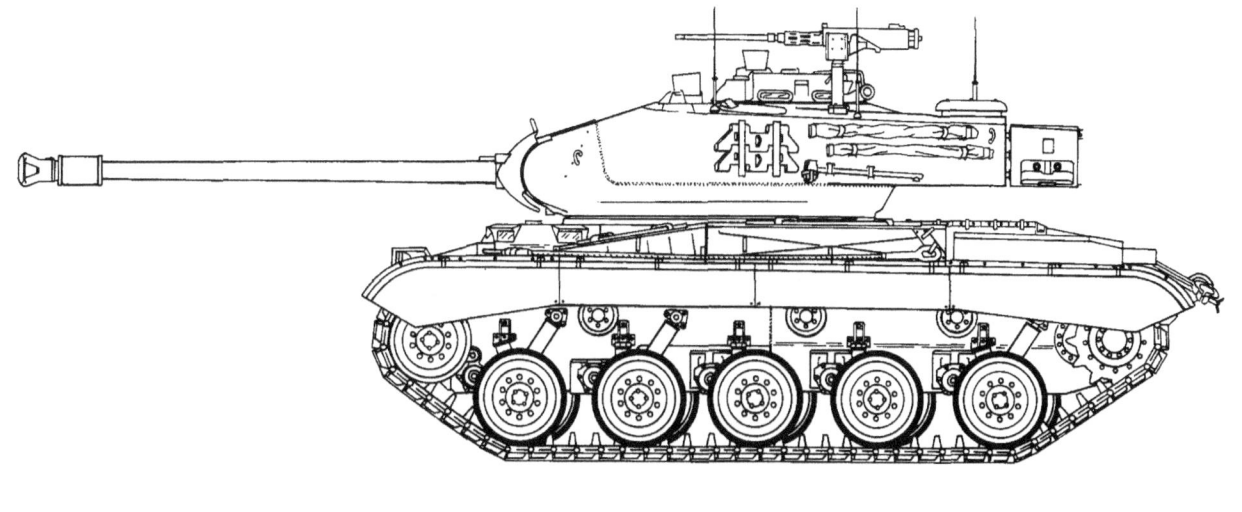

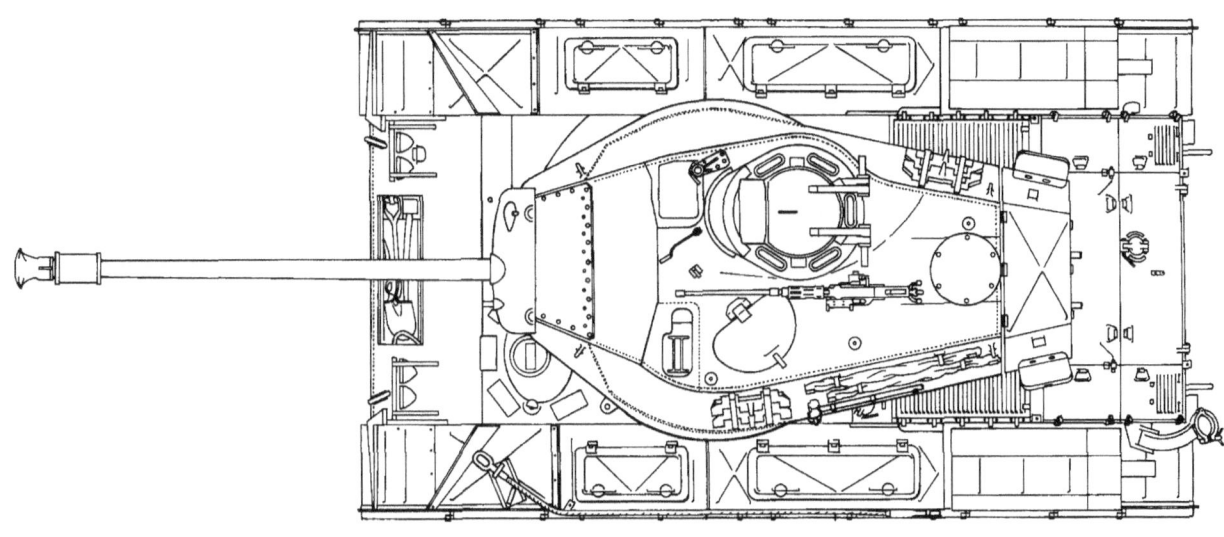

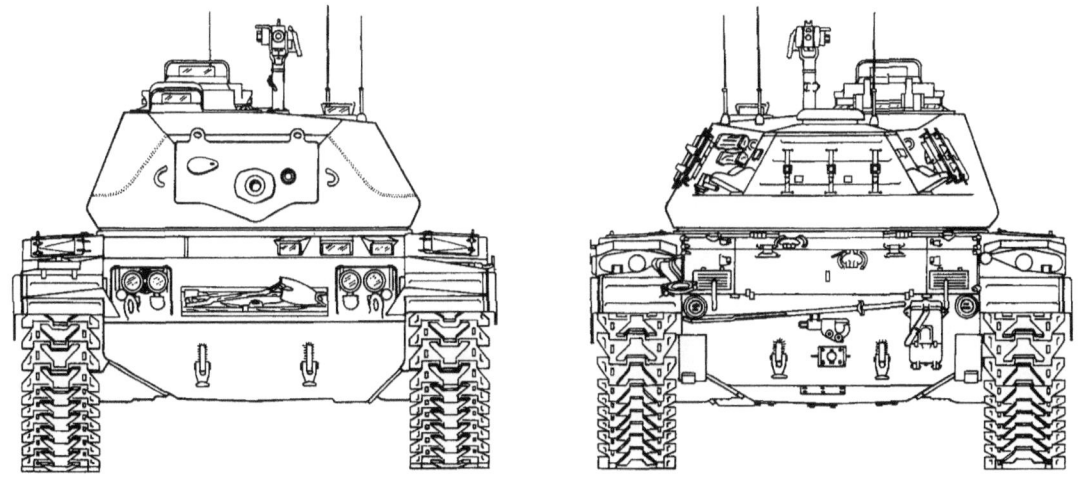

Scale 1:48

©D.P. Dyer

76mm Gun Tank M41, early production

The late production 76mm gun tank M41A1 is shown in the photographs above and below. Note the late turret configuration and muzzle brake. However, the contoured tires on the compensating idler and track support rollers have not yet been installed.

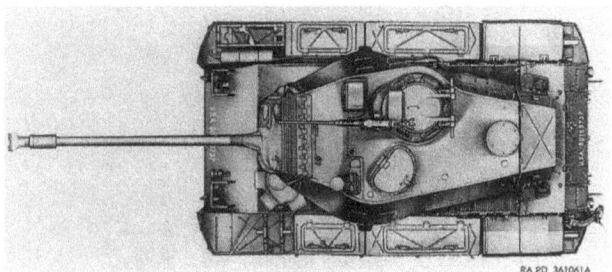

The view of the late production M41A1 at the bottom right can be compared with the earlier vehicle below. The side skirts on the latter have been removed, but it retains the square end fenders.

40

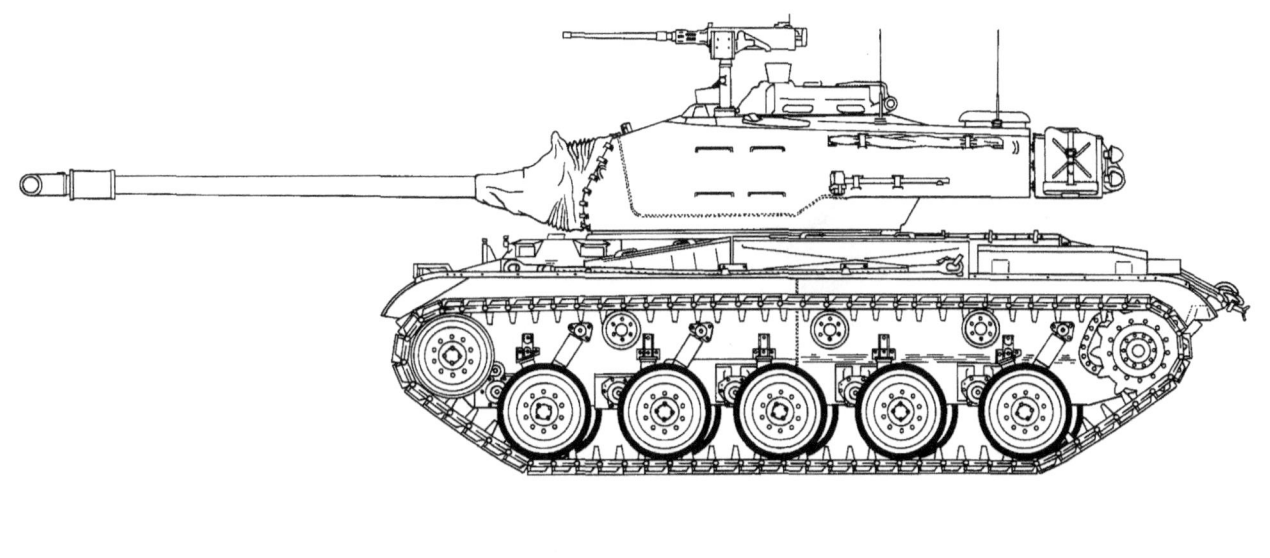

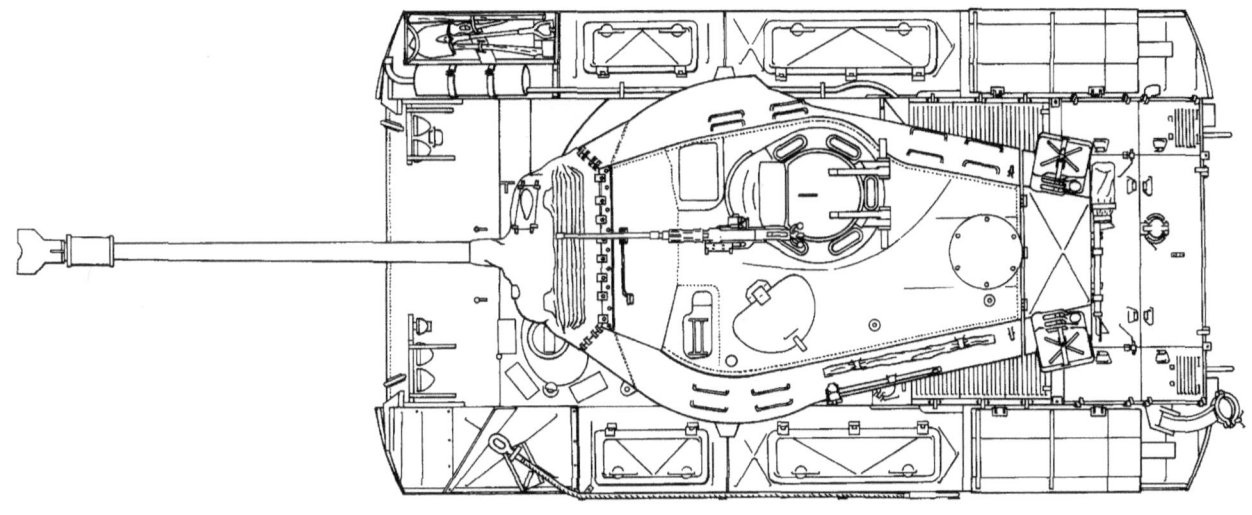

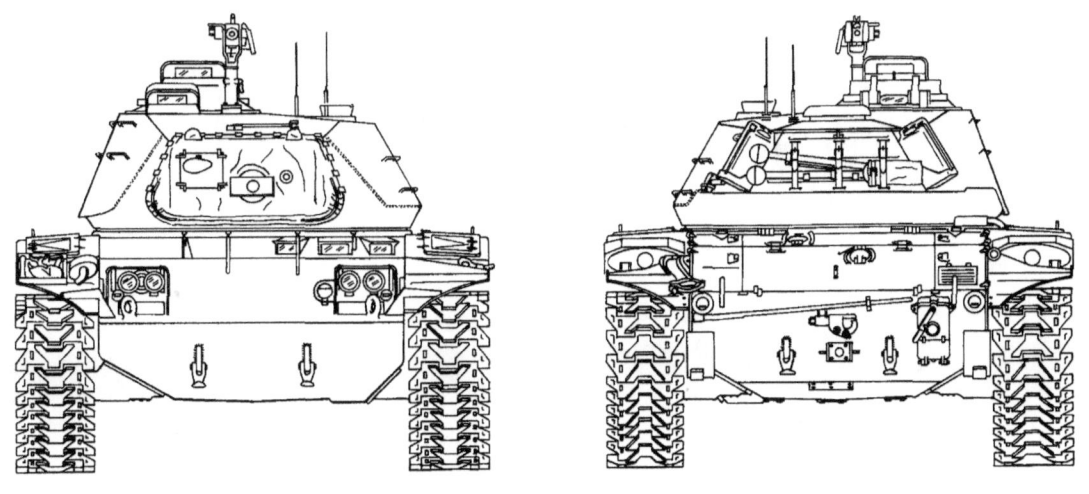

Scale 1:48

©D.P. Dyer

76mm Gun Tank M41A1, late production

The service test conditions imposed upon the M41 are clearly illustrated here, particularly in the mud bath below.

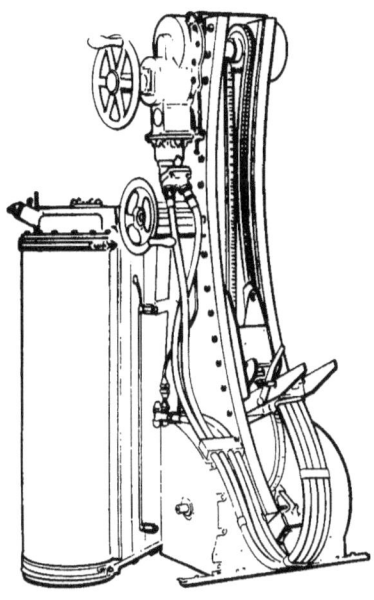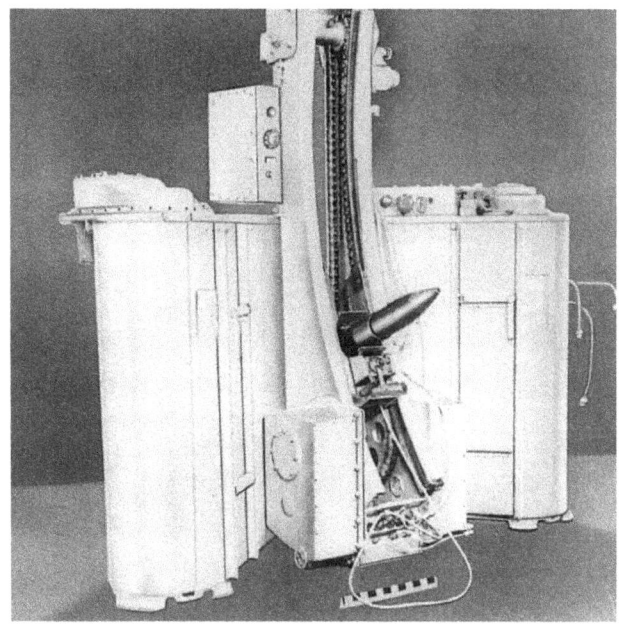

The sketch at the top left shows the single magazine automatic loader developed for installation in the light tank T37. The double magazine version intended for the 76mm gun tank T41E1 appears in the photograph at the top right.

INCREASING THE FIREPOWER

The Phase III turret originally proposed for the light tank T37 specified an automatic loader for the 76mm gun. A contract for the development of such a loader was awarded to the Rheem Manufacturing Company by Detroit Arsenal. The objective of this program was to reduce the crew size by eliminating the loader, increase the rate of fire, and to evaluate the overall performance of the light tank with an automatic loader.

The loader designed by Rheem used the 76mm gun T102. This was essentially the original T94 gun mounted upside down to accommodate the automatic loading equipment. The latter consisted of a 13 round magazine, a transfer mechanism, a hoist, a hoist control arm, and a power supply. The magazine was installed on the turret platform to the right of the hoist. The transfer mechanism moved the vertical round from the magazine into the hoist assembly. The hoist rotated the round from the vertical position until it pointed at the trunnion center line. It was then raised until it was in line with the gun bore and rammed into the chamber. After firing, the empty case was ejected and returned to the magazine. Tests of this automatic loader continued at Aberdeen Proving Ground into 1951. The rate of fire was 18 rounds per minute. Unfortunately, the system was not reliable and several cases of premature ramming occurred resulting in the loading of two rounds, one upon the other. With live ammunition, this would have caused

an explosion. Also, the automatic loader was extremely unreliable if the turret was canted. Under this condition, the round would frequently strike the breech ring. However, the test report concluded that an automatic loader was practical, if these problems could be corrected.

In 1951, a new contract was awarded to the Rheem Manufacturing Company to build an improved version of the automatic loader. This equipment was to be suitable for use with the larger rounds used in the 76mm gun T91E3 mounted in the T41E1 tank then going into production. The new loader incorporated two 13 round magazines and was able to select any one of three types of ammunition from either magazine. Thus the system could handle and select six types of ammunition. The loader was similar in design to the earlier model with the two magazines located one on each side of the hoist. Rows of lights on an electric panel indicated the type of rounds available in the two magazines. A selector switch allowed the gunner or tank commander to control which round would be loaded. The modified cannon, mounted upside down, was designated as the 76mm gun T91E5. Using an on board power supply, the new loader had a maximum firing rate of 26 rounds per minute. Completed in early 1953, the equipment was shipped to Aberdeen for the test program which continued until 1956. Unfortunately, the reliability problems still existed and the loader was considered unsatisfactory for field use.

The 90mm gun tank T49 is shown in these photographs at Aberdeen Proving Ground on 27 May 1954.

Although the 76mm gun T91 was effective against the World War II era Soviet T34/85, the appearance of postwar Soviet tanks such as the T54 indicated that its future usefulness might be limited. On 13 January 1950, the Commanding Officer of Detroit Arsenal recommended the development of a 90mm gun installation for the light tank. Design studies proposed a 90mm smooth bore cannon to be mounted in the light tank T41. This was to be a low pressure gun with a quick change tube firing fin stabilized, shaped charge, projectiles. The new weapon was assigned

The welded steel insert used to increase the height of the turret to accommodate the range finder on the T49 is obvious in these views.

the designation 90mm gun T132. Initial tests with the smooth bore gun showed that the dispersion was excessive and shallow rifling was introduced in an effort to improve the accuracy. This version, designated as the 90mm gun T132E3, was mounted in the modified turrets of two production T41E1

tanks. The converted vehicles were designated as the 90mm gun tanks T49.

The turret roof on the T49 was raised 6¼ inches compared to the T41E1 to provide space for the T41E3 stereocopic range finder. A standard range finder cover plate was

 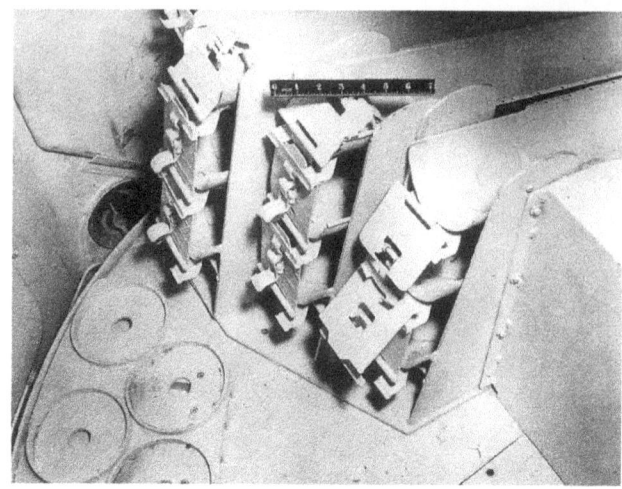

The rear view of the 90mm gun tank T49 appears above at the left. The rack for nine 90mm rounds under the cannon is above at the right and some 90mm ammunition is stowed in the ready rack below.

used from the 90mm gun tank M47. Stowage in the T49 was rearranged to accommodate 46 rounds of 90mm ammunition. The rack in the right front hull contained 29 rounds. Nine rounds were under the gun mount on the turret platform and eight rounds were in the ready rack. The empty cartridge case bin was eliminated to provide stowage space. The 90mm gun T132E3 was installed in the combination gun mount T145 with a coaxial .30 caliber machine gun and a direct sight telescope. An experimental electric amplidyne system controlled the movement of the turret in azimuth and the gun in elevation. The conversion of the T49 tanks was carried out by the United Shoe Machinery Corporation. The installation of the new cannon increased the weight of the T49 by approximately 1200 pounds compared to the original T41E1. Three types of ammunition were provided for the 90mm gun T132E3 which had a maximum rated chamber pressure of 27,500 psi. These rounds were high explosive (HE) T91 (muzzle velocity 2400 feet per second), high explosive plastic (HEP) T142E3 (muzzle velocity 2600 feet per second), and high explosive antitank (HEAT) T108E45 (muzzle velocity 2800 feet per second).

Below, the gunner's station in the T49 is at the left and the range finder can be seen from the loader's position at the right.

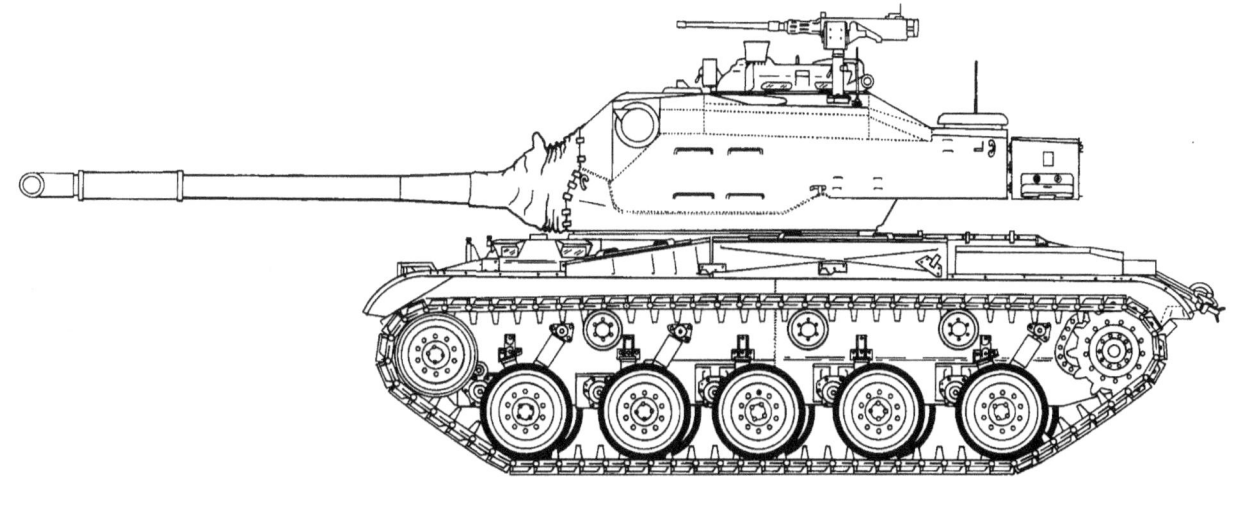

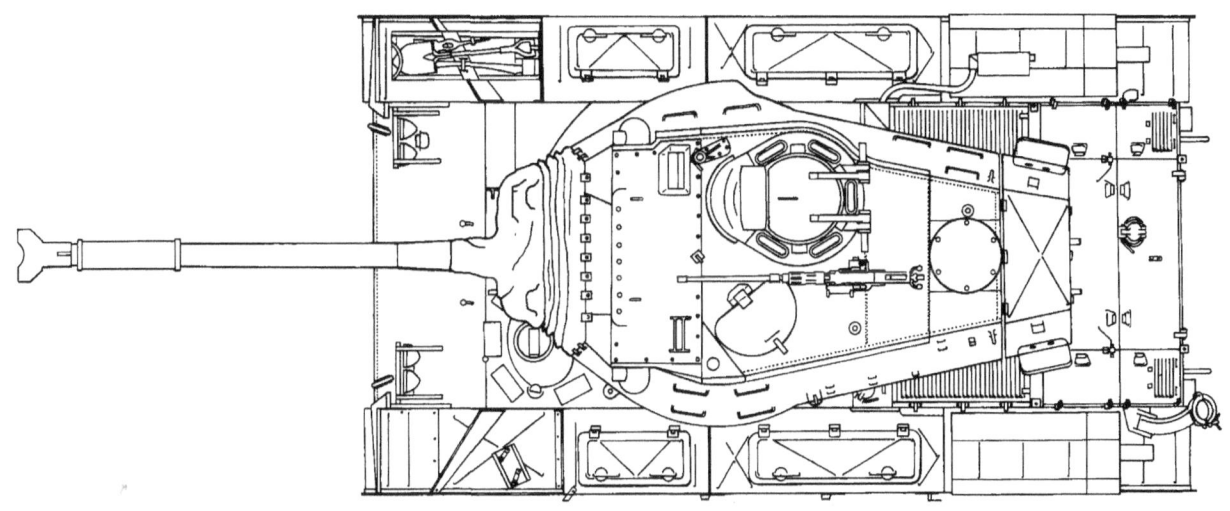

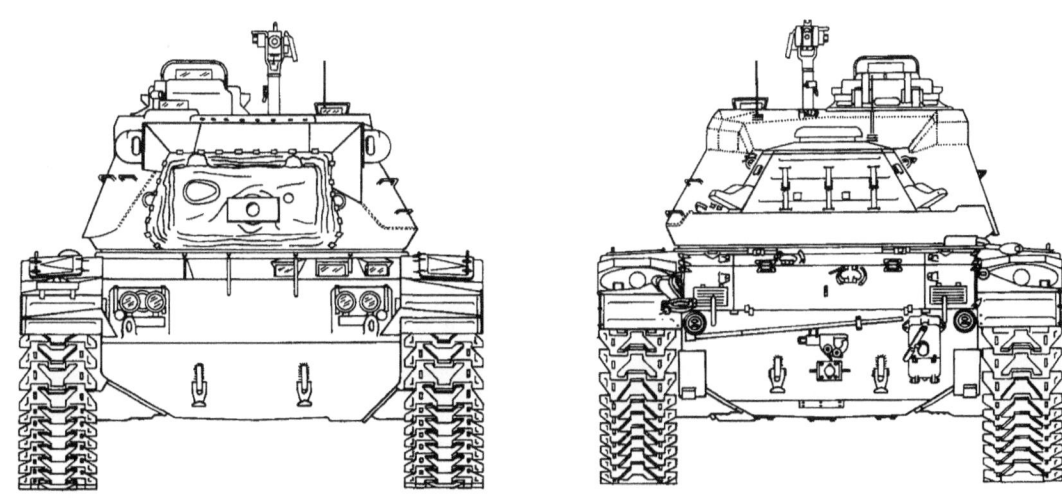

Scale 1:48

©D.P. Dyer

90mm Gun Tank T49

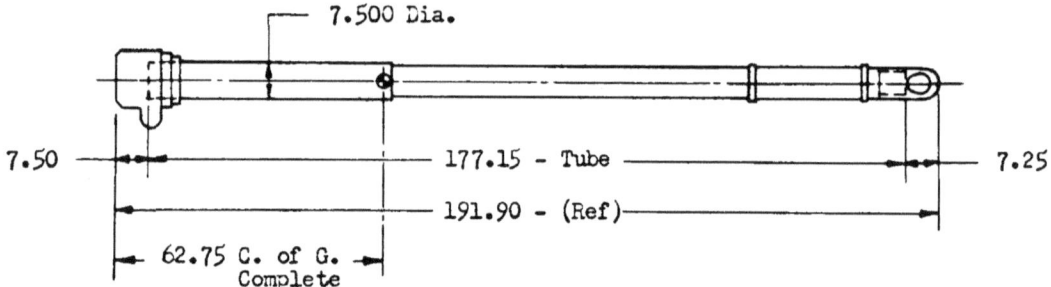

7.500 Dia.

7.50   177.15 - Tube   7.25

191.90 - (Ref)

62.75 C. of G.
Complete

The dimensions of the 90mm gun T132E3 are shown in the sketch above. The open breech of this weapon appears at the right.

The two T49 pilot tanks were shipped to Aberdeen for tests starting on 5 May 1954 and continuing to 10 May 1955. The tests indicated that the 90mm gun installation was satisfactory, but that its performance was limited by the ammunition. Since the HEAT projectile still had excessive dispersion, the HEP round had to be considered as the primary armor defeating ammunition. The amplidyne control system was satisfactory, but the range finder required modification to provide the correct superelevation and to retain the proper boresight. However, the user lost interest in the T49 since other armament options were being considered and the T41E1 itself was regarded as an interim light tank.

The end of the line for the T49. The photograph below was taken by Robert Lessels in the storage yard at Aberdeen Proving Ground during December 1984.

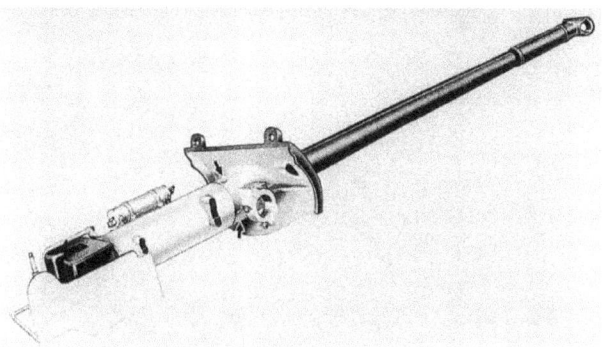

The Cadillac proposal for the M41 rearmed with the 90mm gun M41 appears in the artist's concept above. The new gun and mount are at the top right. The shield modifications required to accommodate the larger weapon are indicated at the right.

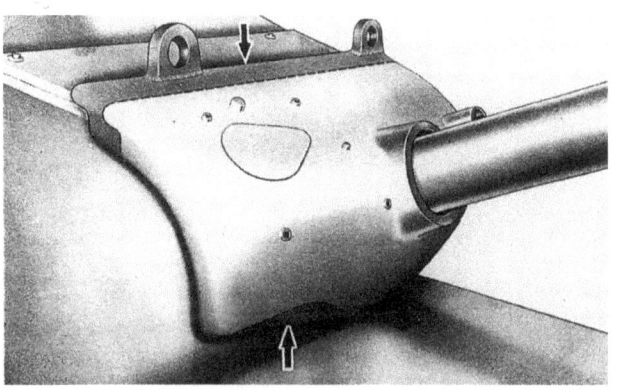

In a report dated 25 November 1958, the Cadillac Motor Car Division proposed rearming the M41A1 tank with the 90mm gun M41 from the M48 tank. The study concluded that this could be accomplished with an increase in combat weight of only about 1500 pounds. The modified tank would carry 44 rounds of 90mm ammunition with 22 in the right front hull, 13 in the turret rack, and nine in the ready rack. Since there was no provision for a range finder, it was suggested that the T156 direct sight telescope be replaced by a spotting rifle to increase the probability of first round hits.

By this time, other design studies were in progress to provide a superior light tank and there was no further interest in upgunning the M41A1, although the Ordnance Committee as early as June 1952 had considered the possibility of installing an unmanned turret on the M41 chassis. This turret was to be armed with a 105mm gun using an automatic loader and provision was to be made for remote stereoscopic vision for the crew in the tank hull.

The new hull stowage rack for 90mm ammunition is at the right and the 90mm turret ready rack appears below. At the bottom right is the new 90mm turret stowage rack.

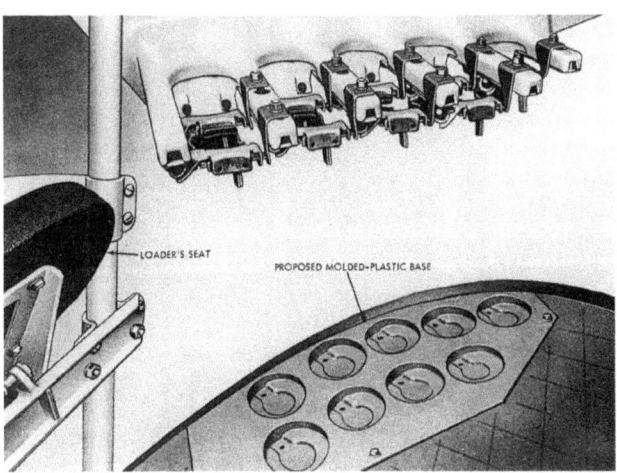

49

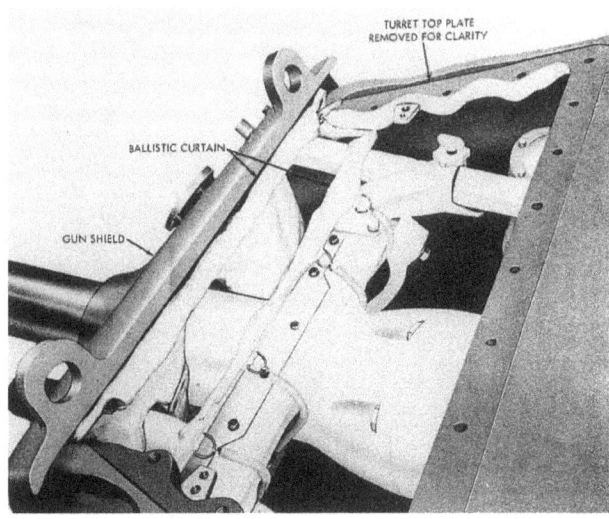

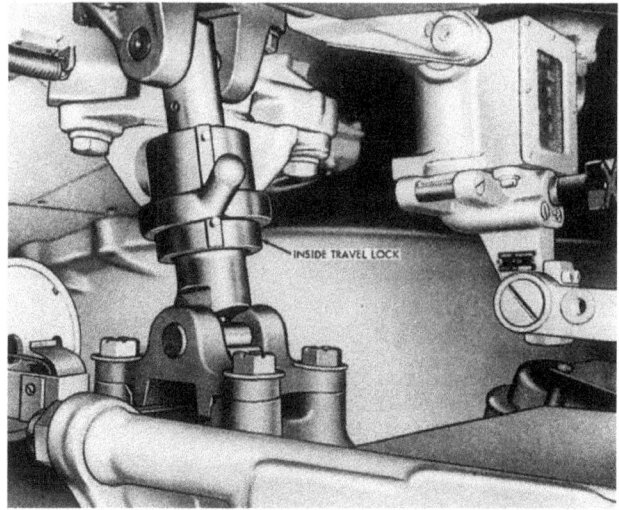

The ballistic curtain and the inside travel lock for the new 90mm gun installation are shown above at the left and right respectively. The modifications required to mount the 90mm gun are indicated in the view below at the left. Below at the right is the relocated turret traverse lock.

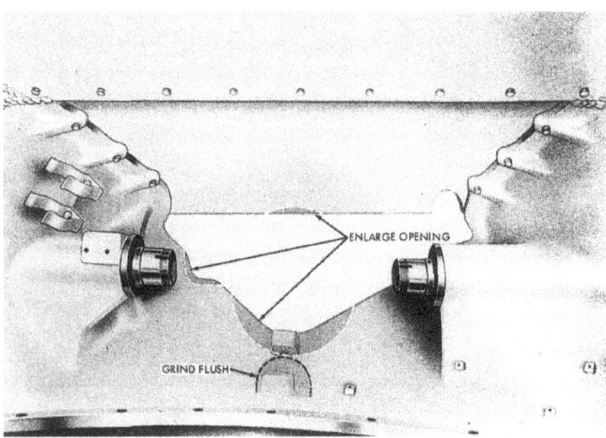

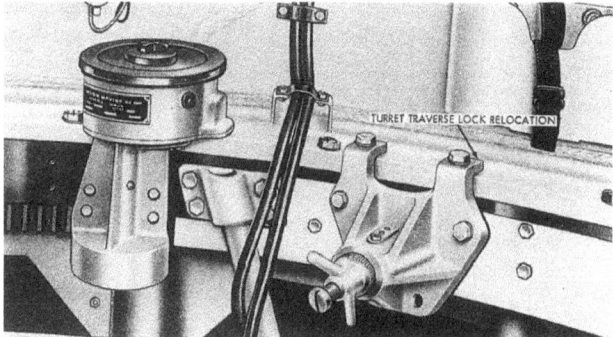

Below is the proposed installation of the spotting rifle. Note that the vehicle retains the M41A1 gun control system. At the right is a sketch of the new fire control linkage independent of the turret roof.

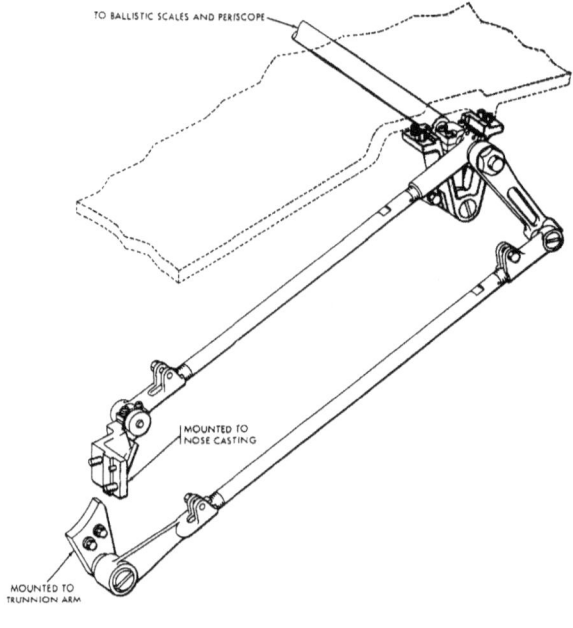

PART II

A FRESH APPROACH TO LIGHT TANK DESIGN

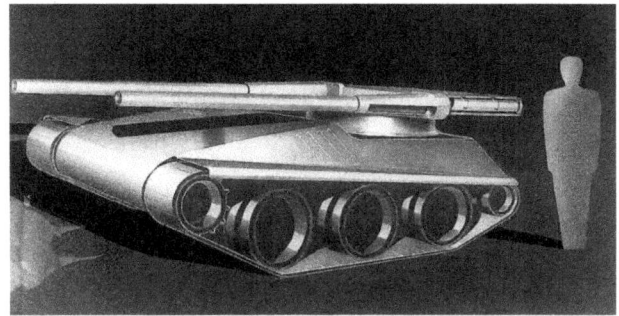

Above, a sectional view of the "primary" tank proposed by the Ballistic Research Laboratory is shown at the left. At the right is a concept drawing of an antitank vehicle armed with automatic loading recoilless rifles.

NEW LIGHT TANK DESIGN CONCEPTS

Although the light tank T41E1 entered production in 1951, it did not completely fulfill the requirements of the Army Field Forces and was regarded as an interim vehicle. In particular, the Army desired a lighter weight, less expensive light tank with a longer cruising range. During this time period, there was considerable discussion over the type of vehicle required to fill the role of the tank. An interesting analysis of the problem was presented by Mr. Floyd Hill of the Ballistic Research Laboratory during the Fourth Tank Vulnerability Conference at Aberdeen Proving Ground in late February 1952. In his presentation, Mr. Hill concluded that the primary role of the tank was to attack and defeat entrenched infantry and that the antitank role was secondary. He proposed that the single all purpose tank be replaced by two less expensive vehicles to handle the two separate tasks. A "primary" tank to perform the assault role would require heavy armor, but it could be armed with an automatic cannon firing high explosive shells. This weapon would be lighter in weight and smaller in caliber than one required for heavy armor penetration. The antitank vehicle would need armament capable of penetrating heavy armor, but it could be lightly armored depending upon high mobility for protection. This, of course, was reintroducing the concept of the highly mobile, lightly armored, tank destroyer that had been rejected as the result of experience during World War II. Mr. Hill also concluded that the Ontos, then under development, would be suitable for the antitank role. This was a lightweight vehicle armed with recoilless rifles. The development of the Ontos will be described in a later section.

In an effort to meet the Army's requirement for a new light tank, a number of design studies were initiated at Detroit Arsenal. These studies included a new light tank weighing about 17 tons with the engine mounted in the right front hull alongside the driver. With the turret installed at the rear, access to the vehicle could be provided through

Below is the mock-up of the proposed 17 ton light tank armed with the 76mm gun. This photograph, taken at Detroit Arsenal, was dated 17 January 1952.

The rear mounted turret on the 17 ton tank mock-up greatly reduced the gun overhang when the weapon was aimed forward.

the back wall of the hull. Armed with the 76mm gun, the 17 ton tank was to be manned by the usual crew of four with the tank commander, gunner, and loader in the turret. As mentioned above, the driver was located in the left front hull. The highly sloped front hull armor extended upward far enough to minimize the shot trap beneath the main gun mount. A mock-up of the new light tank was completed at Detroit Arsenal in January 1952 and this concept provided the basis for six design proposals presented at the Questionmark Conference in April. This was the first of what became a series of conferences between the designers and users to determine the course of future armored vehicle development. These meetings included representatives of the Army Field Forces, the technical design agencies, as well as private industry.

The six light tank proposals presented at the first Questionmark Conference were designated L-1 through L-6 with L-1 being the basic 17 ton design armed with the T91E3 76mm gun and armor equivalent to the 76mm gun tank T41E1. The power train consisted of the AO-536 gasoline engine and the XT-270 transmission. The engine developed 250 gross horsepower providing a power to weight ratio of 14.7 horsepower per ton. The vehicle had an estimated cruising range of 145 miles.

The front armor on the proposed 17 ton tank was extended upward in an effort to minimize the shot trap under the 76mm gun mount. Note the escape hatch for the crew in the rear hull wall.

Concept L-1 from the first Questionmark Conference is illustrated by the drawing and sectional view above. Below is a similar presentation of concept L-2 armed with the 90mm gun M139.

The L-2 proposal replaced the 76mm gun with the 90mm gun T139 using the same 73 inch inside diameter turret ring. Other components remained the same and the weight increased to 19 tons. The estimated cruising range was reduced to 130 miles.

Proposals L-3 and L-4 also were armed with the 90mm gun T139, but the inside diameter of the turret ring was enlarged to 85 inches. Both tanks were to be powered by the new 430 horsepower AX-660 engine using the XT-500 transmission. The two designs were identical except for the front armor which was equivalent to the 76mm gun tank T41E1 on the L-3 and double that on the L-4. This resulted in estimated weights of 22 tons and 25 tons for the L-3 and

L-4 respectively. The estimated cruising range of 150 miles for the L-3 was reduced to 140 miles for the L-4.

More powerful armament was proposed for the L-5 and L-6. This was the 105mm gun T140. On the L-5, the new weapon was mounted in the 85 inch diameter turret ring and the vehicle retained the same power train as the L-3 and L-4. The L-6 was fitted with a 100 inch inside diameter turret ring and powered by the 717 gross horsepower AX-1100 engine using the XT-500 transmission. The armor on both vehicles was equivalent to that on the 76mm gun tank T41E1. The respective weights of the L-5 and L-6 were 26 tons and 30 tons, but both had an estimated cruising range of 140 miles.

Concept L-3 appears in the drawings above. Below is the L-5 proposal armed with the 105mm gun T140.

The L-7 concept presented by Cadillac is shown in the drawings at the left. Note the oscillating turret and the rear mounted engine with the front drive on this design.

A seventh design concept was presented at the Questionmark Conference by the Cadillac Motor Car Division of General Motors Corporation. Designated as the L-7, it was armed with the 76mm gun T91E3 in an oscillating turret. It weighed 17 tons and, like most of the other concepts, its armor was equivalent to that on the 76mm gun tank T41E1. The AO-536 engine was mounted in the rear hull with a drive shaft extending beneath the turret to the XT-270 transmission in the front. The crew was reduced to three with the driver in the left front hull. The tank commander and gunner rode in the oscillating turret with an automatic loader for the 76mm gun.

The seven light tank concepts incorporated new components and design features, some of which were highly experimental. The six vehicles proposed by Detroit Arsenal all featured front mounted engines and final drives. This arrangement reduced the overall length of the tank with the gun in the forward position and increased the frontal protection for the crew. The front drive also reduced track throwing and the rear hull door provided easy access for the crew and simplified the loading of ammunition. A safety feature was the rear mounted external armored fuel tanks that could be jettisoned in an emergency. The trailing idler increased the track ground contact area reducing the ground pressure. The sponson air grills were less vulnerable to attack with napalm than the top deck grills, although they might be clogged by dirt thrown up by the tracks. Another possible problem with the front engine design was the effect of heat wave distortion on the optical fire control equipment. The front armor over the engine also complicated maintenance by reducing accessibility. A new arrangement was proposed for the turret mounted .50 caliber machine gun. The weapon was installed inside the turret with the barrel extending vertically through the turret roof. This barrel was fitted with a movable deflector which was expected to provide fire through an elevation range of −20 to +65 degrees. The weapon was traversed by rotating the turret.

The cooling arrangement for the front mounted engines on the Detroit Arsenal concept studies is illustrated below at the left. The proposed deflector on the vertical .50 caliber machine gun is sketched below at the right.

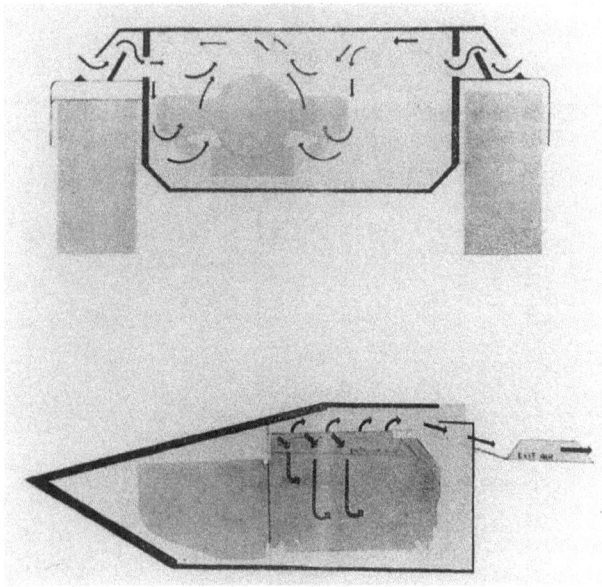

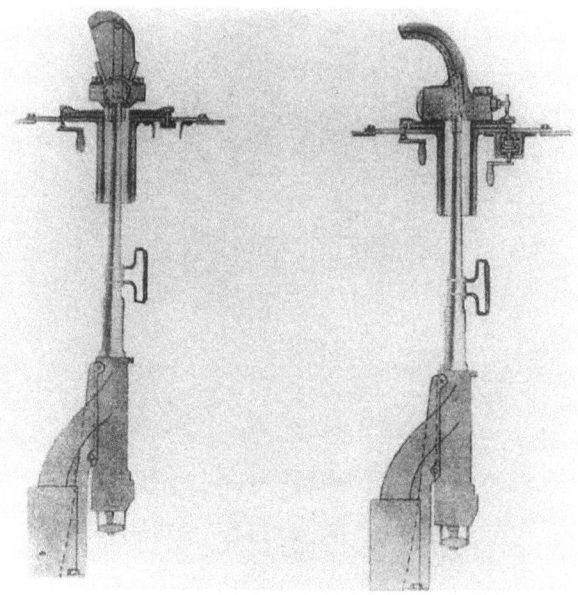

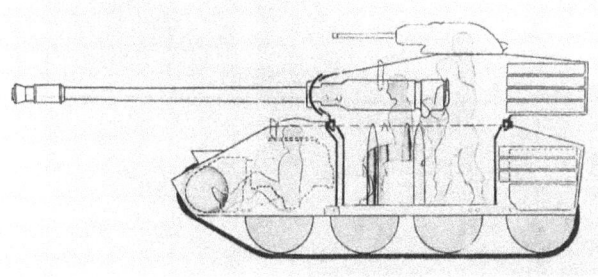

Above and at the left are a model and a sectional drawing of the TS-8 concept presented at the third Questionmark Conference. The TS-10 proposal is illustrated by the views below and at the lower left.

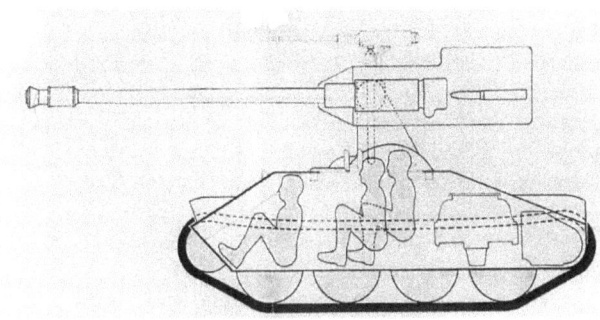

The Questionmark Conference provided a useful interchange of ideas between the designer and the user contributing to a better understanding of each others problems. Other meetings at Detroit Arsenal in February and at Fort Knox in July 1953 eventually resulted in development programs for the light tanks T71 and T92 which are described in detail in the following section. However, the value of the Questionmark studies was recognized and six additional concepts for future light tanks were presented at the third Questionmark Conference in June 1954. The second Questionmark Conference in September 1952 had been devoted to self-propelled artillery and antiaircraft weapons. The six design concepts proposed as replacements for the 76mm gun tank M41A1 were designated as the TS-8, TS-10, TS-26, TS-32, TL-3, and TL-8. The S and L in this nomenclature indicated short and long estimated development times of two and five years respectively.

As mentioned earlier, the limited range and excessive weight were considered to be the major shortcomings of the 76mm gun tank M41. As a result, all of the new designs emphasized increased range and light weight. A strong influence on the latter was the desire to increase the ease with which the vehicles could be transported by air. As a consequence of this policy, the armor protection on all six concepts was drastically reduced to a thickness of ½ inch. This armor was highly sloped in front, but it was vertical on the sides providing only limited protection. The TS-8, TS-10, and TS-32 were armed with a 76mm gun ballistically identical to that in the 76mm gun tank M41. On the TS-8 and the TS-32, the weapon was mounted in a conventional turret and stowage was provided for 55 and 54 rounds of 76mm ammunition respectively. The TS-10 featured a remote control gun in an overhead mount with an automatic loader and a 20 round magazine. A total of 60

A model of the TS-32 concept is at the left and a sectional drawing of the same vehicle is below.

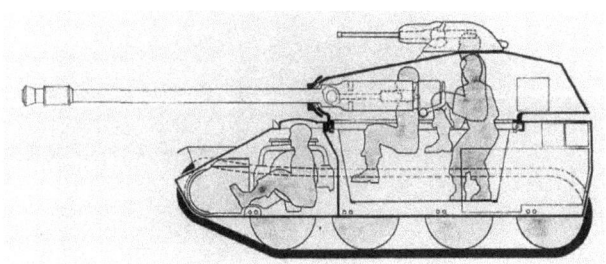

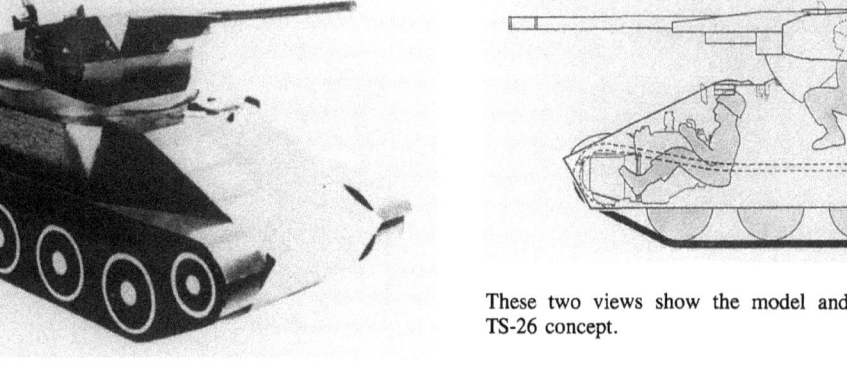

These two views show the model and sectional drawing of the TS-26 concept.

76mm rounds was carried, but the magazine had to be reloaded externally. The crew in the TS-10 was reduced to three compared to four in the TS-8 and the TS-32. The power plants proposed for the TS-8, TS-10, and TS-32 were the AOI-470, the AO-470, and the AOI-628 respectively. All three vehicles used the XT-270 transmission. The respective weights and cruising ranges for the TS-8, TS-10, and TS-32 were 16.5 tons and 128 miles, 15 tons and 160 miles, 17.5 tons and 171 miles.

The TS-26 differed from the other short development time proposals by being armed with the T189 105mm recoilless rifle in an oscillating turret. Stowage was provided for 40 rounds of 105mm ammunition. The vehicle also was lighter than the other short term concepts with an estimated

weight of 11 tons. Powered by a Chrysler VT-350 engine through an XT-90 transmission, it had an estimated cruising range of 170 miles.

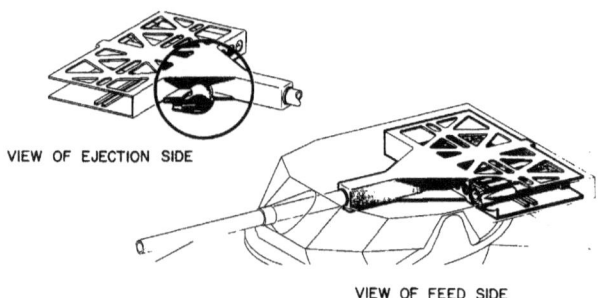

VIEW OF EJECTION SIDE

VIEW OF FEED SIDE

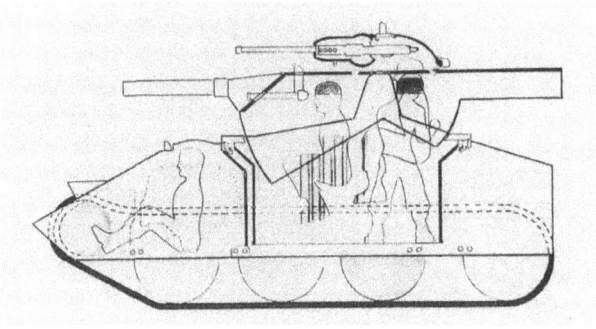

The model and sectional drawing of the TL-3 concept appears above and at the upper left. At the upper right is a sketch of the feed mechanism for the 105mm rocket assisted gun. Below and at the lower left are the model and sectional view of the TL-8 concept.

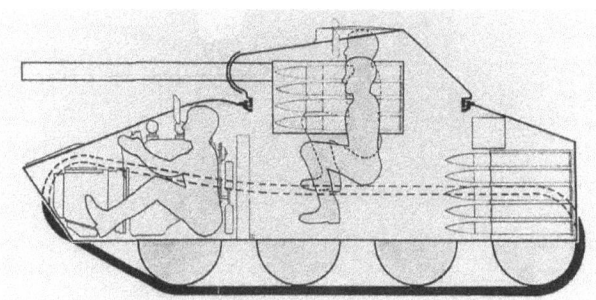

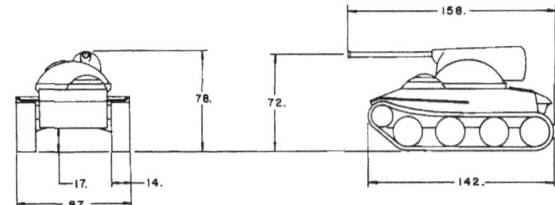

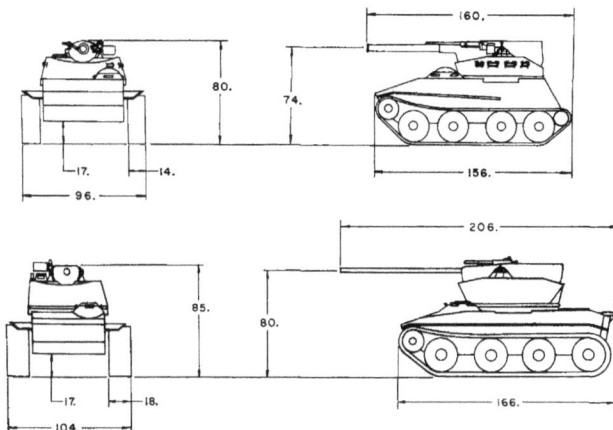

The dimensions of the Cadillac proposed TLAA (above), TLA (top right), and TLB (lower right) are shown in these sketches.

The first of the long term development proposals, the TL-3, weighed 16.5 tons and was armed with a 105mm rocket assisted gun in an oscillating turret. This weapon with a six round feed mechanism was mounted just beneath the turret roof permitting the tank to take the maximum advantage of a defilade position. The gun could fire single shots or a burst of six rounds. Total ammunition stowage was 52 105mm rounds. Manned by a crew of four, the TL-3 was to be powered by the AOI-470 engine with the XT-270 transmission. The estimated cruising range was 128 miles.

The TL-8 was the lightest of the six design concepts with an estimated weight of only eight tons. It was armed with the 105mm rocket assisted gun in a conventional turret and was fitted with an automatic loader reducing the crew to three men. Only 29 rounds of 105mm ammunition were carried and the tank commander transferred these rounds from the stowage rack to the automatic loader. Driven by a Chrysler V-8 engine with an XT-90 transmission, the TL-8 had an estimated cruising range of 200 miles.

All six of the proposed designs were equipped with a flat track suspension eliminating the track support rollers. Except for the TS-10, all of the vehicles had front mounted engines and final drives. On these tanks, the rear road wheels served as trailing idlers. On the TS-10, the power plant was installed in the rear hull with rear mounted drive sprockets.

Further studies of new light tank concepts continued both at Detroit Arsenal and in private industry. In April 1955, the Cadillac Motor Car Division completed a contract for the preliminary study of nine tank concepts weighing from six to 30 tons. These tanks were armed with guns ranging in caliber from 76mm to 105mm. In the Cadillac report, these tanks were assigned the designations TLAA, TLA, TLB, TLC, TLD, TME(c), TME(b), TMF, and TMG.

The TLAA and the TLA weighed a little over six tons and eight tons respectively. They were designed to be dropped by parachute and both were armed with a single 105mm recoilless rifle with a five round repeating magazine. The turret on both tanks was limited in traverse to 220 degrees. Two speed power traverse was included as well as manual traverse and elevation. The TLAA was manned by a crew of two with the driver in the hull and the tank commander-gunner in the turret. The crew of the TLA was increased to three with two in the turret and the driver in the hull. The maximum armor thickness on both tanks was ½ inch. Thus, like the Questionmark proposals, they would have been vulnerable to heavy automatic weapons such as the .50 caliber machine gun.

On the TLB, the weight was increased to over 12 tons with a maximum armor thickness of ¾ inches. The main armament consisted of a 105mm rocket assisted gun in a turret with 360 degree traverse. This weapon was equipped with a five round repeating magazine. The turret had two speed power traverse and manual controls for traverse and elevation. The tank commander and gunner rode in the turret and the driver remained in his usual position in the left front hull.

The TLC weighed less than 14 tons and was armed with a 76mm gun in an oscillating turret with full 360 degree traverse. Power traverse was provided with manual controls for both traverse and elevation. The maximum armor thickness was ¾ inches on both the TLC and the TLD. The weight of the latter was increased to 16 tons and it was armed with a 90mm gun in a fully rotating conventional turret. Power was supplied for both elevation and traverse. Provision was made for the installation of a short base range finder in the TLD turret. Both the TLC and TLD were manned by a crew of three with two in the turret and one in the hull.

Below are the dimensional sketches of the Cadillac TLC (left) and TLD (right) proposals.

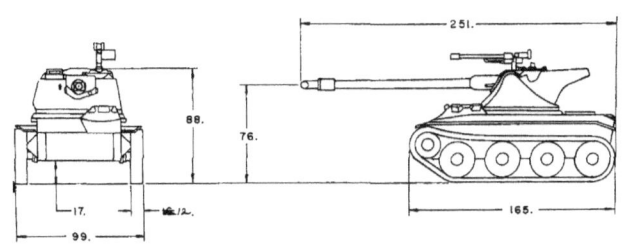

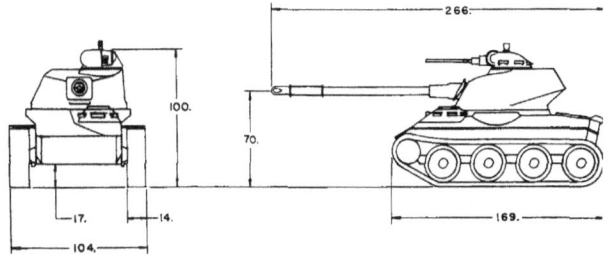

59

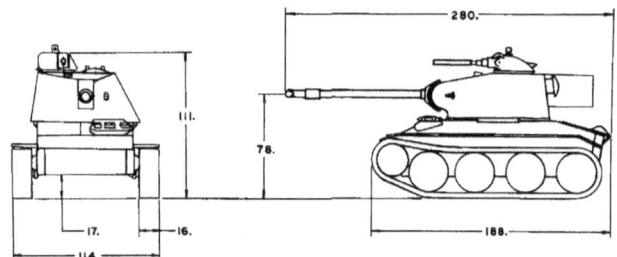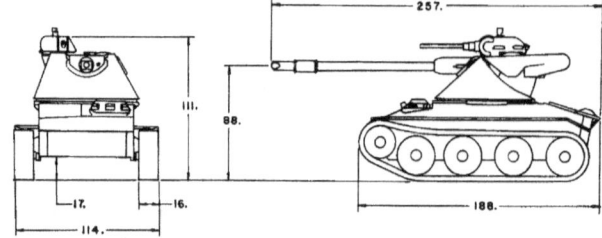

The dimensions for the TME(c) (left) and the TME(b) (right) concepts are shown above. The dimensions of the TMF can be seen below at the right with the TMG at the lower right.

The TME was proposed in two versions, the TME(c) and the TME(b), both of which weighed slightly under 20 tons. The (c) and (b) in the designations referred to a conventional or balancing turret respectively. The latter was the Cadillac terminology for an oscillating turret. The turret was the only major difference between the two vehicles. Both tanks carried a four man crew with three in the turret and one in the hull. The main armament was a 90mm gun and the maximum armor thickness was one inch. Power traverse and elevation were provided and the installation of a radar or Optar range finder was proposed. The latter determined the range by measuring the reflection time of a pulsed light beam from the target.

The TMF weighed 23½ tons and was armed with a rocket assisted 105mm gun in a conventional turret. This weapon was fitted with a five round repeating magazine. The maximum armor thickness was increased to 1¾ inches, the first of the proposed concepts to exceed the protection level of the 76mm gun tank M41A1. Power and manual gun and turret controls were provided in both elevation and traverse. An Optar range finder was proposed as part of the fire control system.

The heaviest of the Cadillac design concepts was the TMG with a weight of 30 tons. It was armed with a 105mm gun in an oscillating turret and provided with automatic ramming. The maximum armor protection was 1¾ inches as on the TMF. The oscillating turret was equipped with full power elevation and traverse as well as manual controls. The Optar range finder also was proposed for installation.

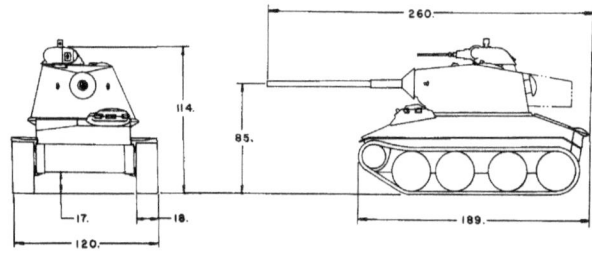

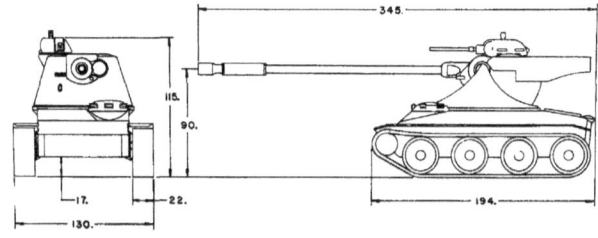

Models of the various Cadillac light tank proposals appear in these photographs. At the left, they are identified as follows: 1. TLAA, 2. TLA, 3. TLB, 4. TLC, 5. TLD, 6. TME(c), 7. TME(b) not shown, 8. TMF, and 9. TMG.

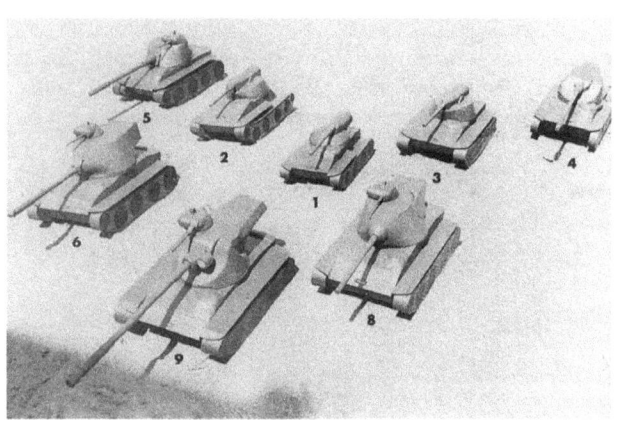

60

DESIGN CONCEPT	TLAA	TLA	TLB	TLC	TLD	TME(c)	TME(b)	TMF	TMG
CREW	2	3	3	3	3	4	4	4	4
WEIGHT (pounds)	12,483	16,650	24,646	27,570	32,000	39,948	39,486	47,000	60,000
GUN	105mm RECOILLESS	105mm RECOILLESS	105mm ROCKET	76mm	90mm	90mm	90mm	105mm ROCKET	105mm
AMMUNITION	25 rounds	30 rounds	46 rounds	60 rounds	50 rounds	52 rounds	58 rounds	60 rounds	49 rounds
ARMOR, HULL	Front 1/2 inches Sides 3/8 inches	Front 1/2 inches Sides 1/2 inches	Front 3/4 inches Sides 3/4 inches	Front 3/4 inches Sides 3/4 inches	Front 3/4 inches Sides 1/2 inches	Front 1 inch Sides 7/8 inches	Front 1 inch Sides 7/8 inches	Front 3/4 inches Sides 7/8 inches	Front 1 3/4 inches Sides 1 inch
TURRET	Front 1/2 inches Top 3/8 inches	Front 1/2 inches Top 3/8 inches	Front 3/4 inches Top 1/2 inches	Front 3/4 inches Top 1/2 inches	Front 3/4 inches Top 1/2 inches	Front 1 inch Top 1/2 inches	Front 1 inch Top 1/2 inches	Front 1 inch Top 1/2 inches	Front 1 3/4 inches Top 1/2 inches
ENGINE	GAS TURBINE 200 hp	AOI-346 165 hp	AOI-470 250 hp	AOI-470 250 hp	AOI-519 275 hp	AOI-692 374 hp	AOI-692 374 hp	AOI-865 469 hp	AOI-865 469 hp
TRANSMISSION	XT-90	XT-90	XT-90	XT-90	XT-90 mod.	XT-300	XT-300	XTG-400	XTG-400
TRACKS	14 inch band	14 inch band	18 inch band	12 inch steel	14 inch steel	16 inch steel	16 inch steel	18 inch steel	22 inch steel
CRUISING RANGE	259 miles	260 miles	255 miles	250 miles	272 miles	270 miles	270 miles	284 miles	260 miles
MAXIMUM SPEED	40 miles/hour	40 miles/hour	40 miles/hour	35 miles/hour	40 miles/hour	40 miles/hour	40 miles/hour	40 miles/hour	35 miles/hour

The characteristics of the various Cadillac light tank proposals are listed in the table above. At the right is a sectional drawing of the (UC)R7 concept presented at the fourth Questionmark Conference.

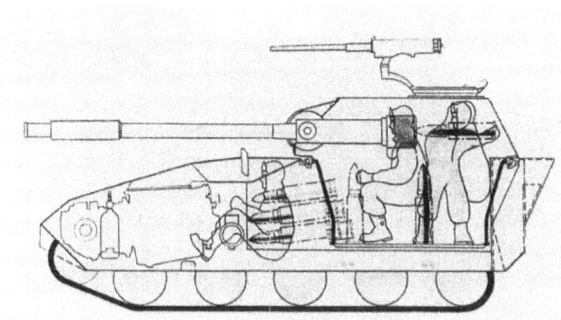

The fourth Questionmark Conference in August 1955 proposed a single design concept as a replacement for the 76mm gun tank M41A1. Designated as the 105mm gun tank (UC) R7, it was a ten ton vehicle armed with a low velocity 105mm gun in a conventional turret. This weapon depended upon a fin stabilized shaped charge (HEAT) round to defeat enemy armor. Powered by a front mounted AOI-470-1 engine with an XT-90 transmission, the R7 had an estimated cruising range of 150 miles. Like many of the earlier proposals, the armor protection was reduced, consisting of ½ inch plate at 60 degrees from the vertical on the front. The driver rode in the left front hull and the tank commander, gunner, and loader were in the rear mounted turret. Thirty-three rounds of ammunition were carried for the 105mm gun. The overall length of the tank including the gun in the forward position was 199 inches.

Although none of these concepts were selected for full development, they served to illustrate the wide variety of design features under considerstion. Some of these features were eventually incorporated into the new development programs.

At the upper right is a mock-up of the original AO-470-1 engine without fuel injection. At the lower right is the XT-90 transmission. A sketch of the early Optar range finder appears below.

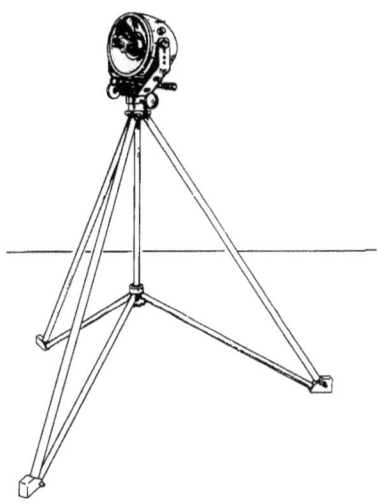

61

Above is a model of the proposed new light tank armed with a 90mm gun. This photograph from Detroit Arsenal was dated 11 February 1953.

76mm GUN TANKS T71 AND T92

In May 1952, Ordnance Committee action outlined the characteristics for a new light tank intended as a replacement for the 76mm gun tank T41E1. Initially, a maximum weight of 20 tons was specified, but this was later reduced to 18 tons. The main armament was to be a 90mm gun. As mentioned previously, numerous design concepts were studied at Detroit Arsenal and subsequently, the requirement for the main armament was changed to a 76mm gun with a quick change tube. In July 1953, a conference at Fort Knox selected three concepts for further evaluation as the new 76mm gun tank T71. The first of these design concepts, proposed by Detroit Arsenal, was an 18 ton tank armed with the 76mm gun in an oscillating turret. Stowage was provided for 60 rounds of 76mm ammunition with 18 available as ready rounds. The main weapon was fitted with an automatic loader reducing the crew to three men. The tank commander and gunner rode in the turret with the driver in the

center front hull. The armor protection was equal to that on the 76mm gun tank T41E1. The vehicle was powered by the AOI-628 engine installed transversely in the rear hull. This engine developed 340 gross horsepower and was fitted to an XT-270 transmission with a concentric planetary final drive and sprockets at the rear of the hull. A torsion bar suspension supported the tank on four road wheels per track.

These photographs show the model of the Detroit Arsenal proposed version of the 76mm gun tank T71. They were dated 3 June 1953.

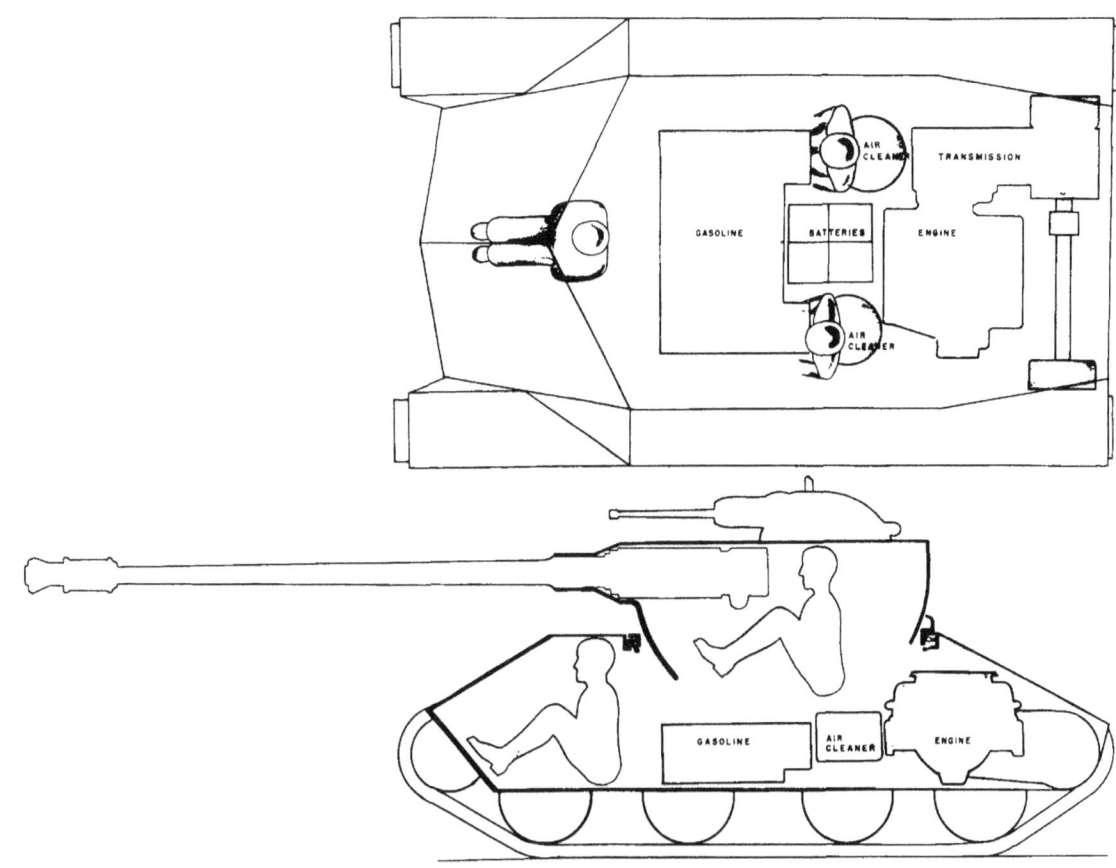

The internal arrangement of the Detroit Arsenal proposal for the 76mm gun tank T71 can be seen in the sectional drawings above. The stowage of the 76mm ammunition in this vehicle is shown below at the left.

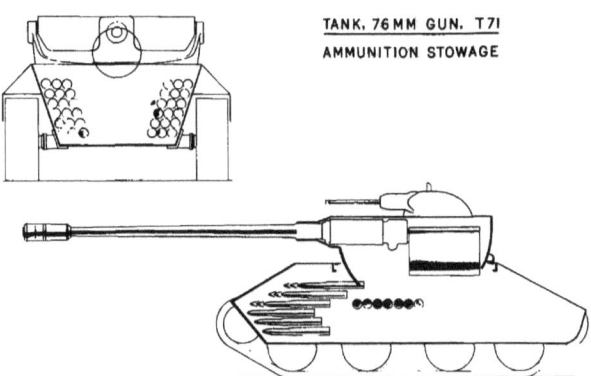

TANK, 76 MM GUN, T71
AMMUNITION STOWAGE

An idler was mounted at the front of each track and the upper track run was carried on three support rollers.

The second concept of the 76mm gun tank T71 was proposed by the Cadillac Motor Car Division of General Motors Corporation. It weighed a little under 18 tons and the 76mm gun T185 was mounted in a conventional turret. This weapon was the same as the 76mm gun M32, except that it was fitted with a quick change tube. The tank was manned by a crew of four with the tank commander, gunner, and loader in the turret. The driver was located in the left front hull. Armor protection was essentially equivalent to that on the 76mm gun tank T41E1. An AOI-628-1 engine powered the vehicle through an XT-300 transmission. The front mounted sprockets drove the 14 inch wide, single pin, center guide tracks. A torsion bar suspension supported the vehicle on four large road wheels per side, the rear pair of which served as trailing idlers. This flat track suspension did not require support rollers for the upper track run. With a fuel capacity of 150 gallons, the estimated cruising range of the tank was 165 miles.

At the left is a front view of the Detroit Arsenal T71 model. Note the double slope on the front armor.

63

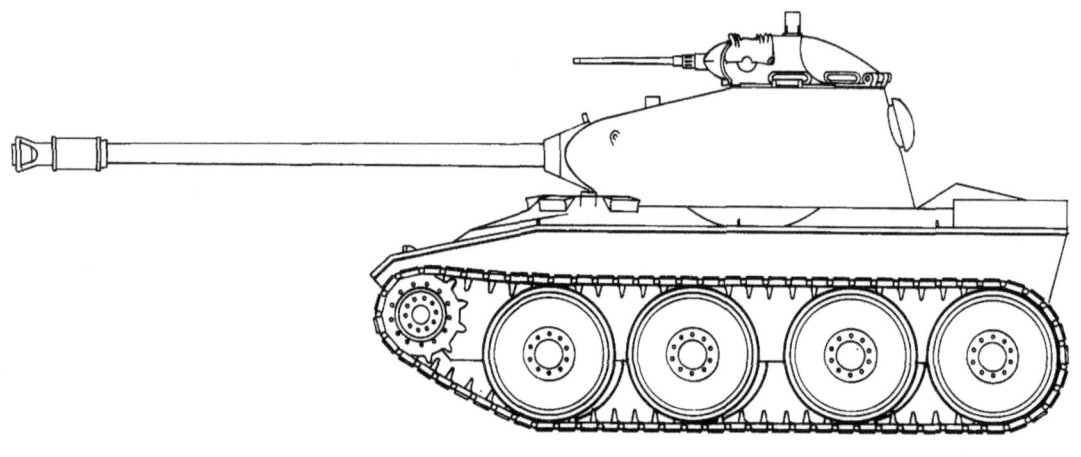

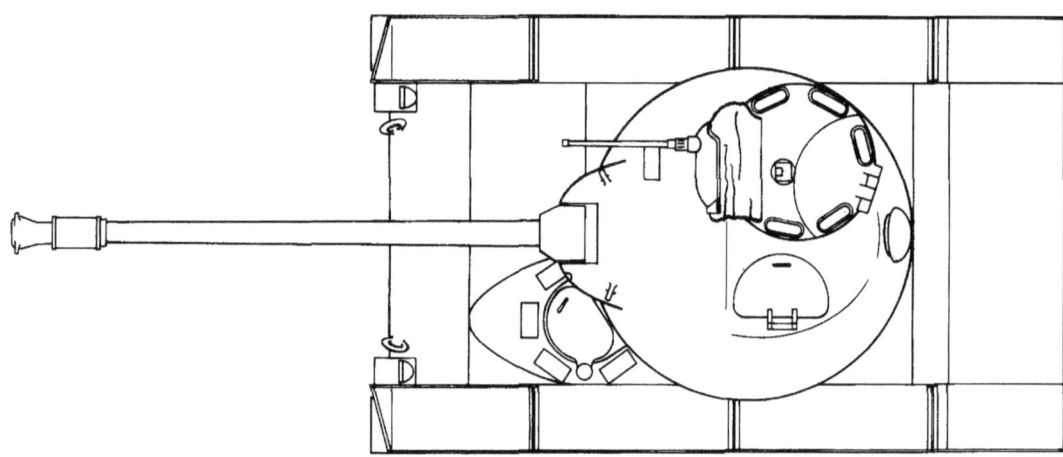

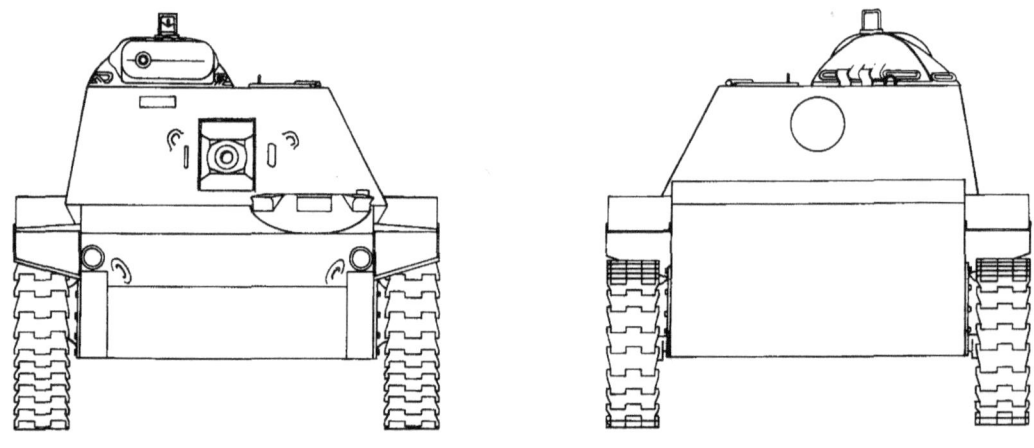

Scale 1:48

©D.P. Dyer

76mm Gun Tank T71

The early mock-up of the 76mm gun tank T92 appears above on 8 October 1954. No track support rollers were included in the design at this stage.

The third light tank concept under consideration was presented by Aircraft Armaments, Incorporated (AAI). Weighing a little over 18 tons, it differed considerably from the other proposals. The 76mm gun was installed in a cleft turret reducing the height of the tank and the tank commander and gunner each rode under a cupola armed with a .50 caliber machine gun. A .30 caliber coaxial machine gun was installed above the 76mm gun. The loader was located in the left rear of the turret where he inserted the 76mm rounds into a semiautomatic loading device. The driver was seated in the left front hull alongside the engine. The power pack consisted of the transverse mounted AOI-628-1 engine with the XT-300 transmission. The front mounted sprockets drove 16 inch wide, band type, tracks on a torsilastic suspension. This suspension consisted of four road wheels per side with the rear road wheels acting as trailing idlers. In the original proposal, no support rollers were installed.

1 DRIVER
2 COMMANDER - GUNNER
3 LOADER
4 CUPOLA - CAL. .50
5 76 MM GUN
6 76 MM READY RACK
7 76 MM AMMUNITION STOWAGE
8 FUEL TANKS
9 BATTERIES
10 BLOWER
11 CONTROL PANEL
12 COOLING AIR INTAKE GRILL
13 PERISCOPES
14 GUN SIGHTS
15 VISION RINGS
16 COAXIAL CAL. .30 M.G.
17 RADIO
18 CAL. .30 AMMUNITION COAXIAL M.G.
19 CAL. .50 AMMUNITION CUPOLA GUN
20 TURRET CONTROLS
21 TORSILASTIC SUSPENSION
22 ENGINE ACCESS HATCH
23 FENDER BOX
24 CUPOLA HATCH
25 BRASS EJECTION TRAP
26 CO₂ BOTTLES

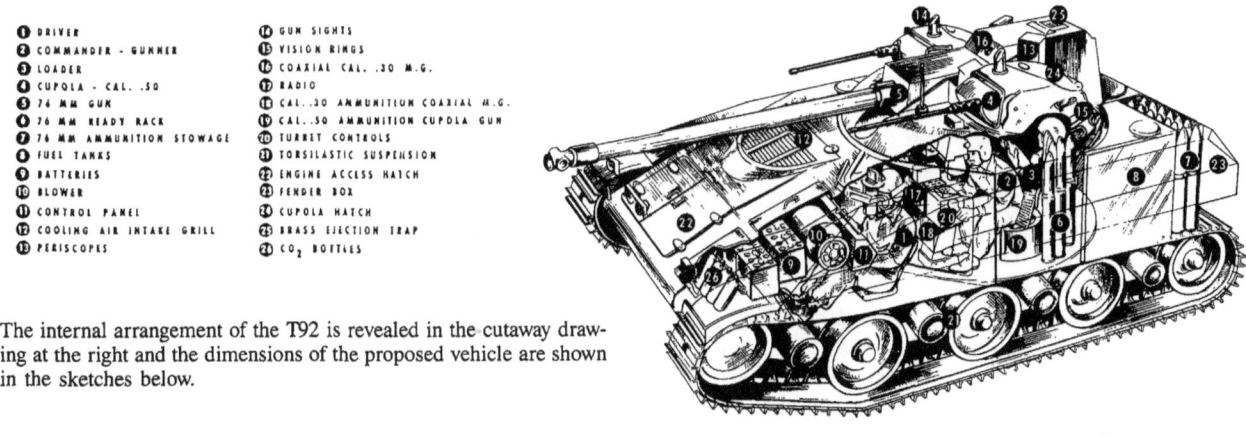

The internal arrangement of the T92 is revealed in the cutaway drawing at the right and the dimensions of the proposed vehicle are shown in the sketches below.

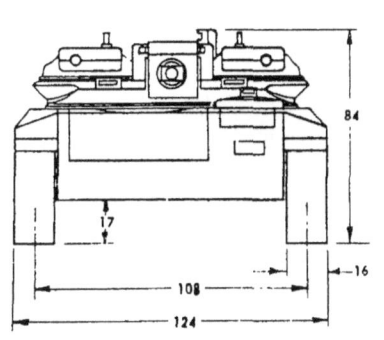

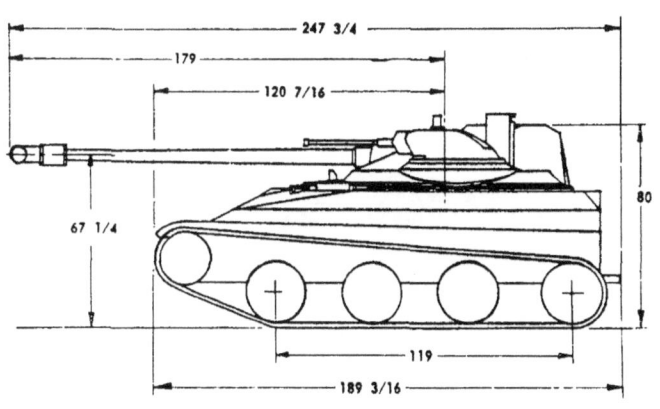

The late mock-up of the 76mm gun tank T92 at the top right now includes two track support rollers on each side. The other photographs show the T92 pilot after completion. Note the hatch arrangement for the crew in the closed (upper) and open (lower) views at the right.

After review of the three concepts, contracts were awarded to the Cadillac Motor Car Division and to Aircraft Armaments, Incorporated to further develop their designs and to construct full scale mock-ups. A mock-up also was completed for the Detroit Arsenal concept. Originally, it had been expected that the final version of the T71 would incorporate features of all three design concepts. However, this was not practical because of the tightly integrated arrangement of the various features and components. The Cadillac design was considered to be a conservative approach following conventional practice which offered a low weight and a long cruising range at minimum risk. The Aircraft Armaments tank was a highly innovative design which had the potential for greatly improved performance, but included new features and components entailing a higher risk development program. In late July 1953, the Chief of the Army Field Forces and the Assistant Chief of Staff G-4 authorized the development of both the Cadillac and Aircraft Armaments designs. Pilots would be built of both tanks and tests would determine the final selection. The Ordnance Technical Committee approved these projects in March 1954 and in May the Cadillac version was designated as the 76mm

gun tank T71. Because of the radical difference in design, the Aircraft Armaments vehicle became the 76mm gun tank T92. Construction was authorized for two pilot tanks of each design. Development work continued on both tanks until January 1956. At that time, the T71 project was cancelled prior to completion of the first pilot. Two factors resulted in the termination of the T71 program. The first was the rapid progress on the T92 and the second was the limited availability of funds at that time.

On 18 June 1954, Aircraft Armaments, Incorporated had received official notification to proceed with the construction of the two T92 pilots. On 5 November 1954 and 27 January 1955, meetings of the Mock-up Board were held which resulted in recommendations for numerous design changes. Some of these were incorporated in the fabrication of the two pilot tanks. The most obvious modification was the addition of two track support rollers (referred to in Aircraft Armaments documents as idlers) on each side of the torsilastic suspension. Also, the .50 caliber machine gun in the gunner's cupola was replaced by a .30 caliber weapon. However, both cupolas were designed so that either weapon could be installed.

The original band type tracks are clearly visible on the T92 pilot shown on this page. Also, note that the gunner's .50 caliber machine gun on the mock-up has been replaced by a .30 caliber weapon.

The first pilot 76mm gun tank T92, registration number 9B1281, was delivered to Aberdeen Proving Ground on 2 November 1956. Due to a subcontractor machining error, the vehicle arrived without the cupolas for the gunner and the tank commander. Weights were added to the tank to bring it up to combat weight for the initial automotive tests before the new cupolas arrived. The second pilot tank, registration number 9B1282, arrived at Aberdeen on 22 July 1957 and was used for the evaluation of the fighting compartment. At that time, it was expected that the T92 would be ready for full scale production by mid 1962.

The hull of the T92 was a welded assembly of armor steel castings and plates. The protection level was essentially the same as on the 76mm gun tank M41, but a lower weight was obtained through the use of high obliquity surfaces and some lighter weight materials. Aluminum alloy doors were installed on the compartments for the power plant, batteries, and auxiliary generator. The fenders were assembled from an aluminum alloy and fiberglass reinforced plastic. The engine compartment in the right front hull was enclosed by welded steel firewalls. The driver in the left front hull was behind the compartments for the auxiliary

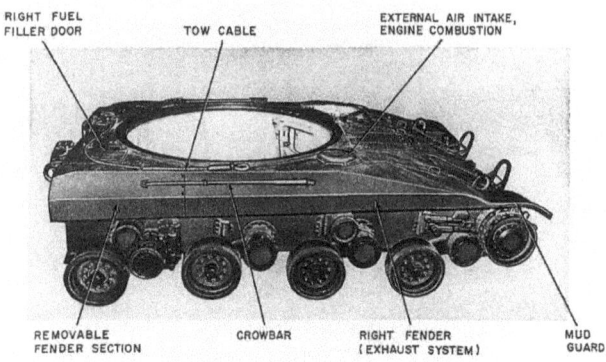

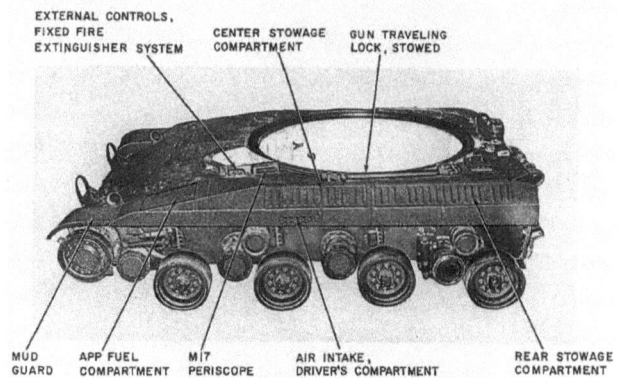

Details of the T92 hull and suspension can be seen above and below. The engine exhaust is concealed beneath the right fender.

power plant (auxiliary generator) and batteries. Two control handles were used to steer and brake the vehicle. Protected vision for the driver was provided by four M17 periscopes around his hatch and the hatch cover was fitted for the installation of an M19 infrared periscope. An escape hatch also was located in the floor of the driving compartment. The transverse mounted AOI-628-1 engine in the right front hull drove the vehicle through the Allison XT-300 transmission which provided six speeds forward and two reverse. The complete power pack was installed and removed as a single unit. The engine combustion air intake was through a mushroom ventilator in the hull roof at the right front of the turret. The intake for the engine cooling air was the grill in the top of the engine compartment. The exhaust passed under the right fender to a grill at the rear of the tank. The fuel supply was carried in two 75 gallon bladder type fuel cells located at the rear of the hull just forward of the main 76mm ammunition stowage compartment. The final drive and sprocket wheel assemblies at the front of the tank drove the T110, band type, tracks. These, steel cable reinforced, rubber tracks were 16 inches wide and consisted of nine sections with a total length of 390.25

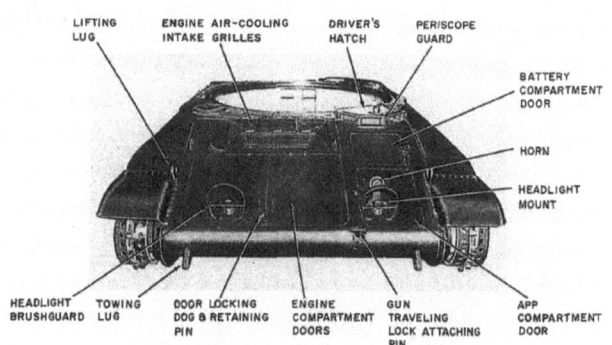

inches. Two spare sections of the track were stowed on the rear of the turret cradle housing. The tank rode on four road wheels per side, each of which was suspended by a torsilastic unit. The rear road wheels served as adjustable trailing idlers. In addition to the cupola and driver's hatches, access to the vehicle was provided through two armor doors at the rear of the hull. Each of these doors was fitted with a vision block.

Two views at slightly different angles looking into the T92 driver's compartment are shown below.

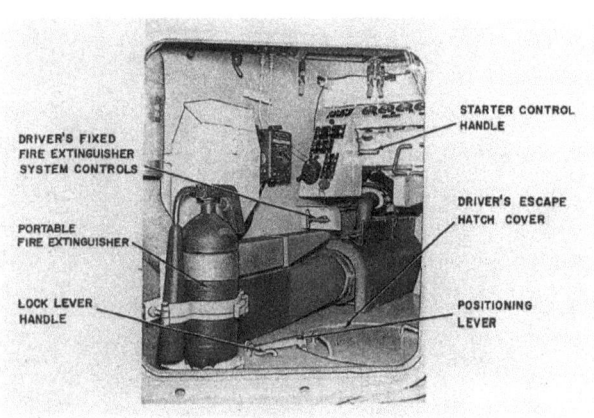

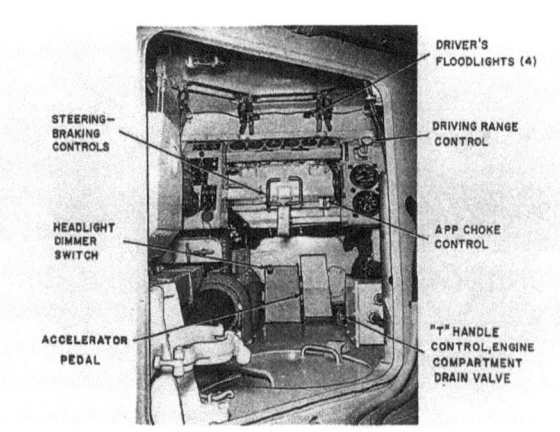

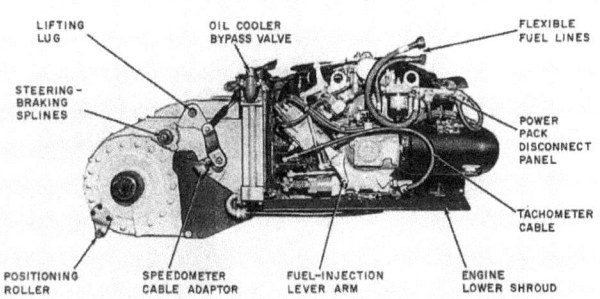

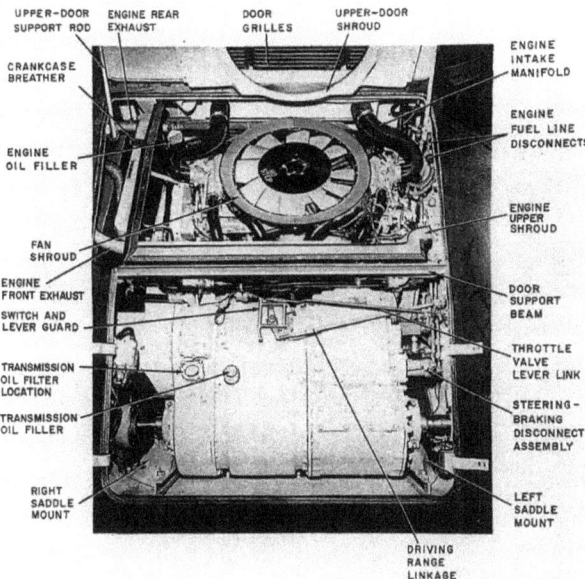

The dismounted power pack appears above. At the right, it is installed in the front hull. Details of the rear hull can be seen below. At the lower right is a sketch of the T92 engine exhaust system.

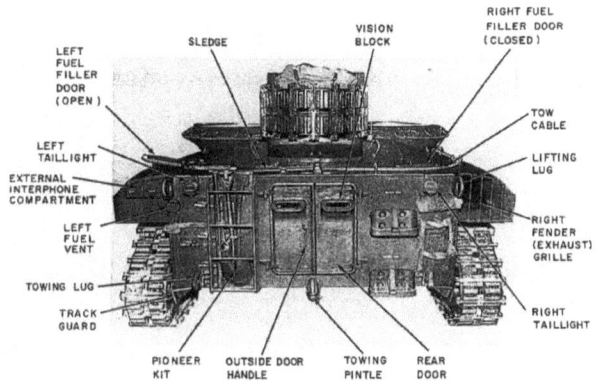

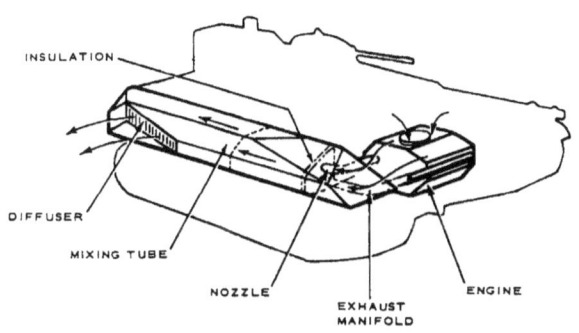

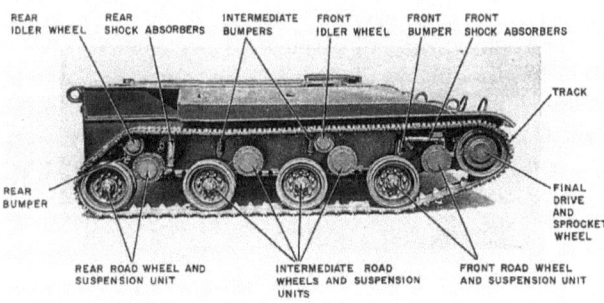

Above, details of the T92 suspension are shown at the left and a section of the original band type track is at the right. Below are inside (left) and outside (right) views of the driver's hatch and periscope installation.

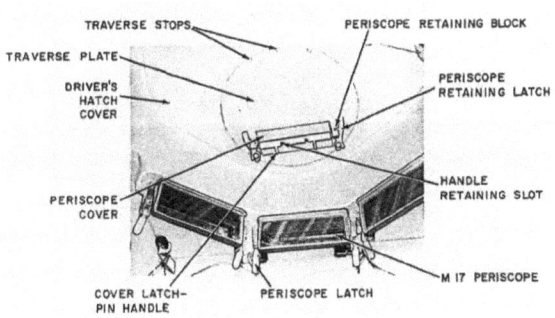

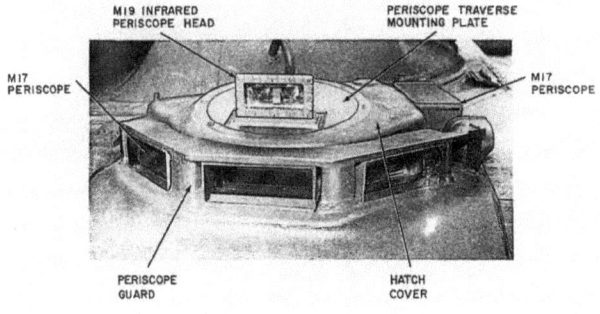

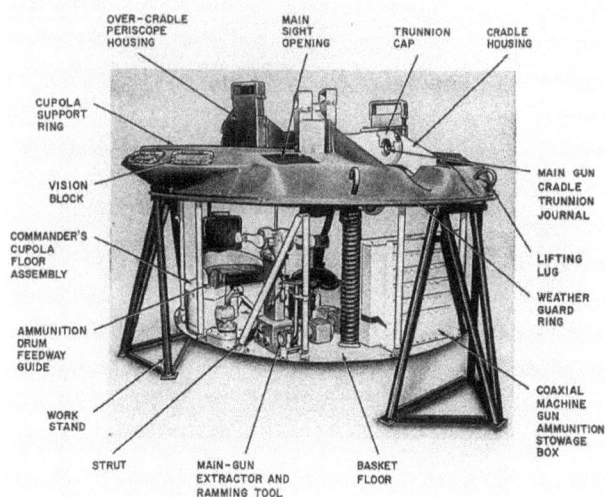

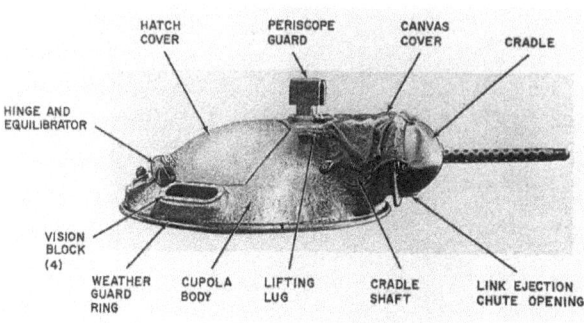

At the left, the T92 turret is removed from the vehicle. The outside and inside of the gunner's cupola can be seen above and below respectively.

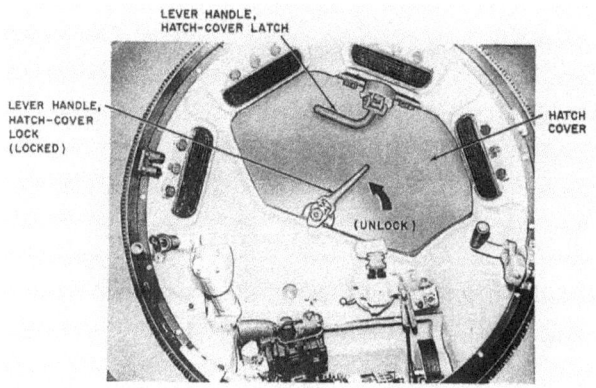

An unusual feature of the T92 was the large cleft turret with an 89 inch inside diameter ring. This turret mounted the 76mm gun T185E1 in a cradle housing between two armored cupolas. The T185E1 was ballistically identical to the 76mm gun M32, but it had a quick change tube and was mounted upside down to accommodate the semiautomatic loader. The .30 caliber coaxial machine gun was installed above the cannon in the cradle. The cupola on the right side was occupied by the tank commander and was armed with the .50 caliber machine gun. The gunner was located in the left cupola armed with the .30 caliber machine gun which replaced the .50 caliber weapon in the original design. A platform was suspended from each cupola upon which was mounted an ammunition drum for the machine gun and a seat for the tank commander or gunner. Each cupola was fitted with four vision blocks and surrounded by six additional vision blocks in the turret. Both the tank commander and the gunner were provided with two periscopes for vision over the main armament cradle housing. In both cupolas, the machine gun was aimed using a periscopic sight in the cupola roof. Since either the gunner or the tank commander could aim and fire the main weapon, each had a modified M16 periscopic sight in the turret directly in front of the cupola. A coaxial telescope also was

included for the gunner. Power and manual traverse and elevation were provided for the gunner and the tank commander and the latter could override the gunner's controls. Each cupola could be traversed 194 degrees, 10 degrees inboard forward and 4 degrees inboard aft. Normally the cupolas were traversed manually, but they also could be brought rapidly into alignment with the main armament using the hydraulic slewing motors. The cupola machine guns were elevated manually through a range of +60 to −10 degrees. The loader, from his position in the left rear of the turret, had access to the main ammunition stowage compartment in the rear hull containing 28 76mm rounds. As required,

Below, the 76mm gun mount is at the left and the .30 caliber coaxial machine gun installation is at the right. Note the inverted position of the 76mm gun.

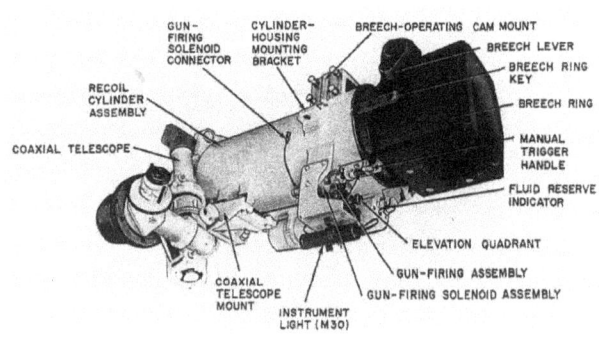

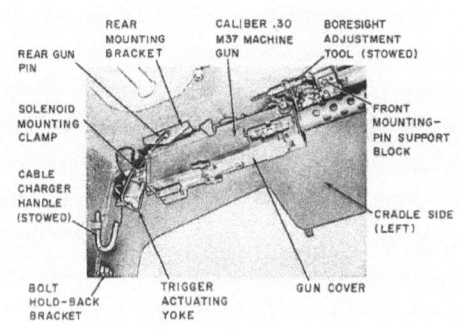

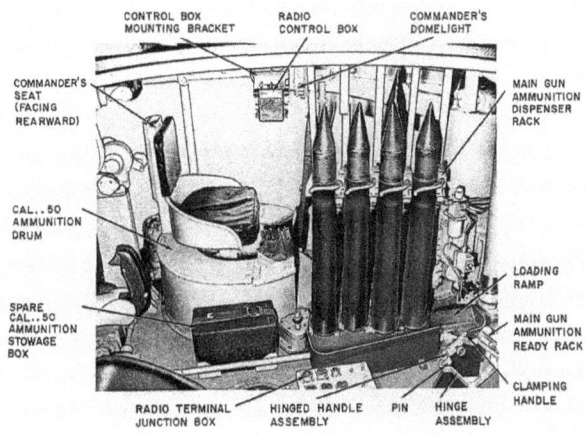

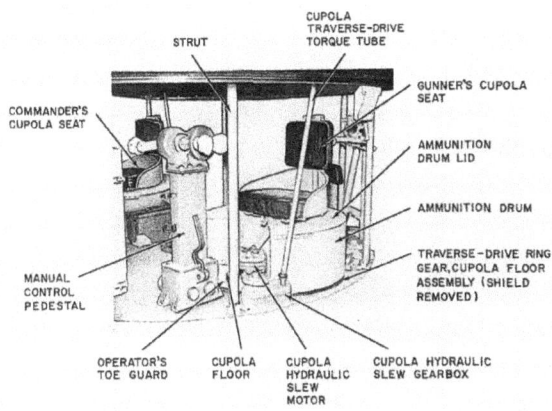

he transferred ammunition to the 24 round ammunition dispenser rack, the seven round ready rack, and to the semiautomatic loader. With one round in the semiautomatic loader, the total 76mm ammunition stowage was 60 rounds. Another new feature of the T92 was the automatic ejection from the tank of the empty 76mm and machine gun cartridge cases after firing.

Above, the commander's seat and the gunner's seat are shown at the left and right respectively. The loader's seat behind the gunner's station can be seen below.

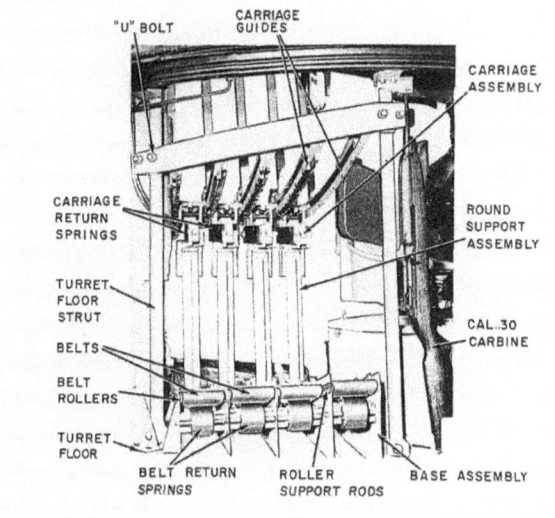

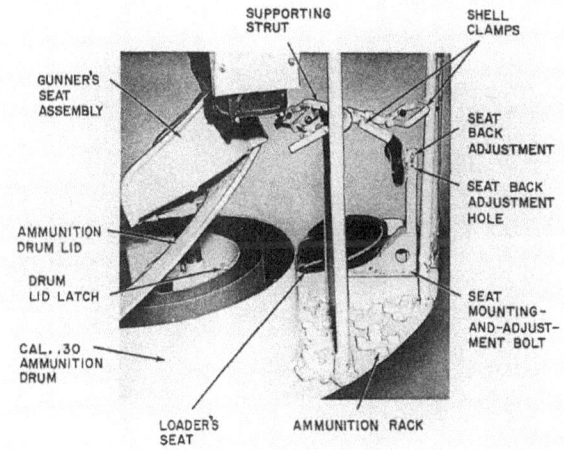

The 76mm ammunition rack appears at the left. Below, a top view of the turret and gun cradle housing is at the left and the semiautomatic loader is at the right.

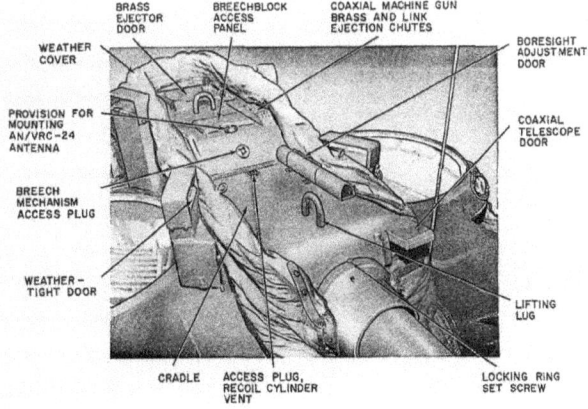

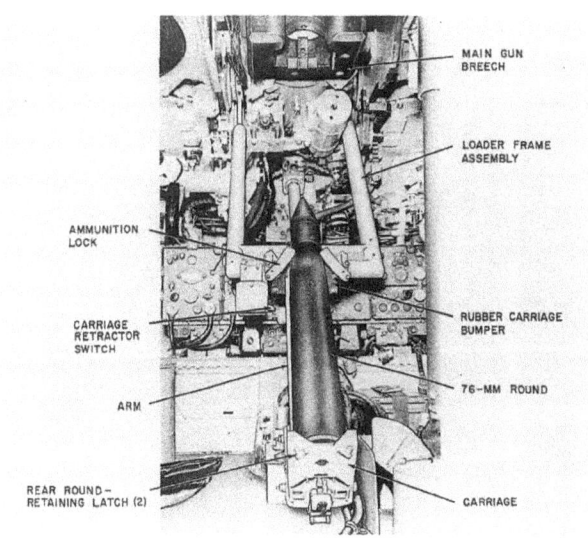

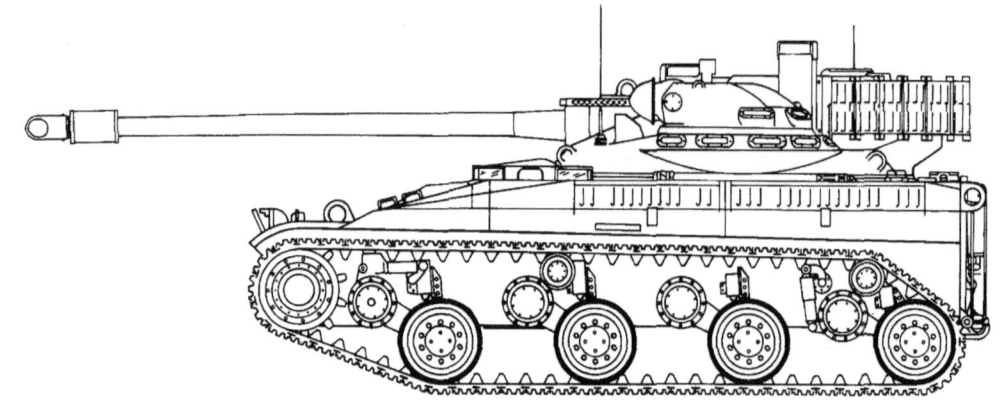

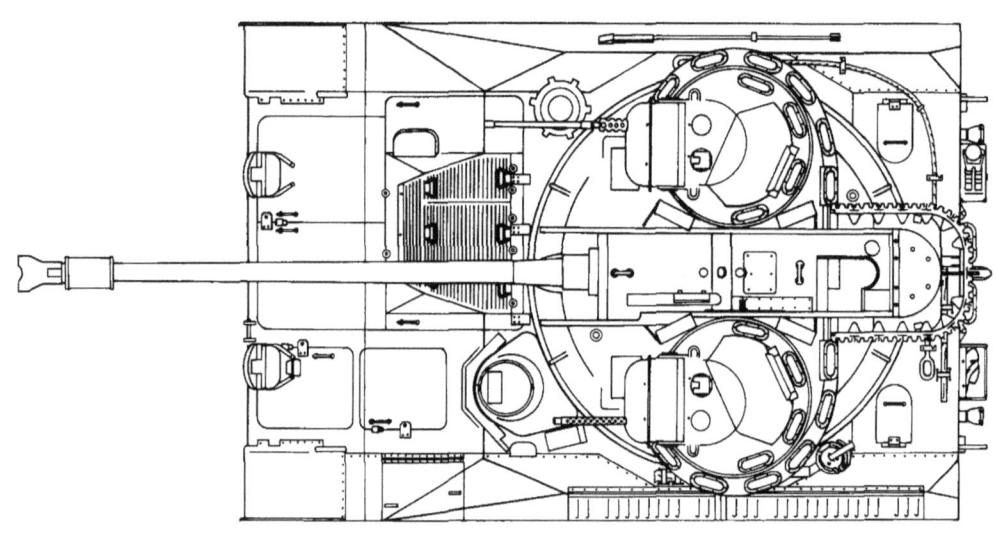

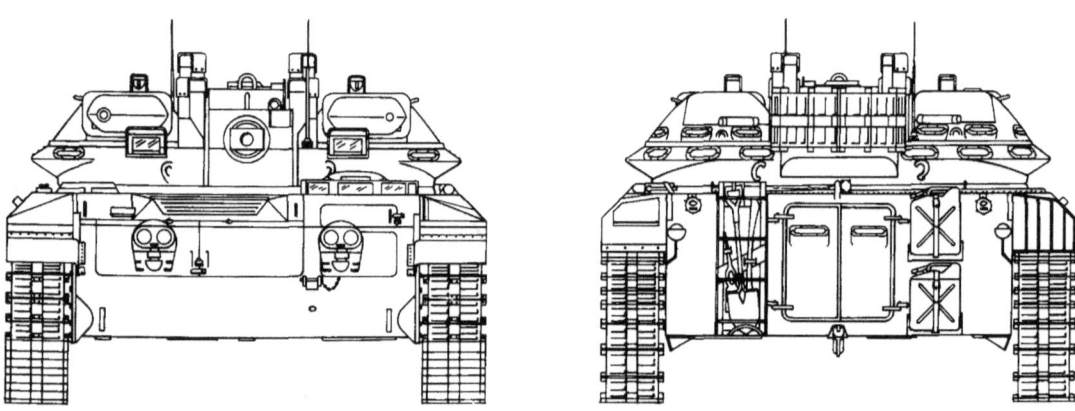

76mm Gun Tank T92

The test program at Aberdeen revealed the need for a number of modifications. The band type track broke frequently and after 202 hours of operation, it was replaced by the 14 inch wide T85E1 double pin track normally used on the light tank M24. Development was initiated on an improved, wire mesh, band type track. Because of a tendency to throw the track, the suspension was modified to include a compensating idler mounted at the rear. Other changes included a stronger final drive assembly, relocation of the personnel heater to the aft right side of the engine compartment, incorporation of an inflatable seal to protect the turret ring bearing, redesign of the 76mm ammunition dispenser to prevent the entrance of dirt into the roller bearings, and the installation of a stronger actuator spring into the 76mm cartridge case ejector. A total of about 50 modifications were proposed.

Below, the first pilot 76mm gun tank T92 is at Aberdeen Proving Ground on 3 May 1957 after installation of the cupolas. At the right is T92 pilot number 2 after arrival at Aberdeen on 22 July 1957.

These photographs show the 76mm gun tank T92 after the suspension was modified to include a rear mounted compensating idler. The vehicle also is fitted with the 14 inch wide T85E1 tracks.

The low silhouette of the T92 compared to the M41 is obvious in the front and side views of the two 76mm gun tanks. The T92 is fitted with the T85E1 tracks, but the compensating idler has not yet been installed.

Funds were allocated during 1957 for the construction of two additional T92 pilots incorporating the changes indicated by the test program. Unlike the first two tanks, numbers three and four were to be ballistic pilots with the full protection specified in the design. Delivery of these two vehicles was expected in mid 1958. However, due to changed circumstances, both were cancelled prior to completion.

In early 1957, a Congressional committee investigating military affairs noted that intelligence reports indicated that the Soviet Union was equipping its forces with an amphibian light tank. Later, this was revealed to be the PT76. Questions to the Army Chief of Staff as to why the United States Army was not provided with such a vehicle resulted in a full scale review of the overall tank program. At that time, the T92 design was evaluated to see if it could be made amphibious. This was not feasible as the original design concept was to make the tank as compact as possible to minimize the size and weight with a maximum level of protection. Modifying the design to obtain the volume required

for flotation was impractical. In June 1958, the Office, Chief of Ordnance directed that the fabrication of T92 pilots three and four be cancelled, but that the engineering and vulnerability tests be continued. Preparation began to develop the characteristics for a combined armored reconnaissance and airborne assault vehicle which would be amphibious. As a result, the light gun tank program was cancelled.

A NEW LIGHT COMBAT VEHICLE

Preliminary concept studies were initiated in January 1959 for a light combat vehicle which would meet the newly specified military characteristics. Designated as the armored reconnaissance/airborne assault vehicle (AR/AAV), it was to be amphibious, have air drop capability, and improved firepower and mobility compared to the earlier light tanks. It also was to have the maximum armor protection possible consistent with the other requirements.

In July 1959, a conference at the Ordnance Tank Automotive Command (OTAC) presented specifications for the AR/AAV and requested proposals from industry for the development and construction of six pilot vehicles. Later, the number of pilot vehicles was increased to twelve. Twelve proposals were received in October for evaluation by OTAC. In December, two of these were selected for further detailed design and the construction of mock-ups. One of the proposals selected was a joint effort of Aircraft Armaments Incorporated (AAI) and the Allis Chalmers Manufacturing Company. The development work was to be carried out at AAI, but Allis Chalmers would handle the production. The second proposal selected was from the Cadillac Motor Car Division of General Motors Corporation. Work continued at both contractors and the competing mock-ups were reviewed in late May 1960. The AAI candidate featured a three man crew with two in the turret and the driver in the hull. This concept was close to the ten ton weight limit of

The drawing below shows the AAI proposal for the new AR/AAV. Note that the gun-launcher is fitted with a muzzle brake. The AR/AAV specification required protection against Soviet 12.7mm armor piercing ammunition at a range of 100 meters and 152mm air bursts at a distance of 60 feet as well as antipersonnel mines. The three man crew consisted of the driver, the commander-gunner, and the loader.

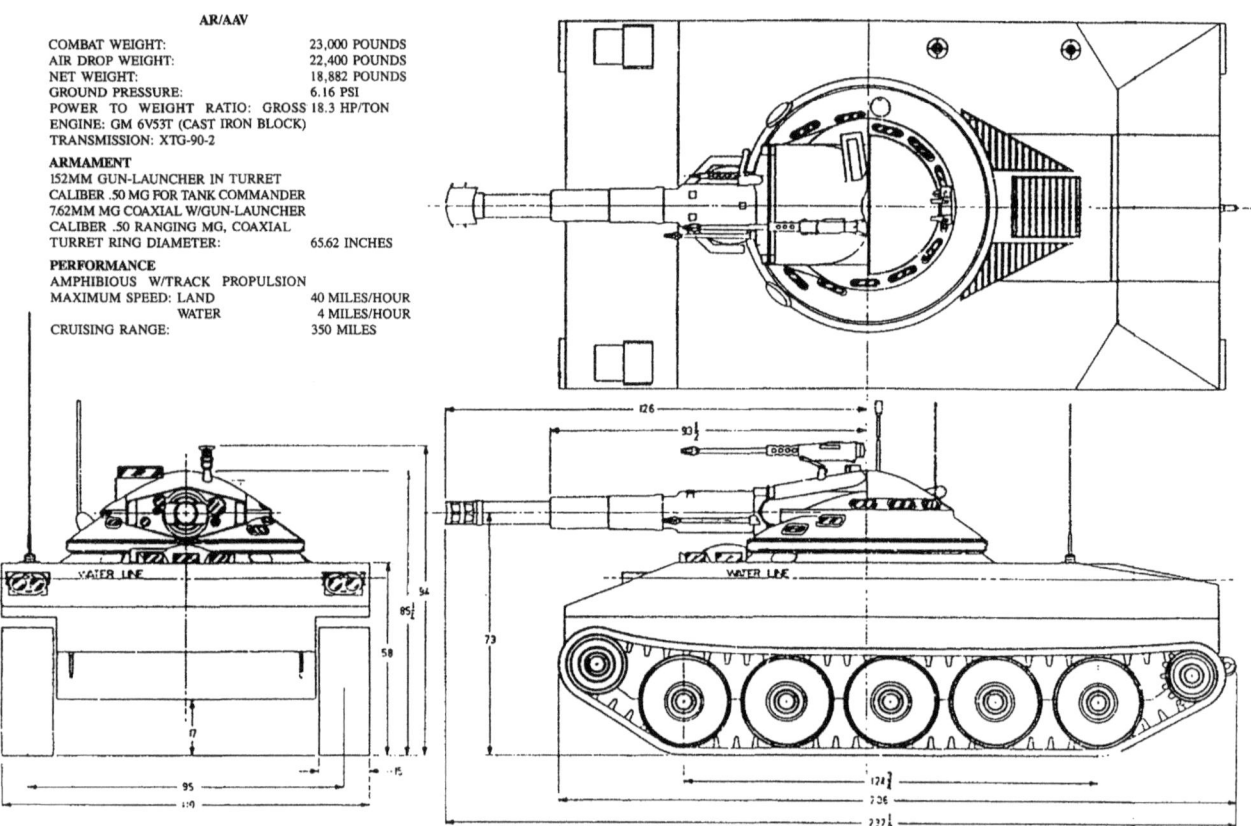

AR/AAV

COMBAT WEIGHT:	23,000 POUNDS
AIR DROP WEIGHT:	22,400 POUNDS
NET WEIGHT:	18,882 POUNDS
GROUND PRESSURE:	6.16 PSI
POWER TO WEIGHT RATIO: GROSS 18.3 HP/TON	
ENGINE: GM 6V53T (CAST IRON BLOCK)	
TRANSMISSION: XTG-90-2	

ARMAMENT
152MM GUN-LAUNCHER IN TURRET
CALIBER .50 MG FOR TANK COMMANDER
7.62MM MG COAXIAL W/GUN-LAUNCHER
CALIBER .50 RANGING MG, COAXIAL

TURRET RING DIAMETER:	65.62 INCHES

PERFORMANCE
AMPHIBIOUS W/TRACK PROPULSION

MAXIMUM SPEED: LAND	40 MILES/HOUR
WATER	4 MILES/HOUR
CRUISING RANGE:	350 MILES

the original specifications. The Cadillac design was somewhat heavier and carried a crew of four with a three man turret. However, this arrangement was preferred by the user despite the increased weight because of the improved fightability of the three man turret. The specifications were then revised to increase the maximum weight to fifteen tons and to improve the armor protection. In June 1960, a contract was signed with the Cadillac Motor Car Division for the further development of their concept. Now designated as the AR/AAV XM551, the name General Sheridan was

The sketch at the top of the page and the drawings at the right illustrate the Cadillac design concept adopted for the new AR/AAV. Unlike the AAI candidate, it is manned by a crew of four. Both the Cadillac and AAI design proposals were intended to be able to swim without any special preparation.

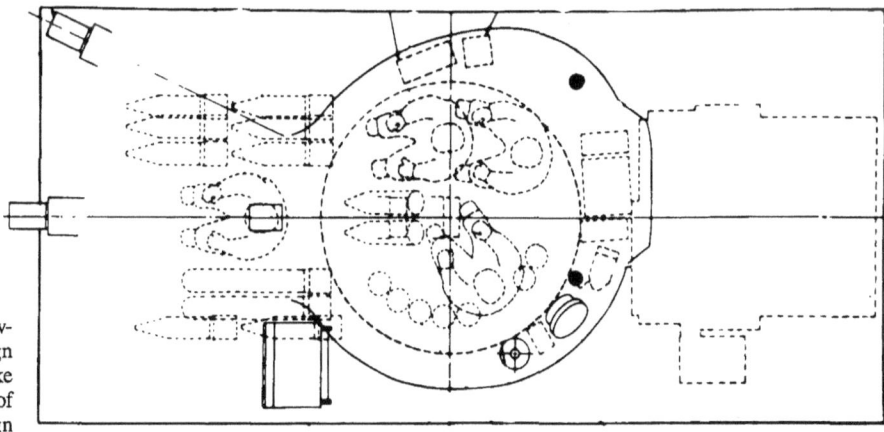

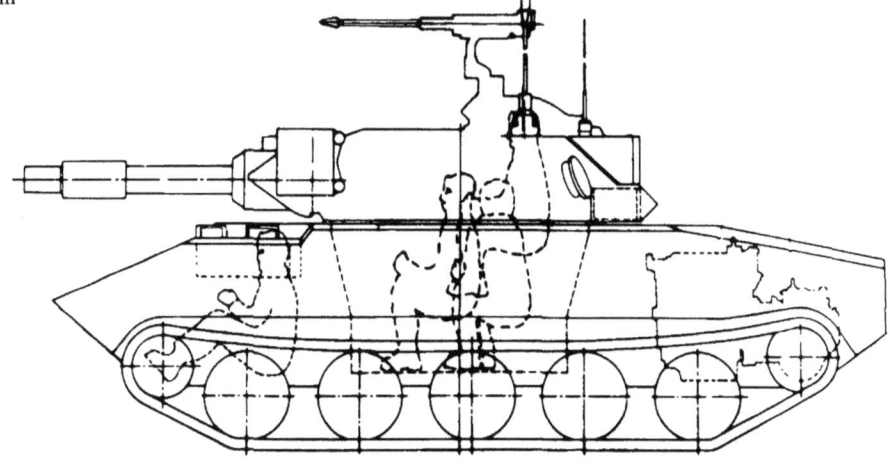

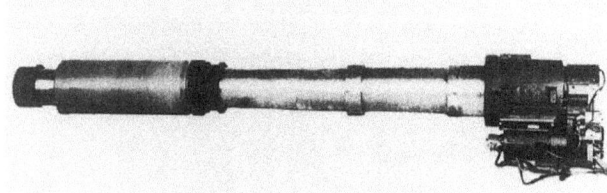

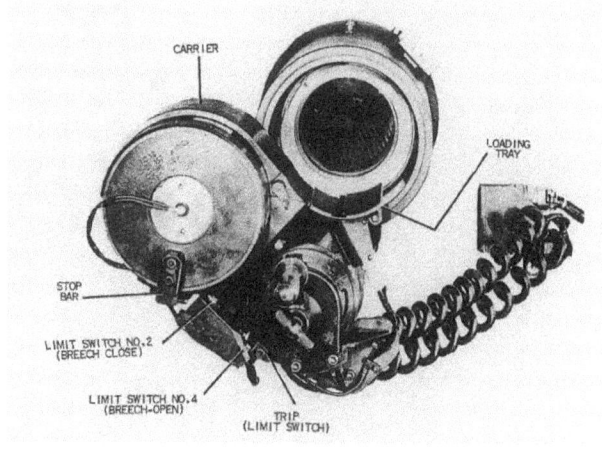

The 152mm XM81E3 gun-launcher appears above. At the right are views of the breech open and closed.

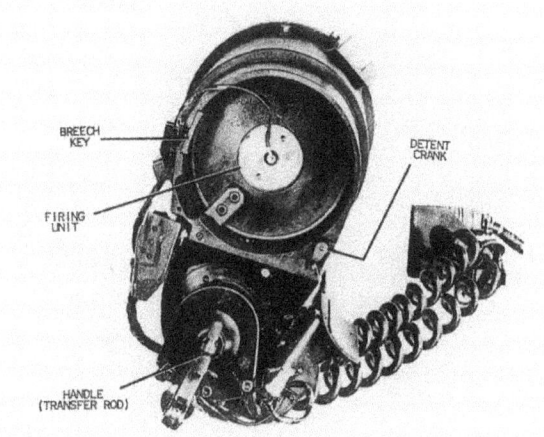

approved by the Secretary of the Army on 14 August 1961 in honor of Major General Philip Sheridan, the Civil War Union cavalry leader. Work continued on the development program and on 12 December 1961, a test bed vehicle began operation at the Cleveland Tank Plant. In service, the Sheridan was intended to replace both the 76mm gun tank M41 and the 90mm self-propelled gun M56.

The new vehicle made use of a radically new main armament system. Dubbed the Combat Vehicle Weapon System (CVWS) Shillelagh, it consisted of the 152mm XM81 gun-launcher which could fire combustible case conventional ammunition or launch the XM13 guided missile. This short barrel cannon weighed approximately half as much as the 105mm gun M68 and thus was particularly attractive for installation where weight was a critical factor. Preliminary firing tests of this weapon began at the Erie Army Depot in late 1961.

The XM81 gun-launcher incorporated a separable chamber breech assembly with a metal seal ring between the two parts providing obturation. Thus it could use a completely combustible cartridge case and primer. In this design, the breech chamber, consisting of the rear 5½ inches of the chamber, separated from the tube by rotating 30 degrees inside the coupling joining the two parts. It then moved to the rear outside of the coupling and pivoted 60 degrees to provide a clear path to the bore. On the original XM81 gun-launcher, the breech mechanism was actuated hydraulically. The XM81E3 gun-launcher, which was designed specifically for the AR/AAV, used electrical breech actuation. The short tube of the gun-launcher was fitted with a bore evacuator and threaded for the installation of a muzzle brake. However, the latter was not installed and the threaded muzzle was fitted with a thread protector. Initially, three types of conventional 152mm ammunition were under development, all of which were spin stabilized by the rifling in the gun-launcher tube. The most important of these was the XM409 HEAT-MP. As its designation indicated, this was a multipurpose round combining the armor piercing performance

of the shaped charge with the fragmentation and blast effect of a high explosive shell. The XM409 had a muzzle velocity of 2260 feet per second and its spin compensated shaped charge was estimated to penetrate seven inches of rolled homogeneous armor at 60 degrees obliquity. In addition to the XM409, the XM410 white phosphorus, and the XM411 training rounds were under development.

At the right are the three types of ammunition originally intended for use with the 152mm gun-launcher.

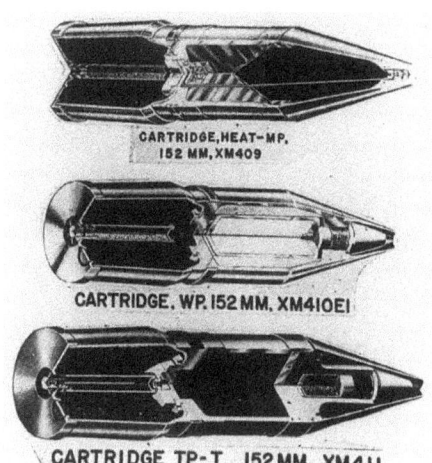

78

SHILLELAGH

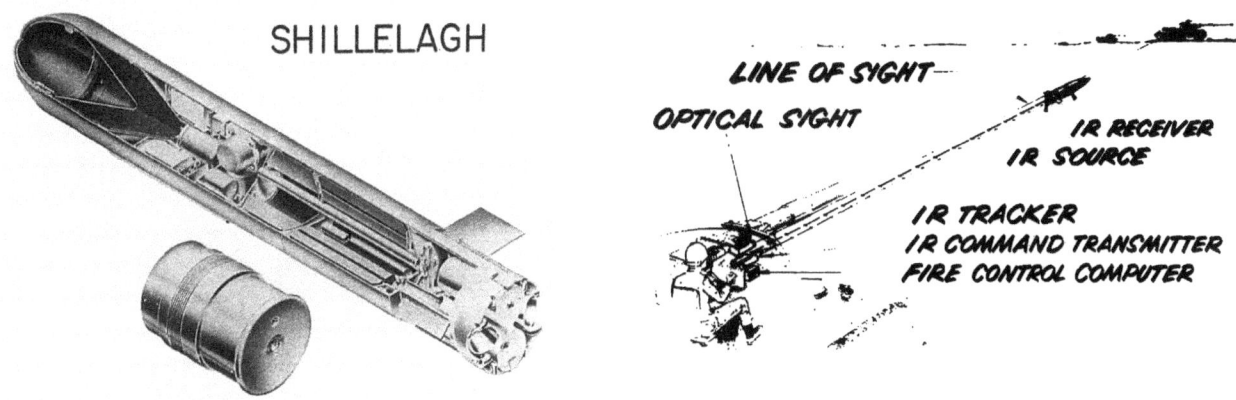

Above, a cutaway drawing of the Shillelagh missile and a sketch illustrating the method of operation are at the left and right respectively.

The XM13 missile had a muzzle velocity of 260 feet per second which was increased to about 1060 feet per second by the solid rocket booster. Rotation of the missile during launch was prevented by a key which engaged a keyway at the six o'clock position in the tube of the gun-launcher. The rear end cap on the missile remained in the gun-launcher and was ejected when the breech opened after firing. The missile flew a line of sight path to the target using an infrared (IR) tracking and command system. The gunner held his sight on the target and the IR tracker measured any deviation from the proper flight path using the IR source in the missile tail. This deviation was processed by the electronic computer and a correction command was transmitted to the missile using a xenon arc lamp. The shaped charge warhead in the missile was approximately six inches in diameter and was lethal to any known armored vehicle at that time.

Unfortunately, work on the XM13 missile encountered various problems resulting in delays in the development schedule and in late 1961, the missile program was reclassified as an applied research project. As a result, there was concern that the Shillelagh system might not be available in time for the AR/AAV. To consider this problem, a meeting was held at the Ordnance Tank Automotive Command on 10 January 1962. After discussion between the various ordnance agencies, seven armament systems were recommended as back-ups for the original Shillelagh system in the event of further delay. At that time, it was expected that a decision on the armament for the AR/AAV would have to be made by April 1962.

The seven back-up systems under consideration were as follows:
1. 76mm conventional gun and ammunition.
2. 90mm conventional gun and ammunition.
3. 105mm conventional gun and ammunition.
4. 152mm conventional gun and ammunition.
5. ENTAC missile with any of the above conventional gun systems.
6. TOW (Tube launched, Optically tracked, Wire guided) missile system
7. POLCAT (POst Launched Correction, AntiTank) system

Review of the seven systems resulted in the following conclusions.

The 76mm gun could easily be installed in the AR/AAV and the ammunition was already in production. However, there would be no improvement in firepower compared to the earlier light tanks.

The drawing at the right shows the XM551 armed with the 76mm gun M32.

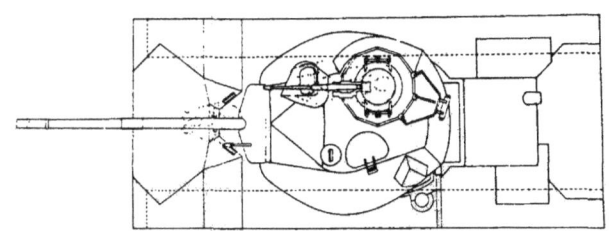

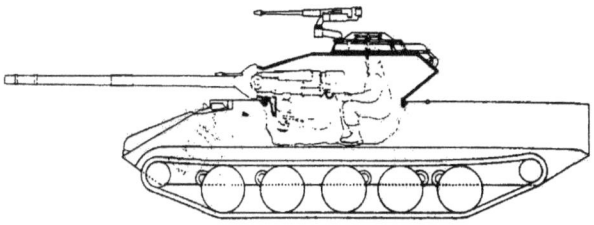

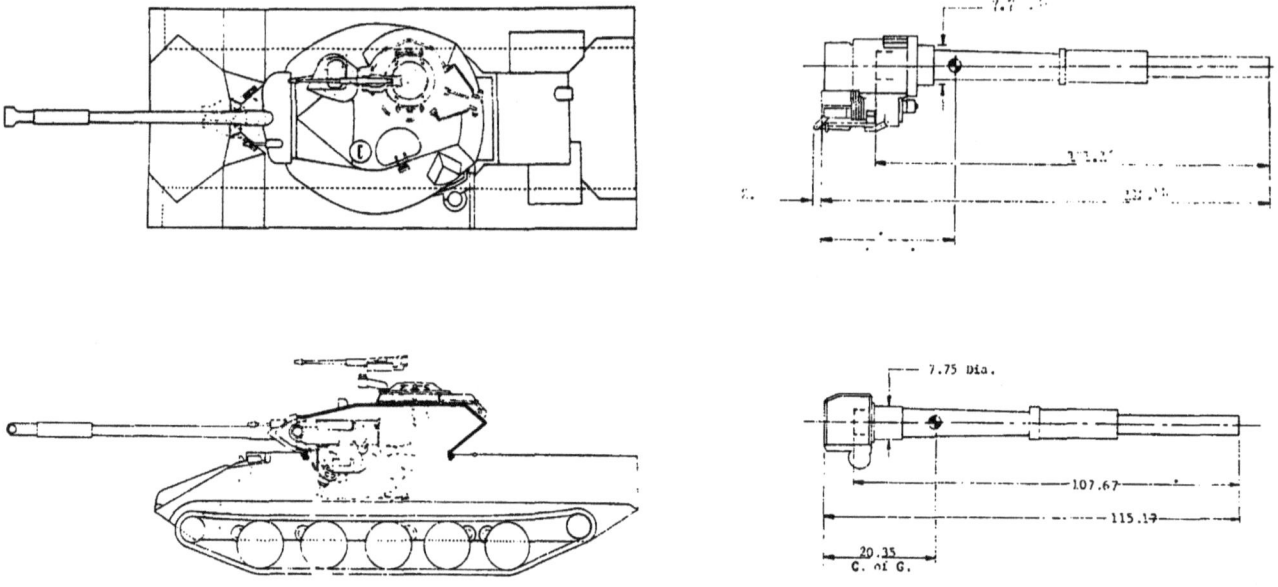

Above at the left, the Sheridan is armed with the 90mm gun M41 using a lightweight tube. At the right are drawings of two versions of lightweight 90mm cannon intended for use with T300 HEAT ammunition. These drawings show the weapon with a separable breech (upper) and a conventional vertical sliding breechblock (lower).

Two 90mm guns were considered, both of which increased the firepower compared to the 76mm gun, but did not equal the 152mm gun-launcher with the conventional HEAT-MP round. The first was the standard M41 90mm gun with its ammunition already in production. It could have been adapted to the AR/AAV in about 15 months. The second was a new lightweight 90mm gun using the T300 HEAT shell as the primary armor piercing round. It was estimated that 27 months would be required to develop this weapon for installation in the AR/AAV.

The 105mm gun proposed was a new lightweight weapon with a combustible cartridge case using a shaped charge projectile as the antiarmor round. The estimated development time for this weapon ranged from 21 to 27 months and it did not have the growth potential of the 152mm gun-launcher with conventional ammunition.

The Sheridan sketched at the left below is armed with the proposed lightweight 105mm gun intended for use with an HEAT round. Versions of this cannon are shown at the right with the separable breech (upper) and a conventional vertical sliding breechblock (lower).

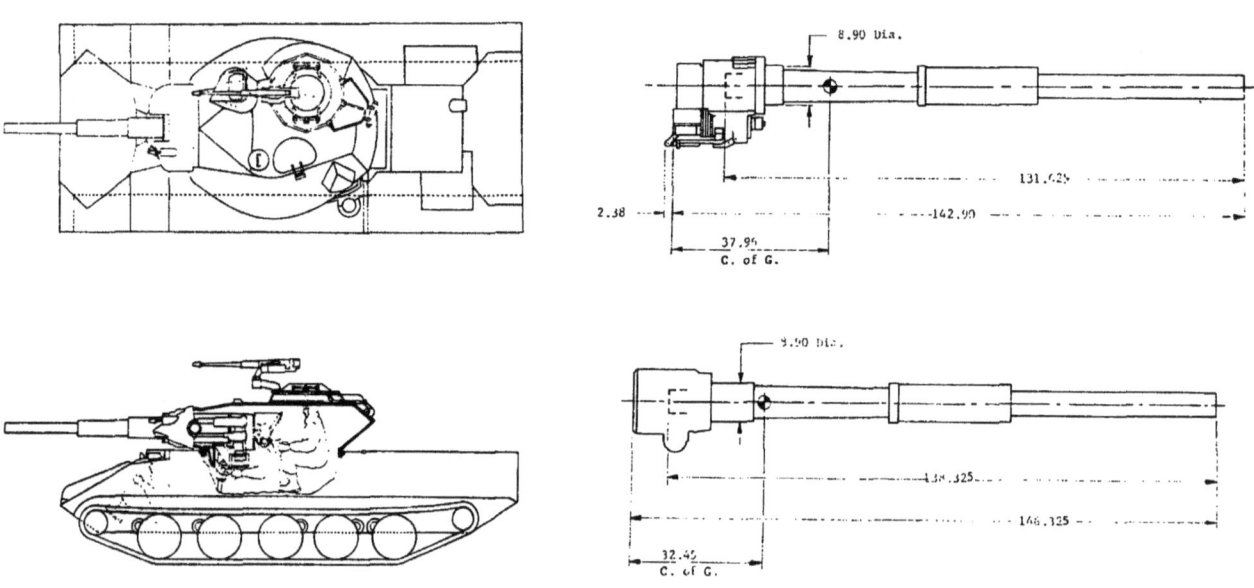

80

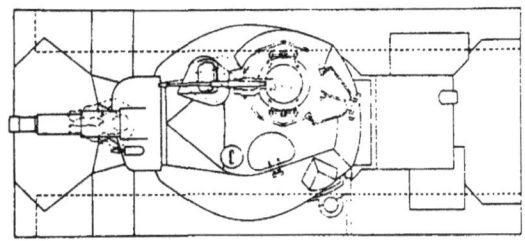

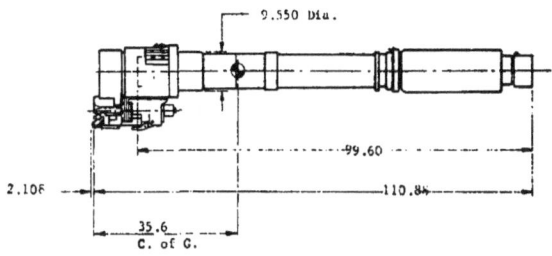

At the left, the Sheridan is armed with the 152mm gun-launcher without the missile system. The sketch above shows the XM81E3 gun-launcher itself.

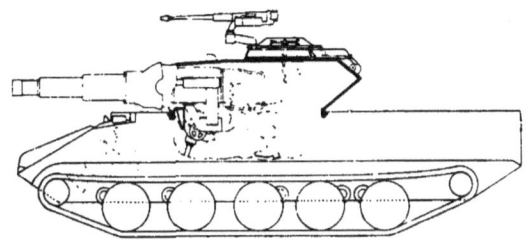

The fourth concept was using the 152mm gun-launcher itself without the missile system, relying on the armor piercing performance of the XM409 HEAT-MP round. This system provided the greatest firepower and had already been designed into the AR/AAV. Thus it could be available in the shortest possible time. The Shillelagh missile then could be added later when it became available.

The ENTAC missile could be installed with any of the gun systems previously described.

If the Shillelagh program failed to reach a satisfactory conclusion, the TOW missile was considered to be a possible substitute. However, to adapt it for closed breech firing from the gun-launcher, a method would have to be developed to disperse the command wire.

The POLCAT was a more remote possibility as a back-up weapon for the AR/AAV. This weapon used a gun which fired a projectile using only conventional sights at the shorter ranges (under 500 yards). At longer ranges, a detector in the projectile nose used a reflected illuminator signal to home in on the target. The path of the rotating projectile was altered by the automatic firing of an impulse cartridge. The original concept featured a 42 pound projectile with a total round weight of 52 pounds.

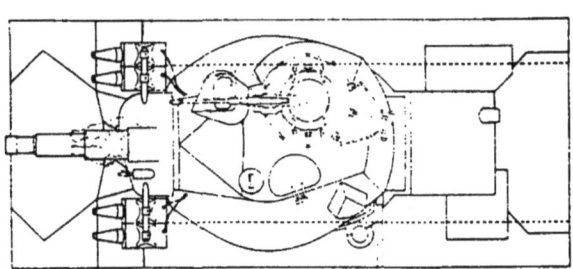

The XM551 appears above at the left armed with the 152mm gun-launcher, but without the missile system. The drawing at the lower left shows this vehicle with the addition of four ENTAC missiles. The sketch below illustrates the operation of the POLCAT system.

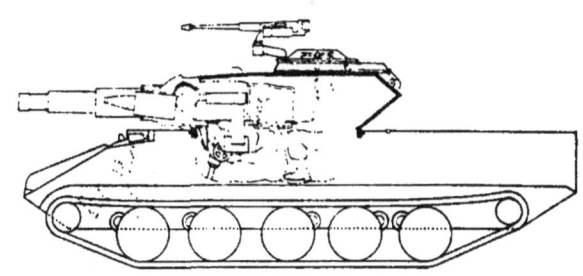

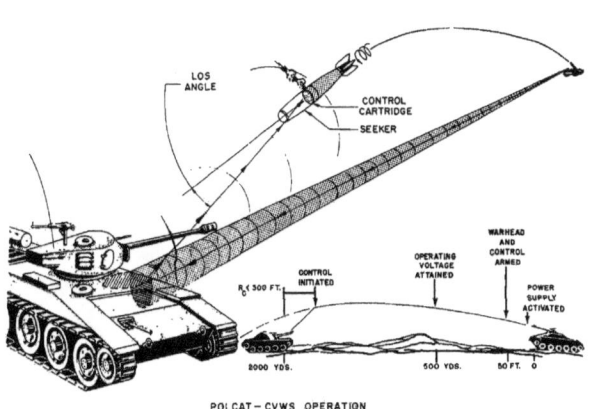

POLCAT - CVWS OPERATION

The XM551 turret is installed on an M41 tank chassis at the top left. At the top right is a view of XM551 pilot number 1. Note that a travel lock has been installed on the rear deck for the 152mm gun-launcher.

After consideration of the various alternate weapon systems, the decision was made to use the 152mm gun-launcher with the conventional combustible case ammunition. The Shillelagh or some other missile would then be introduced later when it reached a satisfactory stage of development. In July 1962, an XM551 test bed turret with the 152mm gun-launcher was installed on an M41 tank chassis to permit the evaluation of the turret and armament. Firing tests began at Aberdeen Proving Ground on 23 August 1962 with 590 152mm rounds being fired during the test program.

Construction of the XM551 pilot vehicles had been proceeding and pilot number 1 was delivered in June 1962. The first three pilots were essentially the same, representating the first generation in the development of the new vehicle.

The hull of the XM551 was a welded assembly of aluminum alloy armor plate designed to present a high obliquity surface to enemy fire. This hull was surrounded by a lightweight structure filled with polystyrene foam to obtain the volume required for flotation. On pilots 1 through 3, this outside structure was rectangular with a vertical rear surface. The front sloped down to track level and was fitted

with a trim vane or surfboard which folded up against the front plate when operating on land. For water borne operation, the surfboard was moved into its forward position and the space between the surfboard and the front plate was filled by two flotation bags to maintain the buoyancy of the front hull. A window in the center of the surfboard between the flotation bags provided forward vision for the driver. The driver was located under a flat hatch in the center front hull with three M27 periscopes.

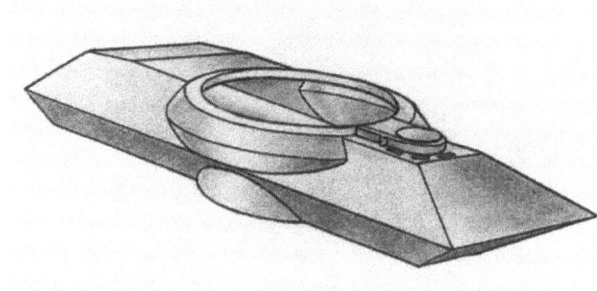

The drawing above shows the configuration of the aluminum armor hull without the flotation enclosure. Below are two photographs of XM551 pilot number 3. Apparently, the travel lock has been omitted from the rear deck.

Details of the flotation enclosure around the hull can be seen in these photographs of XM551 pilot number 1. At the bottom rear, note the two nozzles and the center intake for the water jet propulsion system.

The muzzle of the XM121 spotting rifle can be seen above on XM551 pilot number 3. Also, note the details of the band type track.

Above is XM551 pilot number 3 with its turret rotated to the rear. At the right, a Shillelagh missile is loaded into the vehicle during tests at White Sands Missile Range in New Mexico on 15 April 1964.

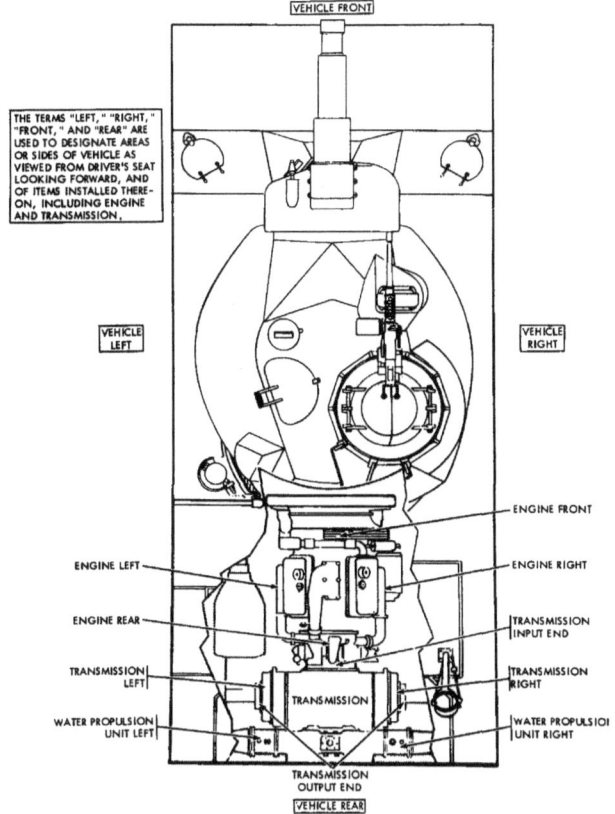

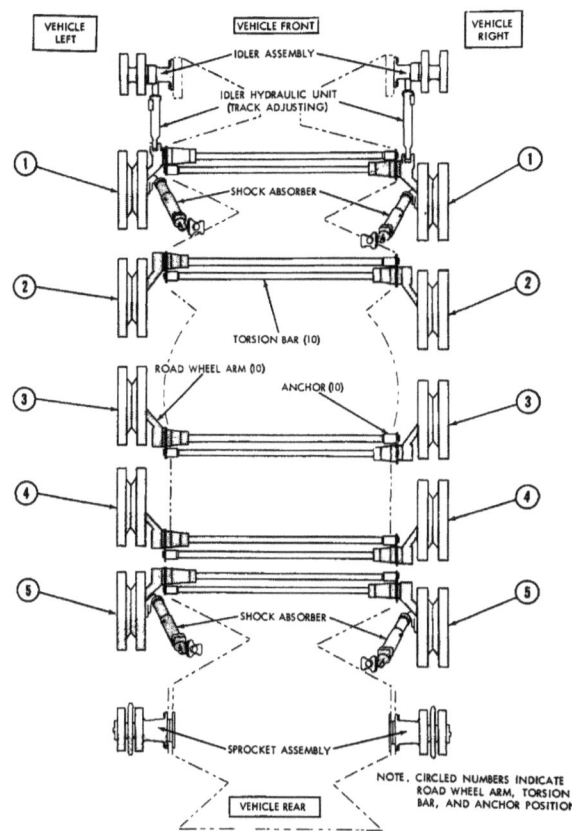

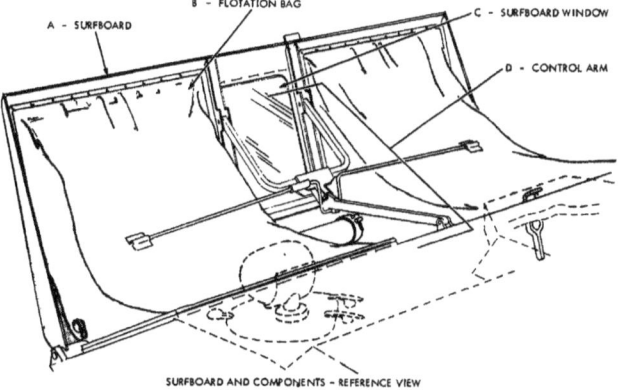

SURFBOARD AND COMPONENTS – REFERENCE VIEW

The sketch at the top left shows the internal arrangement of the engine compartment components in the XM551. The layout of the suspension system can be seen above. At the left, the surfboard is extended for water operation.

The XM551 was powered by a six cylinder General Motors 6V53T, two stroke cycle, diesel engine installed in the rear hull. This turbosupercharged, liquid cooled, engine developed 285 gross horsepower in the early pilots. The engine drove the vehicle through an Allison XTG-250 transmission which had a power take-off to drive the water jet propulsion unit when the vehicle was afloat. The eight tooth, rear mounted, sprockets drove the 19 inch wide, band type, tracks. These tracks, each consisting of 12 sections, contained two rubber, fabric, and steel cable bands connected crosswise by eight pressed steel, rubber faced, grousers per section. The suspension of the first three pilots supported the vehicle on five 24 inch diameter dual road wheels per side using transverse mounted torsion bars. Hydraulically adjusted idlers were mounted at the front of each track.

At the left, the driver's controls are visible through the hatch on XM551 pilot number 3.

85

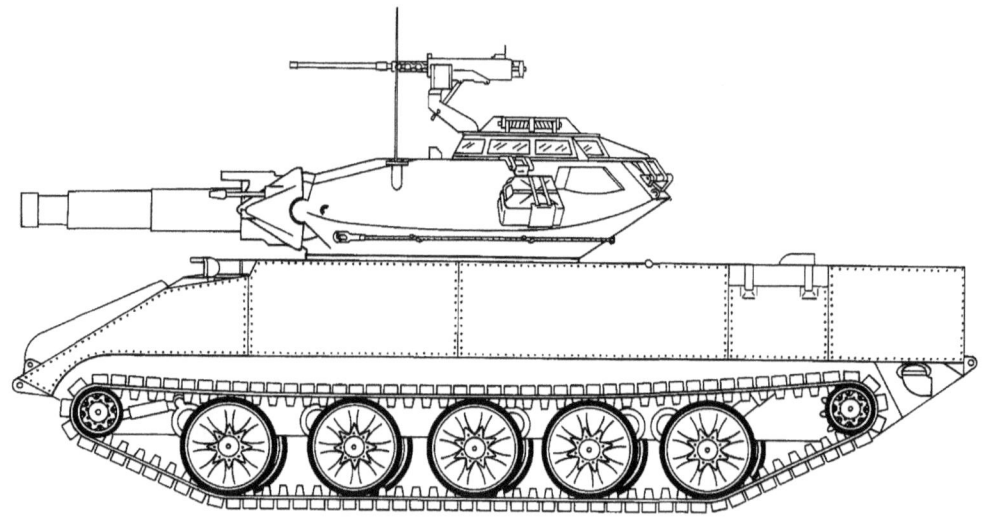

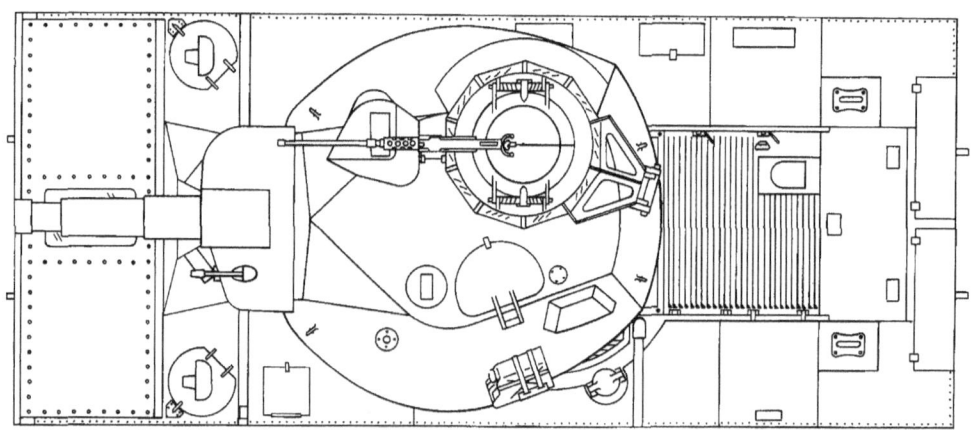

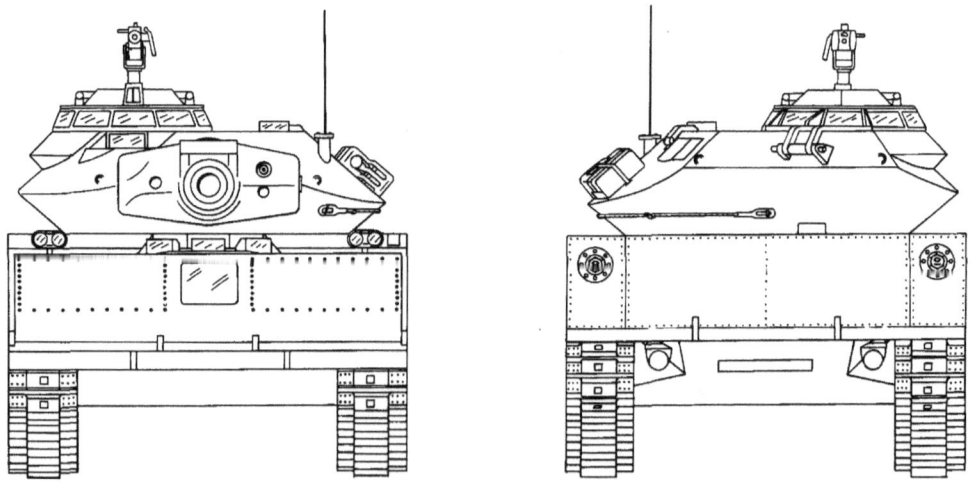

152mm Gun-Launcher AR/AAV XM551, 1st generation pilot

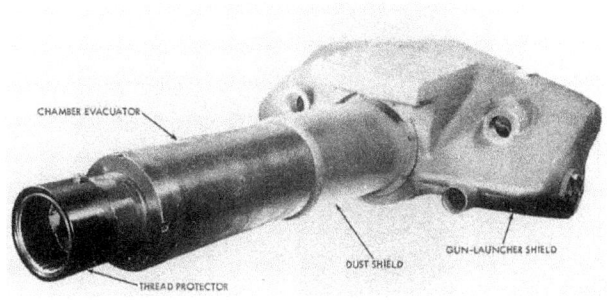

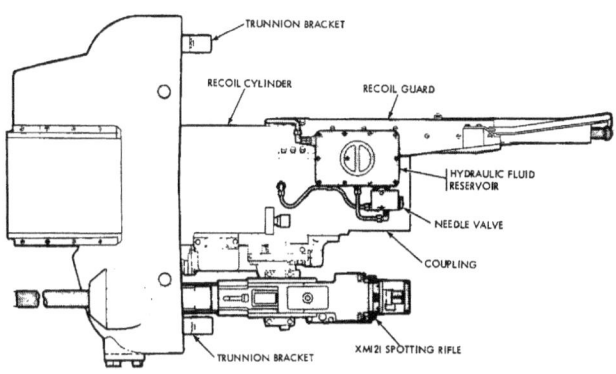

The 152mm gun-launcher XM81E3 is installed in its mount above. Below, the coaxial machine gun and the spotting rifle can be seen in the left rear view. At the right are drawings of the gun-launcher mount.

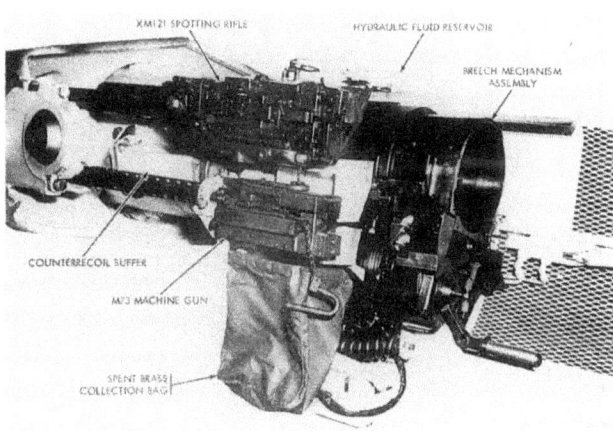

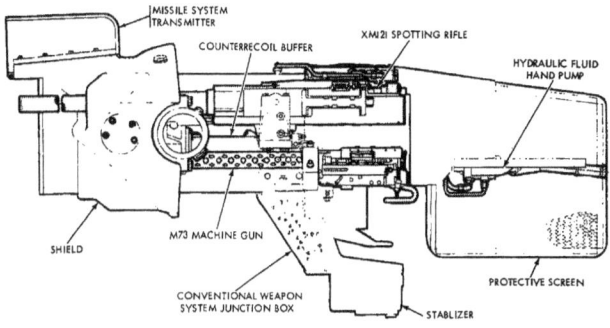

At the right is the commander's hatch assembly from XM551 pilot number 3.

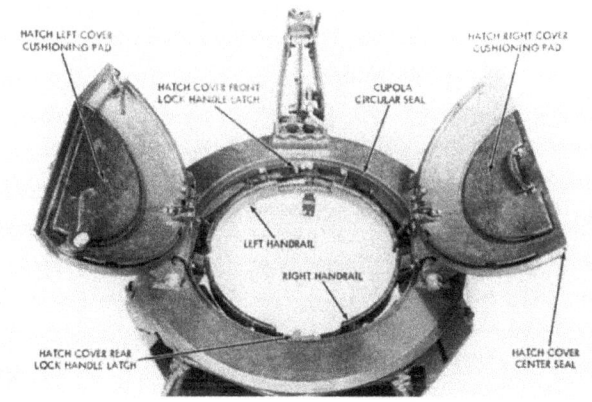

The three man turret crew rode in the turret basket with the gunner and tank commander on the right side of the cannon and the loader on the left. A 7.62mm M73 coaxial machine gun was installed in the mount to the left of the 152mm XM81E3 gun-launcher with an XM112 direct fire telescope on the right. An XM121 .50 caliber spotting rifle was located on the left side of the mount just above the coaxial machine gun. This weapon matched the trajectory of the HEAT-MP XM409 round at ranges up to 1300 yards. The gunner also was provided with an XM38 infrared periscope. The tank commander rode behind the gunner under a cupola fitted with ten vision blocks for a 360 degree view. This cupola mounted a .50 caliber machine gun for use with the hatch open.

The turret interior on pilot number 3 appears at the right. The gunner's seat and controls can be seen in this photograph.

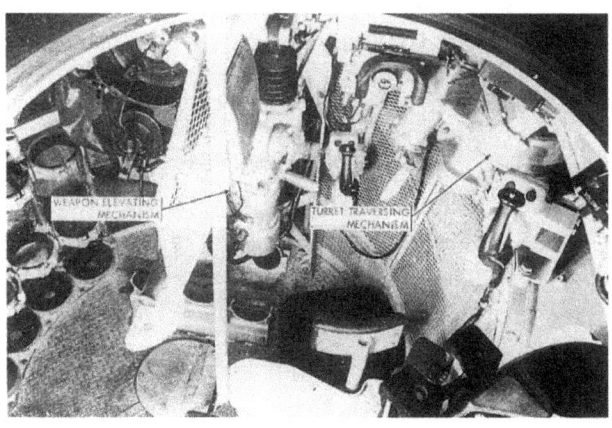

Above is XM551 pilot number 5. Note the absence of a window in the surfboard and that the muzzle of the gun-launcher is no longer threaded for a muzzle brake. Below, pilot number 4 is operating with the flotation screen raised. Obviously, the turret has been removed for this phase of the program.

As mentioned previously, the first of the 12 XM551 engineering pilots was delivered in June 1962 and testing began shortly thereafter. These tests revealed the need for numerous modifications.

The second generation of vehicles was introduced with pilot number 4. It replaced the band type track with a single pin, link type, track and it rode on new design road wheels 24.5 inches in diameter. The water jet propulsion was retained for amphibious operation, but the flotation cells on the hull were redesigned and some were eliminated. The highy sloped aluminum armor rear hull was now exposed with a removable cover permitting easy access to the transmission. The front hull and driver's hatch remained the same as on the first three pilots, but the window was eliminated from the surfboard and the flotation bags were modified. The early ventilator cover on the left rear turret wall was replaced by a new circular cover. The Shillelagh missile also had made satisfactory progress and the components of the missile system were installed on the vehicle. XM551 pilot number 4 is on display by the Patton Museum at Fort Knox, Kentucky. Pilot number 5 had a similar configuration.

In these photographs, pilot number 4 is on display by the Patton Museum at Fort Knox, Kentucky. A closeup of the new turret ventilator on the second generation of pilots can be seen at the right.

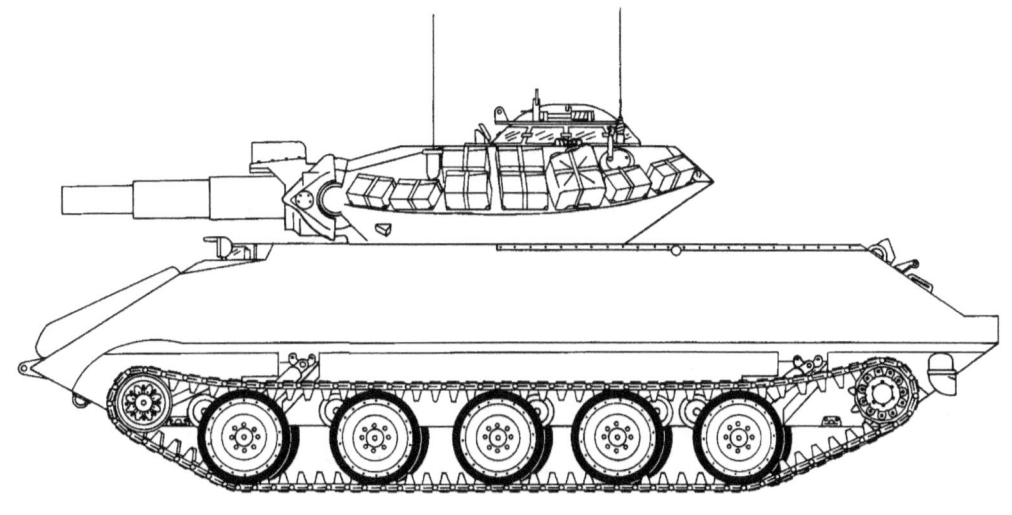

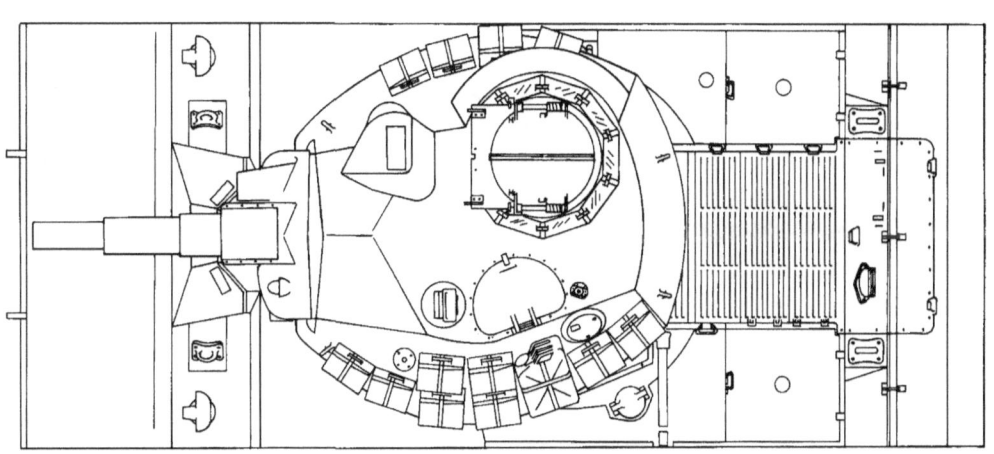

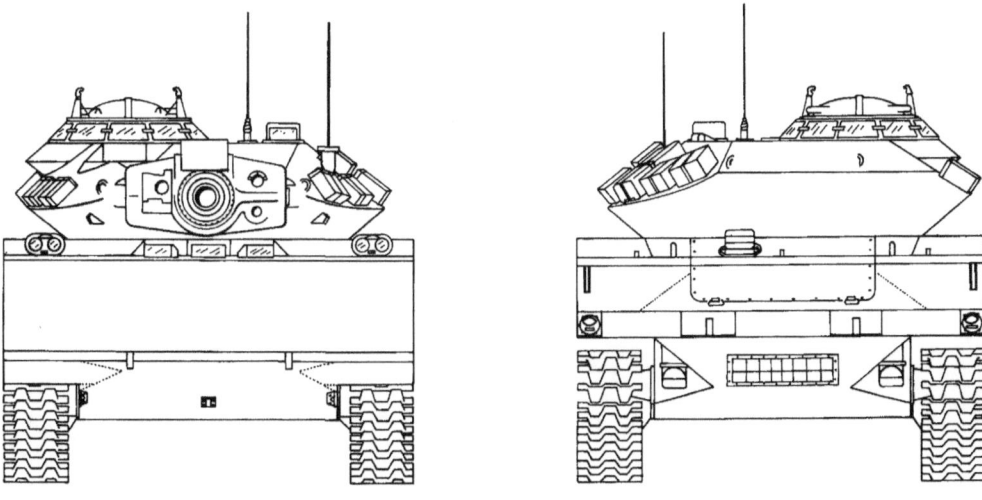

Scale 1:48

©**D.P. Dyer**

152mm Gun-Launcher AR/AAV XM551, 2nd generation pilot

The new rotating driver's hatch on pilot number 7 is obvious in the views above and below. Note the bulbous covers attached to the road wheels in an attempt to improve the buoyancy.

By the end of 1963, the first eight pilot vehicles had been delivered. Number 7 represented the third generation of development. The front hull and driver's station were drastically modified. The conventional flat hatch and periscope arrangement on the early vehicles was replaced by a rotating hatch assembly fitted with three XM47 periscopes. The center one of these could be replaced by an XM48 infrared periscope for use at night. With this design, the entire hatch assembly rotated behind the driver for open hatch operation and he did not have to raise or lower his seat. Also, his head was never in any danger from the rotating gun mount when the hatch was open. A new design surfboard was installed and bulbous covers were added to the road wheels to increase the displacement volume and improve the buoyancy. The water jet propulsion was eliminated and the vehicle was propelled in the water by its tracks. A 15mm XM122 spotting rifle was specified for pilot number 7. This rifle was expected to match the trajectory of the XM409 round up to 2000 yards. During 1963, three fire control systems were evaluated for use with the gun-launcher. They used the spotting rifle, a laser range finder, or a simple optical sight. Pilot number 8 had a modified aluminum hull with a different level of protection than pilot number 7.

Above at the left, pilot number 7 emerges from the water with the flotation screen erected. The long barrel 15mm spotting rifle is installed in the view at the left below. Note that the water jet propulsion system has been eliminated on the rear of pilot number 7 at the bottom right and two of the wheel covers have been removed.

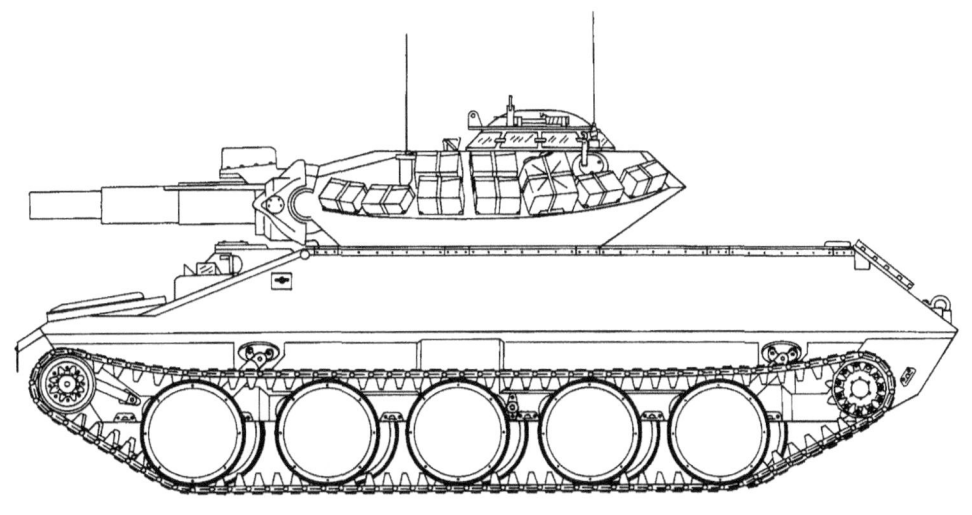

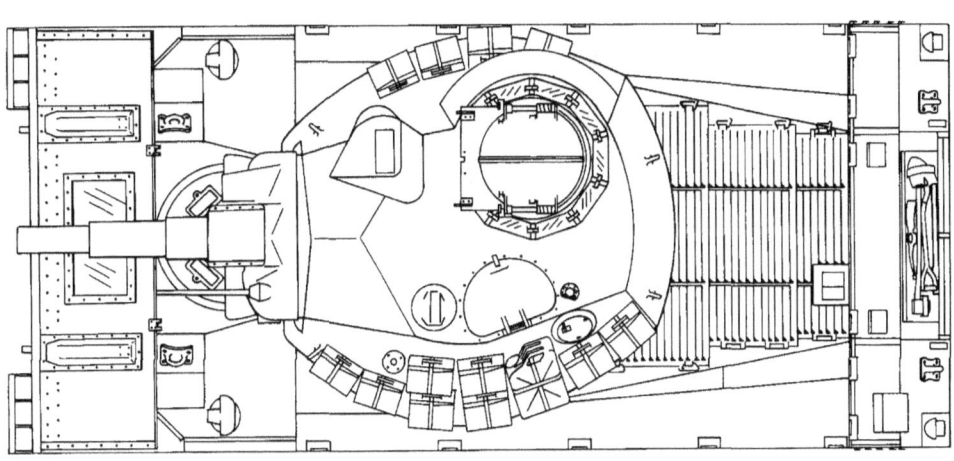

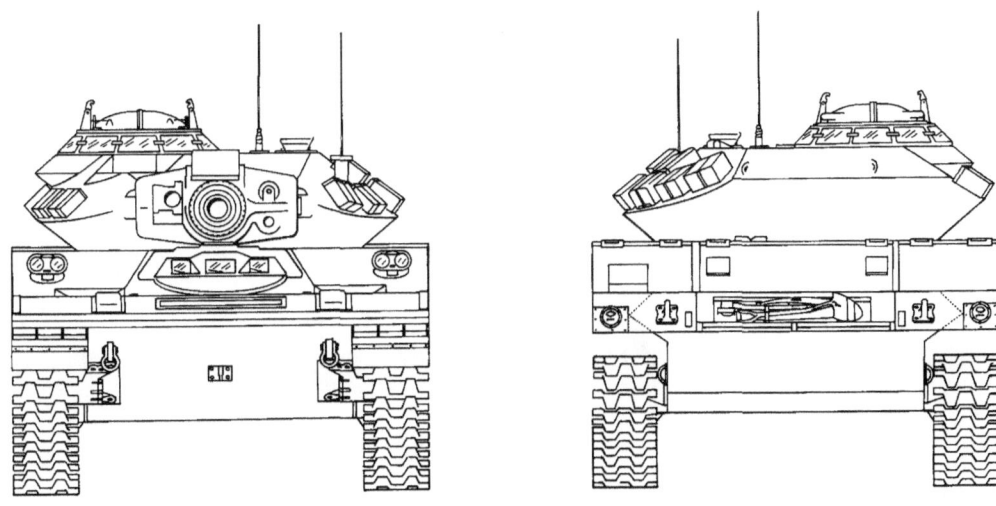

152mm Gun-Launcher AR/AAV XM551, 3rd generation pilot

XM551 pilot number 9 is shown above and below during the test program at Aberdeen Proving Ground. The spotting rifle port in the gun shield has been sealed off and the bulbous wheel covers have been eliminated.

Pilot vehicles 9, 10, and 11 were delivered during 1964. The engineering test/service test program for the system was initiated in October and during the same month, the Department of the Army approved the fielding of the Sheridan using the simple optical sight fire control system. Development of the spotting rifle was terminated, but work continued on the laser range finder for future application. Pilots 9, 10, and 11 were similar in appearance to number 7, but guards were installed to protect the headlights and the bulbous flotation covers were eliminated from the 28 inch diameter road wheels. Pilot number 9 is on display by the Ordnance Museum at Aberdeen Proving Ground, Maryland.

The rotating driver's hatch introduced on pilot number 7 can be seen below in two different positions. Note that this vehicle is equipped with the spotting rifle.

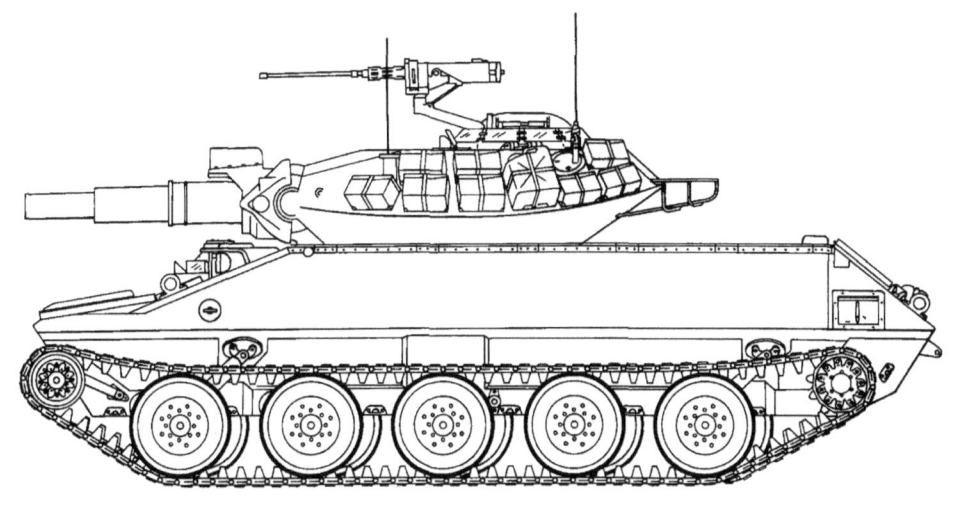

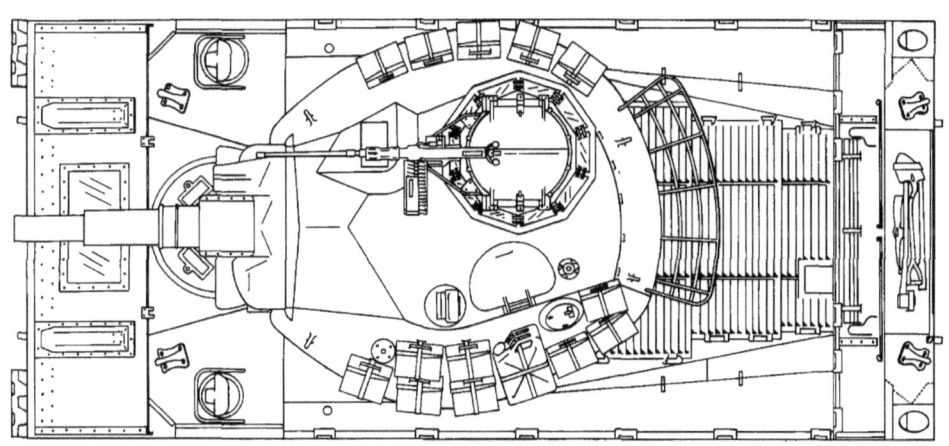

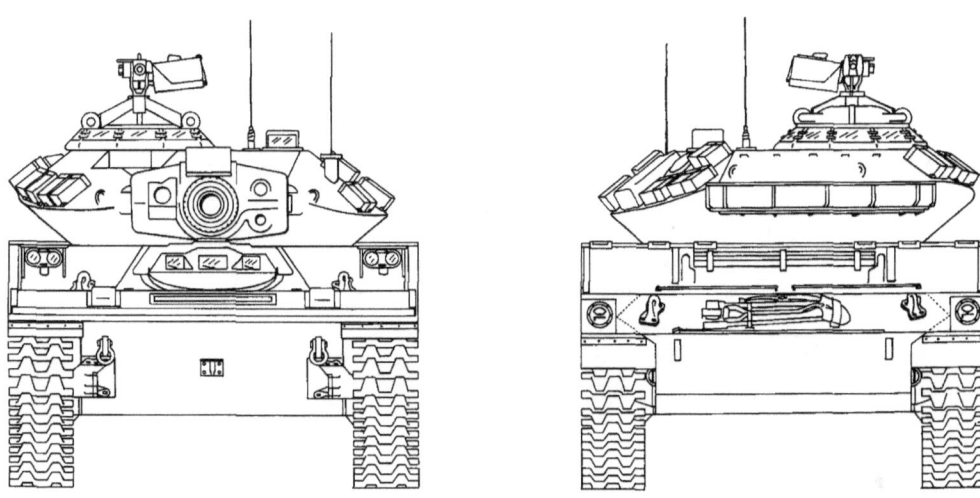

Scale 1:48

©D.P. Dyer

152mm Gun-Launcher AR/AAV XM551, 4th generation pilot

XM551 pilots number 9 (above) and 10 (below) are shown here during tests at Aberdeen Proving Ground.

The flotation screen is erected on pilot number 9 above. Below, pilot number 10 is fitted with test equipment on the turret and rear deck during the evaluation program.

The stowage covers for the new flotation screen can be seen along the sides and rear of the hull on XM551 pilot number 12 in these two photographs.

All of the photographs on this page show XM551 pilot number 12 during the engineering and service tests. Below at the right is a night launch of a Shillelagh missile.

XM551, pilot number 12, was delivered in February 1965. This vehicle represented the proposed production configuration and it was subjected to extensive engineering/service tests. The most obvious change on pilot number 12 was the new two piece folding surfboard and the flexible flotation barrier stowed along the top side corners and rear of the hull. The installation of this new flotation barrier increased the freeboard of the vehicle to about 24 inches when afloat.

Below, the vehicle is operating cross-country and at the bottom right, the new two piece folding surfboard is clearly visible.

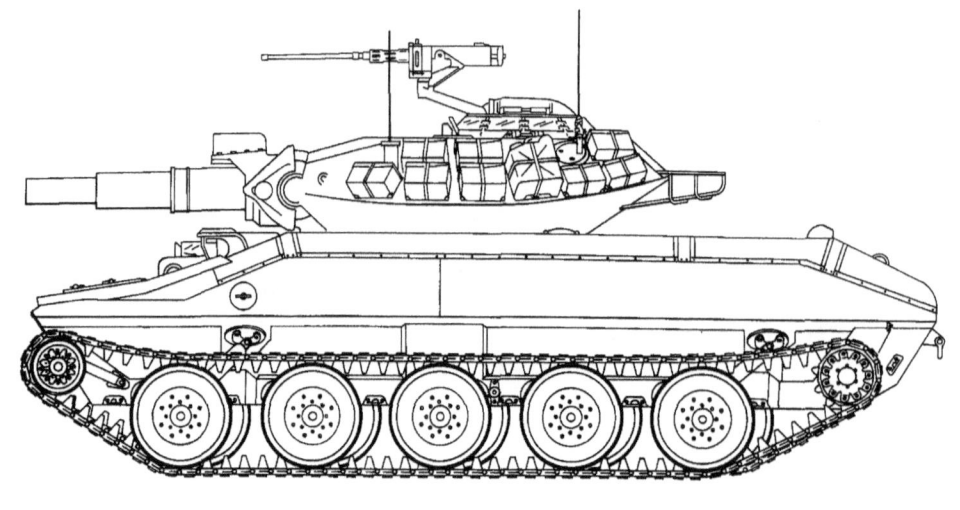

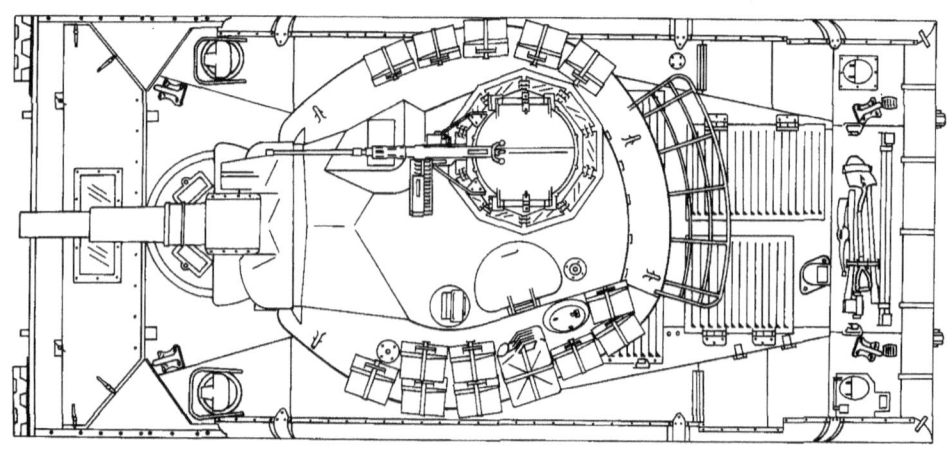

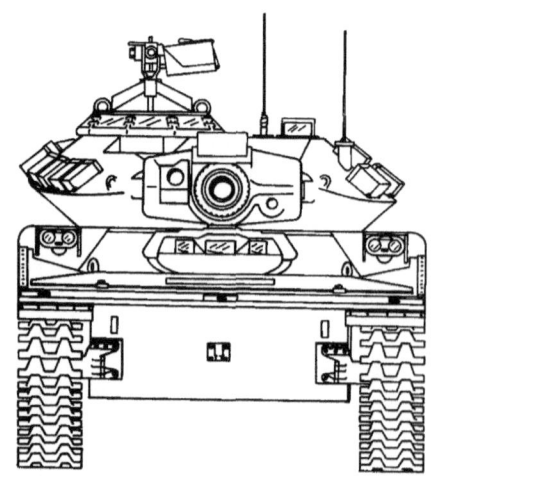 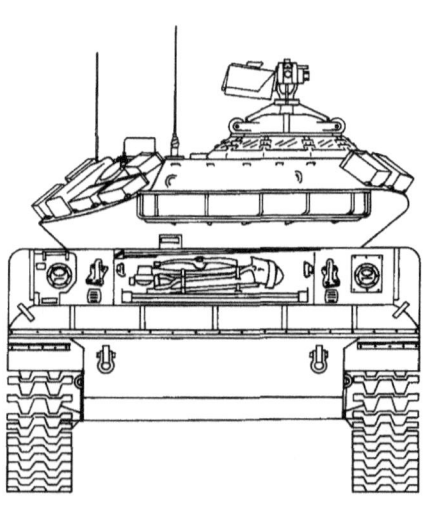

Scale 1:48   ©D.P. Dyer

152mm Gun-Launcher AR/AAV XM551, 5th generation pilot

The new higher flotation screen with the two piece folding surfboard can be seen in these two views of pilot number 12.

These two photographs of an early production Sheridan show the flotation screen fully erected. Note the small folding step just above the registration number in the upper view. Compare with pilot number 12 which does not have this step.

Above, an early production M551 is being evaluated by the Armor and Engineer Board at Fort Knox in October 1967.

THE SHERIDAN IN PRODUCTION

The first increment (Fiscal Year 1966) of a multiyear production contract was awarded to the Allison Division of General Motors Corporation on 12 April 1965 to produce the Sheridan at the Cleveland Army Tank Automotive Plant. Classification of the vehicle as a Limited Production type was approved in May. Watervliet Arsenal was assigned the responsibility for the manufacture of the gun-launcher, however, the breech mechanism was to be produced by Allison. Because of numerous problems with the combustible case ammunition, limited quantities were being produced in-house by the U. S. Army Munitions Command to meet training requirements. Work continued in an effort to solve the problems with this ammunition. Also, limited quantities of the fire control equipment were produced at Frankford Arsenal to meet the early production schedule.

The Army Vice Chief of Staff approved the classification of the Sheridan as Standard A in May 1966, although several waivers were required for items that still did not meet specification requirements. On 29 June 1966, the first two production vehicles were completed and they were delivered the following month to the U. S. Army Test and Evaluation Command for durability testing. Although production continued in Cleveland, the test program indicated the need for further development, particularly on the weapon system. As a result in December 1967, the Secretary of the Army approved a stretch-out of the program which was finally completed with the acceptance of the last vehicle by the government inspector on 2 November 1970. A total of 1662 M551s were produced. The overall cost of the program exceeded 1.3 billion dollars for the 1662 vehicles including missiles, ammunition, and other support items. That cost was a 23 per cent increase from the initial planning figure which was expected to cover 2426 vehicles plus missiles and ammunition. However, the program did produce a vehicle with capabilities unmatched by any other in the Army inventory.

Above, a Shillelagh missile is launched from a Sheridan during the test program in October 1967. Below, the track action of an M551 is being photographed.

Front and side views of a Shillelagh missile launch from the Sheridan appear at the right and below respectively.

Above, an M551 is fording Otter Creek at Fort Knox. At the left below, six G-11A parachutes are installed on this Sheridan during air drop experiments. Later, the use of eight parachutes was standardized for the low velocity air drop of the M551.

The M551s above and below at the right are operating during the cross-country tests. Both of these vehicles are fitted with the early searchlight. At the bottom left, the commander's armor enclosure is installed on this Sheridan at Aberdeen Proving Ground.

Above, a Sheridan is operating at Fort Richardson, Alaska on 15 February 1968 and at the right, another M551 is crossing a frozen lake at the Arctic Test Center. Below, a Sheridan swims across Kyle Lake at Fort Campbell, Kentucky on 23 May 1968.

The bore evacuator has been eliminated and the tube muzzle reinforced on the gun-launchers installed in these two Sheridans. Both have the turret ventilator on the front of the turret roof and are fitted with the commander's front armor shields. The late model searchlight is mounted on the vehicle below and the commander's .50 caliber machine gun has a flash hider installed.

The gun-launcher with the bore evacuator on the early production M551 in the top photograph can be compared with the later version in the Sheridan below. The latter vehicle has been fitted with an additional shield in front of the commander's armor enclosure. It also features a protective frame around the searchlight.

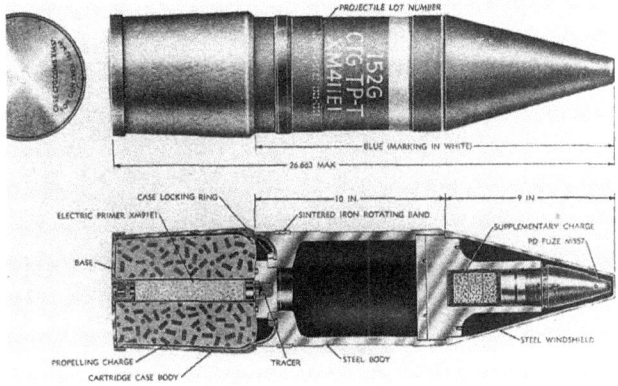

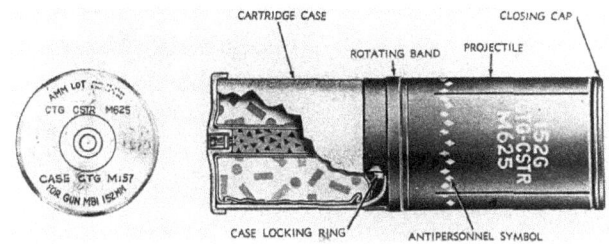

The XM411E1 training round with the XM157 combustible cartridge case appears at the left and the M625 canister round is above. The latter is assembled with the later M157 combustible cartridge case.

Tests of both the pilot and production vehicles had revealed numerous problems. A major one concerned the combustible case conventional ammunition which proved to be fragile and easily damaged by rough handling. If exposed to moisture, the cases would swell or distort making it difficult or impossible to load them into the chamber of the gun-launcher. A particularly serious problem was the incomplete combustion of the moisture laden cases which left a smoldering residue in the chamber after firing. Such residue could ignite the combustible case of the next round loaded with disastrous results.

The original 152mm conventional rounds were assembled with the XM157 combustible cartridge case. This case consisted of a three piece body with tapered side walls and a flat base. A later two piece version, standardized as the M157, had straight side walls and a raised housing for the primer in the center of the base. To solve the moisture absorption problem, both types were enclosed in a neoprene barrier bag that was to be removed just before loading into the cannon. Later, a new combustible cartridge case, standardized as the M205, was adopted. This was the same configuration as the M157, but it was formed from high density felted nitrocellulose, inert fibers, and resin. It was harder and stronger than the earlier cases and less susceptible to mechanical damage. An elastomeric barrier bag was provided to protect the M205 case from moisture.

Later, ballistic protective covers were provided in the vehicle for the combustible case ammunition. These covers were removed when the round was placed upon the loading tray and the elastomeric barrier bag was stripped off just before insertion into the chamber. Needless to say, the time required to remove these items reduced the rate of fire. When not in use, the ballistic protective covers were stowed beneath the hull ammunition racks.

At the right is the open breech carbon dioxide scavenging system installed to blow any residue out of the chamber. Note the mirror fitted to permit a view of the breech section of the chamber.

Although the barrier bags greatly reduced the moisture absorption problem and the resulting chamber residue, they did not eliminate it completely. To remove any residue left after firing, two jets were installed to blow carbon dioxide gas into both parts of the chamber after the breech was opened. A mirror also was mounted so that the loader and the tank commander could see into the open breech chamber and determine if it was clear of residue. Unfortunately, if there was residue, this open breech scavenging system blew any smoldering particles around inside the turret. This was a dangerous situation in the presence of combustible case ammunition. An experimental compressed air system also was evaluated.

A better solution to the residue problem was a closed breech scavenging system (CBSS). The development of such a system continued at the Allison Division through the end of 1967 and in March 1968, it was submitted for service tests. The following July, the Department of the Army approved a program to incorporate the CBSS into production beginning with vehicle number 700. Two months later, a retrofit program was approved to install the CBSS on previously produced vehicles. The CBSS consisted of a four stage air compressor, two compressed air cylinders, a pressure gage, a regulator and fittings. It supplied the high pressure air to jets installed in the gun-launcher. This system blew any remaining residue out of the weapon before the breech opened.

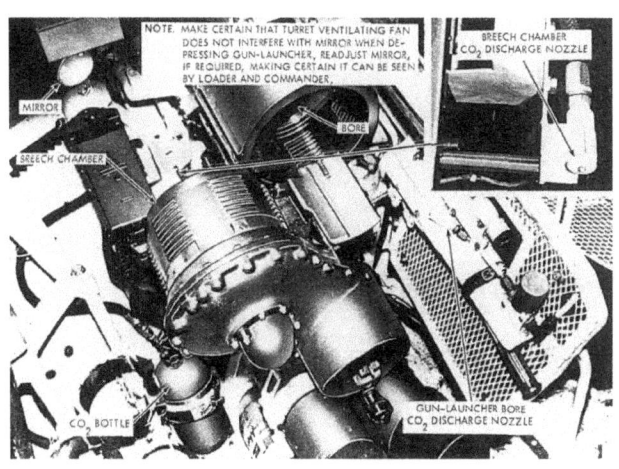

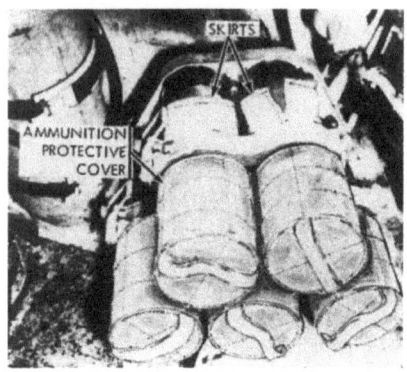  

Above, from left to right, are the turret horizontal ammunition rack, the turret three round vertical rack, and the turret single round vertical rack with a missile installed. Note the compressed air cylinder which replaced a second round in the latter when the CBSS was installed.

Prior to vehicle number 700, stowage space was provided for 30 rounds of ammunition which usually consisted of 20 combustible case conventional rounds and 10 missiles. In the hull, 13 conventional rounds were located in two racks to the left of the driver and seven missiles were to his right. In the turret, five conventional rounds were in a horizontal rack under the gun-launcher. Either conventional rounds or missiles could be stowed in three vertical racks to the left of the gun-launcher. An additional pair of vertical racks at the rear of the turret basket could be stowed with either conventional rounds or missiles. One of the latter racks was replaced by a compressed air cylinder when the closed breech scavenging system was installed reducing the total ammunition stowage to 29 rounds.

As described earlier, three types of conventional combustible case ammunition were under development. These were the XM409 HEAT-T-MP, the XM410-WP (smoke), and the XM411-TP training round. However, because of premature detonations during the development program, the smoke round was not released for production. The standardized M409 and M409A1 HEAT-T-MP became the principal conventional rounds for the gun-launcher. They differed only in the cartridge case, with the M409A1 being fitted with

the later M205 cartridge case. Although the M409 multipurpose round was used for blast and fragmentation effect, an experimental XM657E2 high explosive round was under development, but it was never standardized. The war in Vietnam resulted in the development and production of the M625 and M625A1 canister cartridge. This round contained approximately 10,000 13 grain flechettes which were discharged in a conical pattern when the projectile ruptured after leaving the muzzle of the gun-launcher. This not only had a devastating effect upon enemy personnel, but it also could be used to clear brush for a field of fire. It was frequently referred to as the "beehive" round.

At the upper right an XM409E5 HEAT-MP projectile is in flight. Below is the M409A1 HEAT-MP round with the M205 combustible cartridge case. The XM657E2 HE round appears at the lower right.

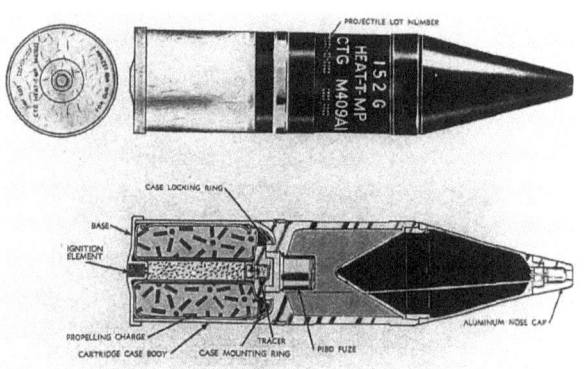 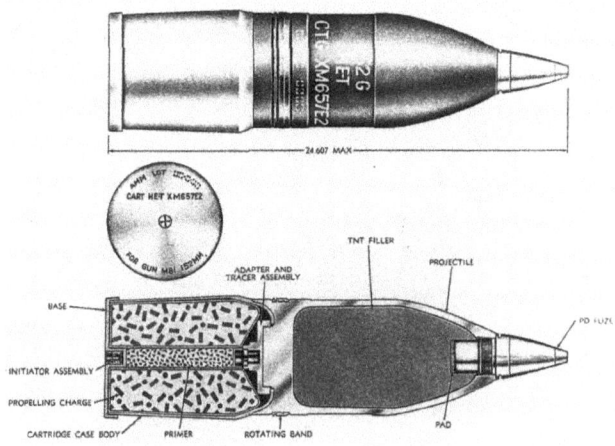

108

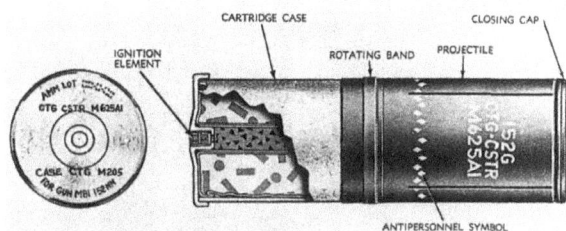

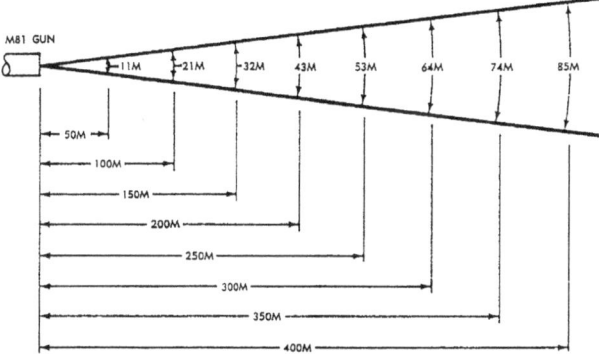

Above is the M625A1 canister round with the M205 combustible cartridge case. The lethal zone covered by the 10,000 flechettes can be seen at the right.

Concern about the problems with the combustible case ammunition resulted, once again, in the consideration of other weapons as the main armament of the Sheridan. During March 1967, the 105mm howitzer XM103E7 was installed in the turret of an M551 at Rock Island Arsenal. Later that month, the howitzer was replaced in the same vehicle by a 76mm gun. Both installations were successful, but they would have reduced the firepower of the Sheridan and were not adopted.

M551, registration number 13C503, is shown at the right and below armed with the 105mm howitzer XM103E7. These photographs from Rock Island Arsenal were dated 10 March 1967.

Here the 76mm gun is installed in M551, registration number 13C503, on 22 March 1967 at Rock Island Arsenal. Note that the 76mm gun had to be fully elevated when the flotation screen was erected.

Above is an early production M551. The gun-launcher is fitted with a bore evacuator and the turret ventilator has not been relocated to the left front on the turret roof.

When the XM551 entered production, it was armed with the 152mm gun-launcher XM81E12. After standardization, the XM81E12 became the 152mm gun-launcher M81. This was the weapon fitted with the open breech carbon dioxide scavenging system. It was equipped with a bore evacuator to help sweep the fumes out of the turret. After the installation of the CBSS, the bore evacuator no longer had any effect and it was eliminated from the later production gun-launchers. When the CBSS was installed on the original M81 gun-launcher, it was referred to as the 152mm gun-launcher M81 modified and the bore evacuator was retained.

The Shillelagh missile, designated as the antitank guided missile MGM 51A, weighed 60.0 pounds and was 43.70 inches in length with a diameter of 5.95 inches. It carried a shaped charge warhead containing eight pounds of Octol explosive and had a maximum range of about 2000 meters.

A training missile, designated as the MTM 51A, was identical in configuration and weight, but was fitted with an inert dummy warhead. An extended range version of the Shillelagh was the MGM 51B in which the weight was increased to 61.3 pounds and the length extended to 45.40 inches. The maximum range was increased to approximately 3000 meters. A similar training missile was the MTM 51B. An XM29 dummy missile could be converted to simulate either the MGM 51A or the MGM 51B. It was used to train the crews to handle and load the missile in the confined space within the turret.

Tests with the 152mm gun-launcher revealed that the tube life was extremely limited when firing the conventional ammunition because of fatigue cracks originating at the muzzle end of the missile keyway. This keyway in the bottom of the bore was required to orient the missile gyros and to prevent rotation of the missile as it moved down the tube. After some modification, the safe tube life was set at 200 conventional rounds. However, experimentation revealed that the depth of the keyway could be reduced until it was no deeper than the rifling grooves thus reducing the stress concentration and extending the tube life to 600 conventional rounds. The tube wall thickness also was increased at the muzzle. The gun-launcher with the shallow keyway was designated as the M81E1. To fit the new keyway, the height of the missile key was reduced from 0.130 inches to 0.075 inches. When fitted with the shorter key, the missiles were redesignated as the MGM 51C and the MTM 51C.

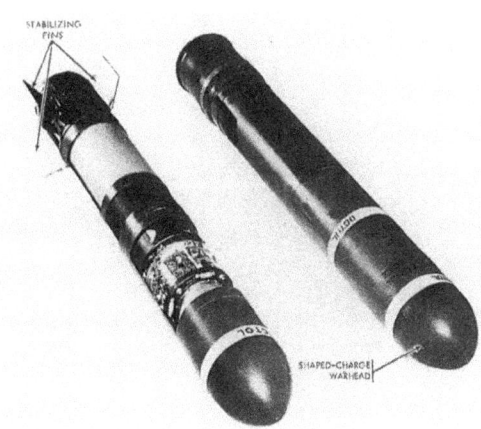

The MGM51A Shillelagh missile is shown at the left complete and partially disassembled.

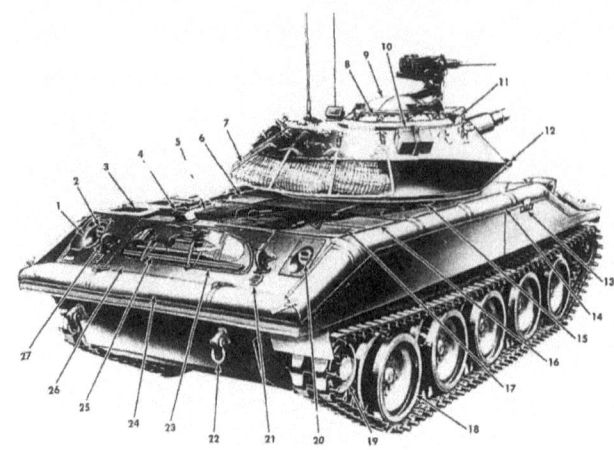

1.	FRONT LIFTING EYE (2)	14.	LEFT SIDE FLOTATION BARRIER COVER
2.	RIGHT FRONT HEADLIGHT ASSEMBLY	15.	XM19 GRENADE PROJECTOR
3.	PERSONNEL HEATER EXHAUST OUTLET	16.	FLOTATION BARRIER STEP
4.	MUZZLE PLUG	17.	FRONT BILGE PUMP OUTLET
5.	152MM GUN-LAUNCHER XM81E12	18.	FIXED FIRE EXTINGUISHER EXTERIOR
6.	7.62MM COAXIAL MACHINE GUN M73		ACTUATING HANDLE
7.	MISSILE SUBSYSTEM TRANSMITTER DOOR	19.	LEFT FRONT HEADLIGHT ASSEMBLY
8.	COMMANDER'S VANE SIGHT	20.	FLOTATION SURFBOARD
9.	CAL..50 MACHINE GUN M2	21.	DUAL IDLER WHEEL (2)
10.	LOADER'S M37 PERISCOPE	22.	M47 PERISCOPE (3) OR M48 PERISCOPE (1)
11.	LOADER'S HATCH COVER		IN CENTER POSITION FOR NIGHT VISION
12.	WATER CAN	23.	DRIVER'S ROTATABLE HATCH COVER
13.	CAL..50 MACHINE GUN AMMUNITION	24.	FRONT TOW SHACKLE (2)

1.	C-2296/VRC INTERCOM SET ACCESS DOOR	15.	FUEL FILLER CAP COVER (2)
2.	LEFT REAR TAILLIGHT ASSEMBLY	16.	BATTERY ACCESS DOOR COVER
3.	ENGINE AIR CLEANER ACCESS DOOR COVER	17.	BATTERY ACCESS DOOR COVER LOCKING PIN (2)
4.	ENGINE EXHAUST OUTLET	18.	DUAL ROAD WHEEL (10)
5.	ENGINE COMPARTMENT EXHAUST GRILL (2)	19.	DUAL DRIVE SPROCKET WHEEL (2)
6.	ENGINE COMPARTMENT AIR INTAKE GRILL	20.	RIGHT REAR TAILLIGHT ASSEMBLY
7.	TURRET STOWAGE RACK	21.	ENGINE COMPARTMENT BILGE PUMP OUTLET (2)
8.	VISION BLOCK (10)	22.	REAR TOWING SHACKLE (2)
9.	COMMANDER'S SPLIT HATCH COVER	23.	TOW CABLE
10.	CAL..50 AMMUNITION	24.	REAR FLOTATION BARRIER COVER
11.	XM44 PERISCOPE BALLISTIC COVER	25.	PIONEER TOOLS
12.	XM19 GRENADE PROJECTOR	26.	ENGINE COMPARTMENT ACCESS COVER
13.	FLOTATION BARRIER STEP	27.	REAR LIFTING EYE (2)
14.	RIGHT SIDE FLOTATION BARRIER COVER		

The various exterior components of the early production M551 are identified in the photographs above. This vehicle has not been fitted with the commander's armor shields and the turret ventilator remains on the left rear turret side wall.

The production configuration of the Sheridan turret followed that of the late pilot vehicles. A welded assembly of rolled homogeneous steel armor, it was mounted on a 76 inch inside diameter turret ring. The port for the spotting rifle, which had been blanked out on the late pilots, was eliminated completely on the production vehicle giving a smooth contour to the upper left of the gun shield. On the early production Sheridans, the turret ventilating blower remained on the left rear side wall. Later, it was relocated to the top on the left front. The turret crew arrangement was the same as on the pilots with the tank commander on the right side behind the gunner. His cupola with its ten vision blocks could be rotated manually or by using the electric power assist which gave it a rotation rate of about nine revolutions per minute. The electric power assist could be operated from a switch on the control box inside the cupola or by pressing the buttons on the spade grips of the .50 caliber machine gun. The latter was mounted on a special pintle support attached to the cupola. To operate the .50 caliber machine gun, the tank commander was required to expose himself in the open hatch, although the split hatch covers provided some protection at each side. When the Sheridan was

deployed to Vietnam, the troops frequently added the armor shield used with the .50 caliber machine gun on the armored cavalry assault vehicle (ACAV) converted from the M113 personnel carrier. Subsequently, a commander's ballistic shield kit was developed which provided protection from the front and improved the protection at the sides. This kit was installed on production vehicles 140 through 233 and also was available for application in the field. Later, the ballistic shield kit was modified to include a rear enclosure to provide protection from the rear. The ungainly appearance of the complete ballistic shield assembly resulted in its being

Details of the turret and stowage can be seen at the right. The AN/PVS-2 night vision sight appears in the lower left of this figure attached to the .50 caliber machine gun.

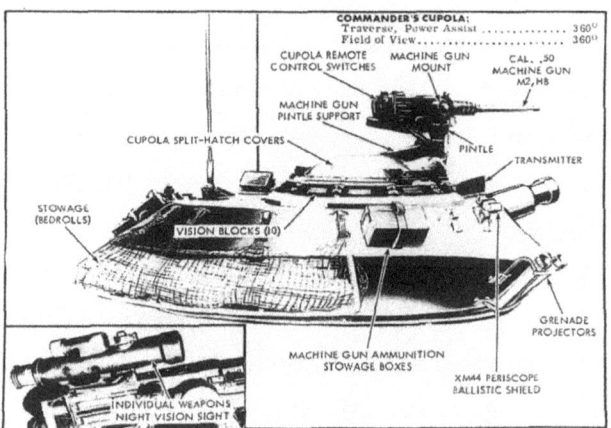

112

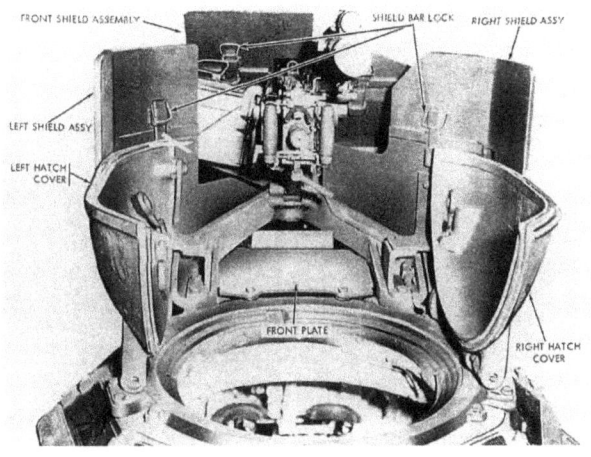

At the left and above are views of the commander's front armor shields installed on the turret. Below, the armor rear enclosure is mounted on this turret trainer vehicle at Fort Knox. The latter photograph was taken by R. P. Vaughan.

referred to by the troops as the "birdcage". Road vibration and firing shock also tended to loosen the screws attaching the ballistic shield assembly to the cupola. On at least one occasion, the "birdcage" fell off while firing on the range.

The electric drive control system provided power to rotate the turret and elevate the cannon with or without gyroscopic stabilization. An electronic servo system provided a constant rate of traverse or elevation proportional to the gunner's or commander's control handle deflection when operating without the stabilizer. When in use, the stabilization system maintained the weapon and line of sight orientation when the vehicle was in motion. The turret could be traversed at maximum rates of six revolutions per minute without the stabilizer and eight revolutions per minute with the stabilizer. The maximum elevation rate was 4

degrees per second. The commander could always override the gunner's controls.

The M73 7.62mm coaxial machine gun in the early production vehicles was succeeded first by the M73E1 and later by the 7.62mm M219 machine gun. Smoke screen protection was provided by eight XM176 grenade launchers installed in two racks of four. One rack was mounted on each side at the front of the turret. The mount design was modified during the production run eliminating the support bars at the front of the assembly. Each grenade launcher contained one M34-WP grenade and one M8-HC grenade. The launchers could be fired one at a time, four at a time, or all at once. The range was 100 to 150 feet.

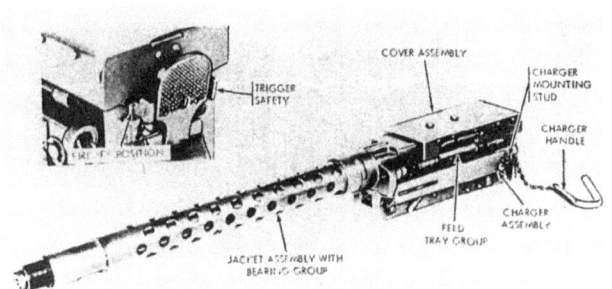

Above is the 7.62mm M73 machine gun used as the coaxial weapon on the M551. Below are the early (left) and late (right) smoke grenade launcher installations.

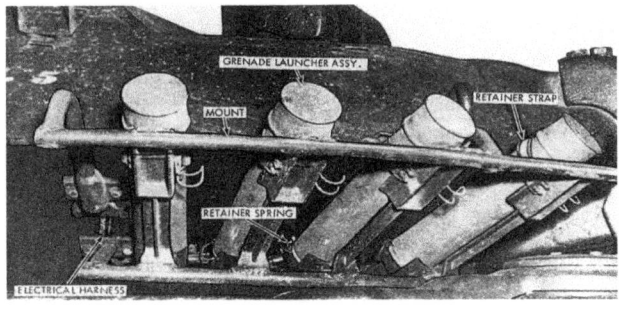

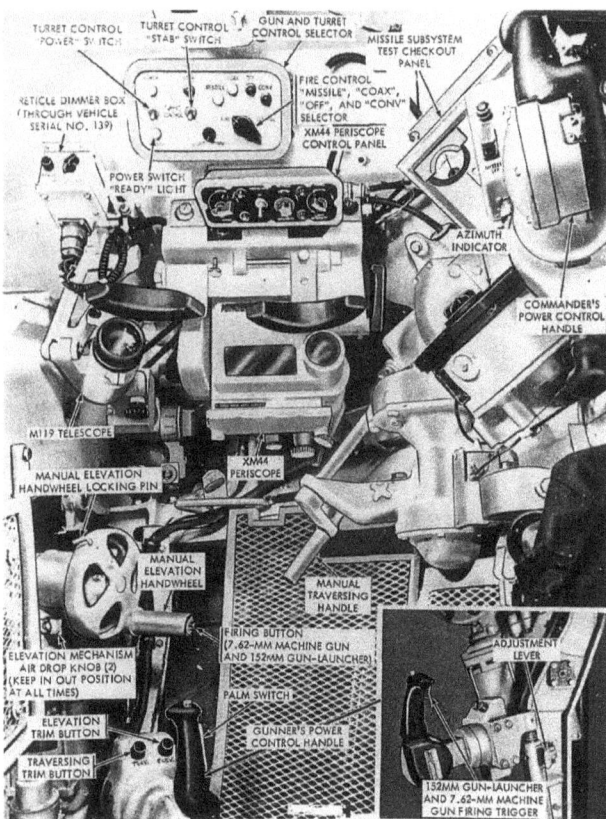

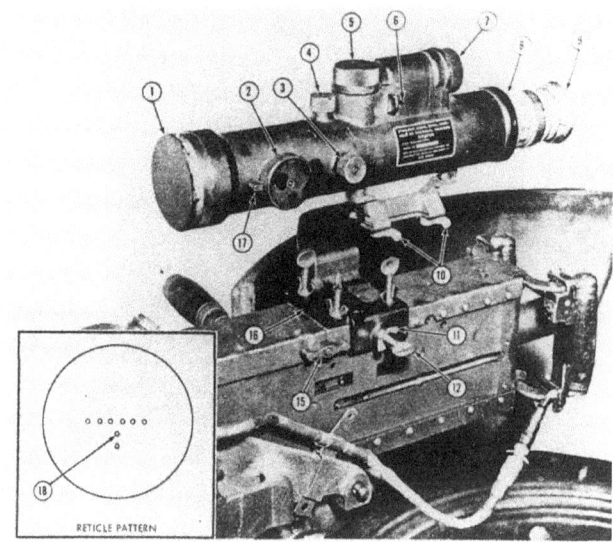

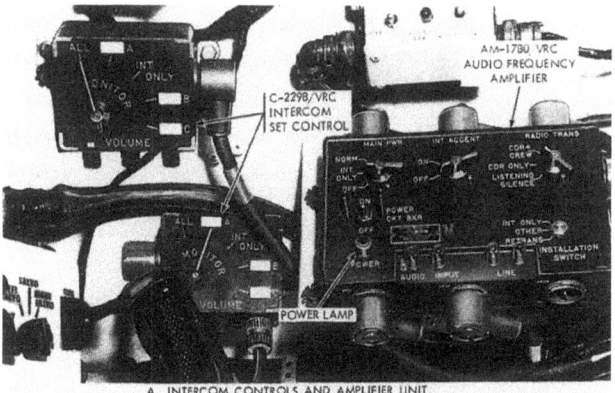

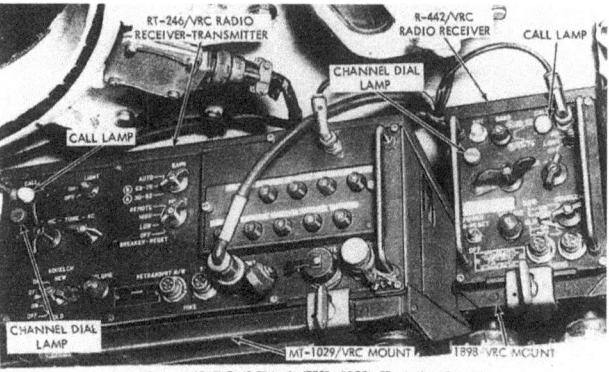

1. LENS CAP
2. OBJECTIVE LENS FOCUS KNOB
3. RETICLE ADJUSTMENT KNOB - AZIMUTH
4. RETICLE ADJUSTMENT KNOB - ELEVATION
5. POWER SUPPLY CAP
6. RETICLE LAMP/IMAGE TUBE ILLUMINATION SWITCH
7. BATTERY CAP
8. FOCUS RING
9. EYESHIELD
10. SIGHT-TO-ADAPTER BRACKET LOCK LEVER
11. WINGNUT (4), WASHER (4)
12. THUMBSCREW (4)
13. SIGHT RETAINING STRAP
14. ADAPTER BRACKET STOWAGE MOUNTING BRACKET
15. MACHINE GUN COVER LATCH
16. SIGHT ADAPTER BRACKET
17. FOCUS KNOB LOCKING DEVICE
18. AIMING REFERENCE DOT

The gunner's controls are at the left and details of the commander's AN/PVS-2 night vision sight can be seen above mounted on the .50 caliber machine gun.

A. INTERCOM CONTROLS AND AMPLIFIER UNIT.

B. COMMUNICATIONS TRANSMITTER, RECEIVER, AND MOUNTS.

The M119 telescope on the right side of the gun mount was the direct fire control sight when using both the missile and the conventional ammunition. This telescope had a single 8x magnification and it was later replaced by the dual power M127 telescope which had both 8x and 12x magnification. The telescopes were installed in the M149 mount along with the optical tracker for the missile. The XM44 periscope was the night sight for use with the conventional ammunition. Later versions included the XM44E1 and XM44E2. All of these were interchangeable as complete units. These periscopes had an image intensifier tube for night use at a magnification of 9x. A unity power optical system also was included for direct "daylight" viewing. An individual night vision sight AN/PVS-2 was provided for use by the tank commander. It could be installed on the .50 caliber machine gun and had a field of view of 185 mils at a magnification of 4X. Later, the AN/PVS-2 was replaced by the AN/TVS-2 with 7x magnification.

The radio installation and intercom controls are above. The external telephone appears at the left.

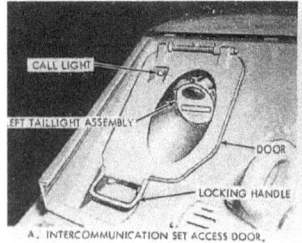

A. INTERCOMMUNICATION SET ACCESS DOOR.

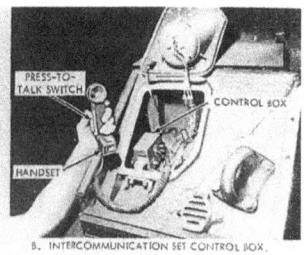

B. INTERCOMMUNICATION SET CONTROL BOX.

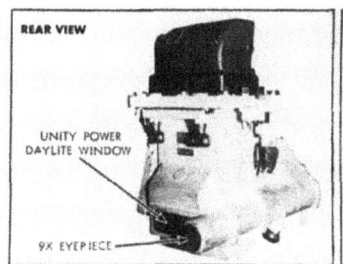
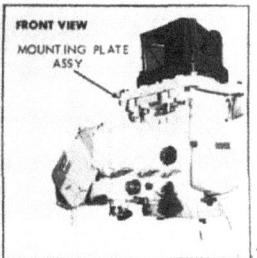
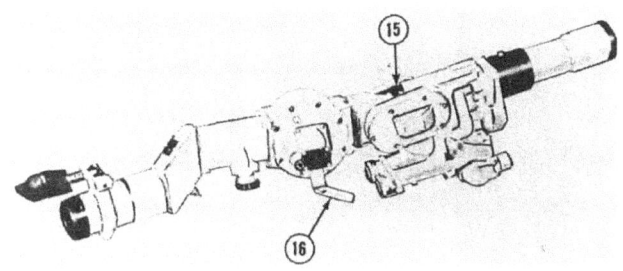

Above, the XM44 periscope can be seen at the left and the M127 telescope is at the right. The numbers 15 and 16 refer to the parallax adjustment and the magnification power lever respectively. The location of the various fire control components is illustrated in the sketch below.

The guidance and control system for the Shillelagh missile consisted of the following:
1. The optical tracker mounted just above and aligned with the gunner's telescope. This was the eye of the system which determined any deviation of the missile from the line of sight and transmitted that information to the signal data converter.
2. The rate sensor measured the rates of turret traverse and gun elevation as the gunner tracked the target. These data were then transmitted to the signal data converter.
3. The signal data converter computed the necessary corrections to keep the missile on the line of sight to the target.
4. The modulator converted the output from the signal data converter to the high current needed to operate the transmitter.
5. The optical transmitter transmitted the infrared guidance command signals to the missile.
6. The power supply converted power from the vehicle system to operating power for the guidance and control system.
7. A test/checkout panel indicated the operational condition of each unit during a test of the system.

When the Sheridan was deployed to Vietnam in early 1969, the missile subsystem was omitted and the main weapon ammunition stowage was limited to the conventional combustible case rounds. A less missile system kit was provided for the removal of some missile subsystem items which were not required and the installation of additional components to permit turret and weapon stabilization. Also, production vehicles 140 through 223 and 740 through 885 were manufactured without the missile subsystem.

On vehicles not equipped with the missile subsystem, the M149 mount was replaced by the telescope mount M165. The stowage of machine gun ammunition and smoke grenades also was increased using the space previously occupied by the missile subsystem components.

A searchlight kit also was provided for installation on the M551 beginning with vehicle number 140. Initially, this was the AEG XSW30U infrared/whitelight unit which was rectangular in shape with a double door cover and was mounted on the left side of the gun shield. Later, the new design AN/VSS-3 or AN/VSS-3A searchlight kit was provided for installation in the same location. The configuration of the stowage rack on the rear of the turret also was modified during the production run.

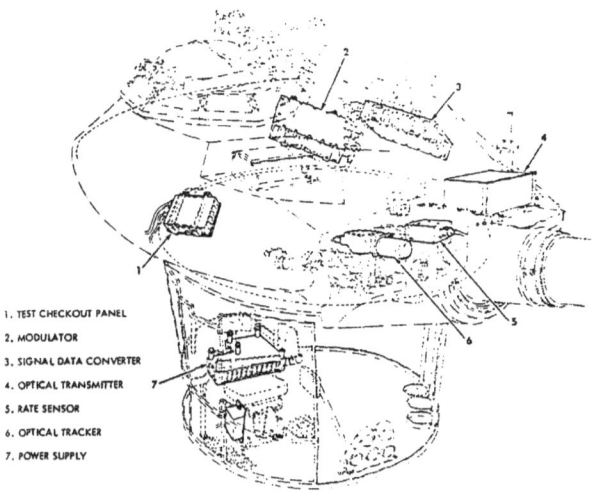

1. TEST CHECKOUT PANEL
2. MODULATOR
3. SIGNAL DATA CONVERTER
4. OPTICAL TRANSMITTER
5. RATE SENSOR
6. OPTICAL TRACKER
7. POWER SUPPLY

GUN SHIELD

The early AEG XSW30U searchlight above can be compared with the later AN/VSS-3 installation below.

115

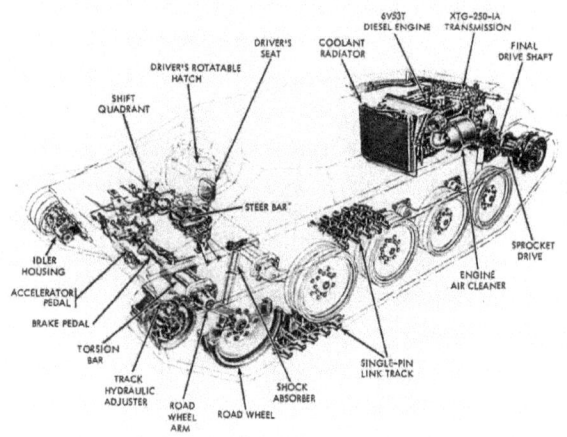

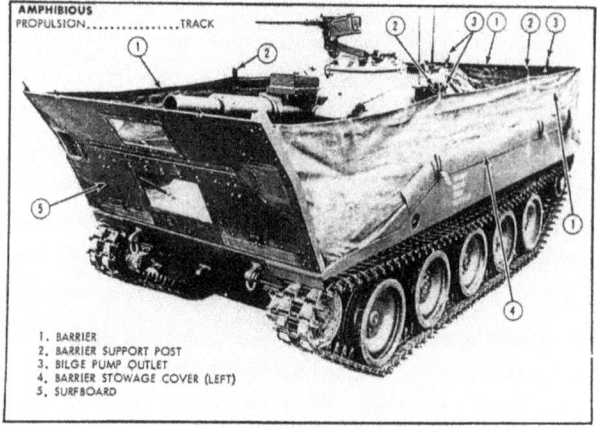

1. BARRIER
2. BARRIER SUPPORT POST
3. BILGE PUMP OUTLET
4. BARRIER STOWAGE COVER (LEFT)
5. SURFBOARD

The illustration at the top left shows the location of various hull mounted components in the M551. At the top right, the flotation screen is fully erected on XM551 pilot number 12. Note the absence of the folding step on the side of the vehicle.

The high obliquity hull on the production M551 was a welded assembly of 7039 aluminum alloy armor plate. As described previously, it was surrounded by a lightweight structure filled with closed cell polystyrene foam and several removable flotation boxes to increase the volume. The collapsible, neoprene coated, nylon flotation barrier was stowed around the top sides and rear of this structure. When erected and attached to the folding surfboard in front, it provided sufficient displacement to float the vehicle with a freeboard to the barrier top of approximately 24 inches. The static freeboard just to the top of the hull was −2½ inches in front and +2½ inches in the rear. When in motion, this changed to −½ inch in front and +½ inch in the rear. Two bilge pumps were installed. One of these pumped water from the crew compartment or the space on top of the front hull behind the surfboard, depending upon the control setting. The other pump removed water from the engine compartment.

Originally, the cannon could not be fired with the surfboard and flotation barrier erected. Starting with vehicle

1. PERISCOPE WASHER LIQUID RESERVOIR
2. PERISCOPE WASHER PUMP
3. HATCH COVER HOLD-OPEN HOOK
4. STEER BAR
5. INDICATOR PANEL
6. PARKING BRAKE LOCK HANDLE
7. HEADLIGHT DIMMER SWITCH
8. BRAKE PEDAL
9. DRIVER'S SEAT
10. CONVENTIONAL AMMUNITION STOWAGE RACK
11. M47 PERISCOPE (3)
12. HATCH COVER HANDLE GRIP
13. HATCH COVER LOCKING LEVER
14. TRANSMISSION SHIFT LEVER
15. DRIVER'S SWITCH PANEL
16. WATER STEER LEVER
17. HAND THROTTLE CONTROL KNOB
18. C-2297/VRC INTERCOM SET
19. ACCELERATOR PEDAL
20. MISSILE STOWAGE RACK

The controls and stowage items in the M551 driver's compartment are identified above.

number 140, a quick release was provided for the surfboard which permitted the weapon to be used when aimed forward. Folding steps were added on the production vehicle for use when the flotation barrier was erected. Mounted on each side of the hull, they could be unfolded to allow a crew member to step over the barrier.

The driver's station in the center front hull of the production vehicle was essentially the same as on the late pilots. The driver was provided with three M47 periscopes

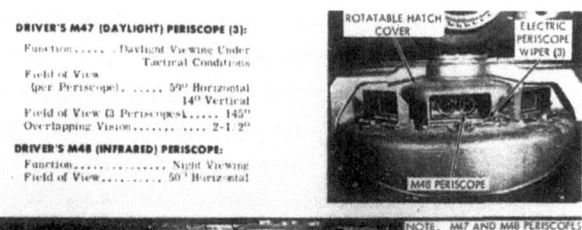

DRIVER'S M47 (DAYLIGHT) PERISCOPE (3):
Function . . . . . Daylight Viewing Under Tactical Conditions
Field of View (per Periscope) . . . . . 59° Horizontal 14° Vertical
Field of View (3 Periscopes) . . . . 145°
Overlapping Vision . . . . 2-1.2°

DRIVER'S M48 (INFRARED) PERISCOPE:
Function . . . . Night Viewing
Field of View . . . . 50° Horizontal

NOTE, M47 AND M48 PERISCOPES ARE INTERCHANGEABLE IN CENTER POSITION.

At the left is the periscope installation in the driver's rotating hatch. Note that the M48 infrared periscope can be substituted for the center M47 periscope.

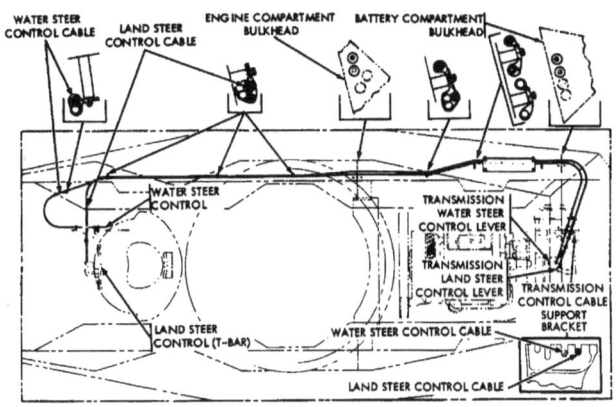

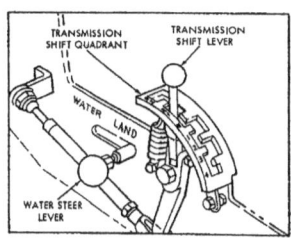

The controls for land and water operation of the M551 are sketched at the left and the transmission and water steer levers are illustrated above.

in his rotating hatch, the center one of which could be replaced by an M48 infrared night vision periscope. These periscopes were equipped with washers and wipers. The driver's adjustable seat could be flipped up to expose the floor escape hatch. Originally, the floor escape hatch cover was aluminum, but it was replaced on later vehicles by a steel cover. The driver had a T-bar steering control for use on both land and water. However, the steer shift control had to be set for either land or water operation.

The production M551 was powered by a version of the General Motors 6V53T diesel engine with an aluminum block to minimize weight. This turbosupercharged, liquid cooled, engine developed 300 gross horsepower at 2800 revolutions per minute. The engine was coupled to the Allison XTG-250-1A transmission which featured four speed ranges forward and two reverse. For land operation, pivot steering was available in first and both reverse ranges. Geared steering was provided in second, third, and fourth. In water, pivot steering was available in all ranges except fourth. The 17½ inch wide, single pin, tracks were driven by the rear mounted, 11 tooth, center drive sprockets. The flat track torsion bar suspension supported the vehicle on five road wheels per side. The diameter of these rubber tired road wheels was 28 inches compared to the 24 inch diameter wheels on the early pilots. An hydraulically adjustable idler was at the front of each track. On the early production vehicles, this idler was attached by eight screws. Effective with vehicle number 178, a redesigned idler was installed attached by 11 screws.

The two photographs at the right side of this page show the right (upper) and left (lower) sides of the power pack consisting of the General Motors 6V53T diesel engine and the XTG-250-1A transmission. The components are identified in each view.

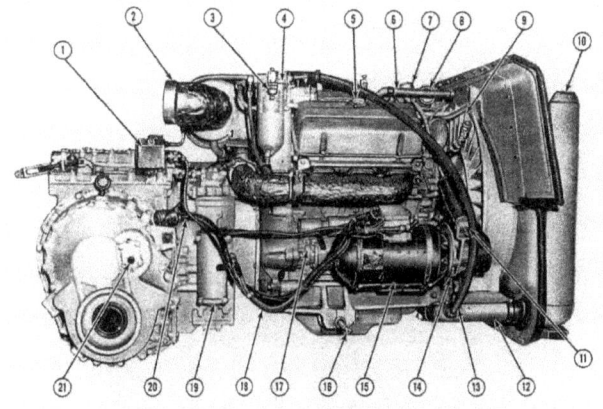

1. STARTER RELAY BOX
2. ENGINE EXHAUST ELBOW
3. MAIN FUEL HOSE QUICK DISCONNECT
4. PRIMARY FUEL FILTER (STRAINER)
5. ENGINE OIL FILLER CAP
6. CRANKCASE BREATHER HOSE
7. AIR BOX ACCUMULATOR GAGE
8. AIR BOX ACCUMULATOR
9. RADIATOR COOLING FAN
10. RADIATOR
11. GENERATOR BELT TENSIONER
12. RADIATOR OUTLET TUBE ASSEMBLY
13. WINTERIZATION COOLANT HEATER HOSE (WHEN INSTALLED)
14. GENERATOR DRIVE ASSEMBLY
15. GENERATOR
16. ENGINE MOUNT SCREW (RIGHT SIDE)
17. ENGINE STARTER MOTOR
18. POWER PLANT GROUND CABLE
19. TRANSMISSION OIL FILTER
20. POWER PLANT ELECTRICAL HARNESS
21. SPEEDOMETER ADAPTER CONNECTION

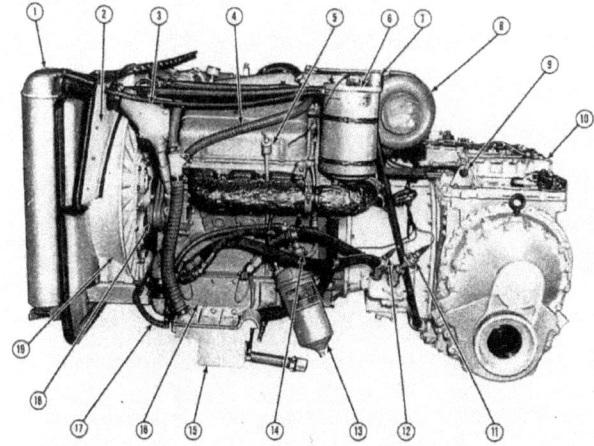

1. RADIATOR
2. RADIATOR COOLING FAN SHROUD
3. INLET THERMOSTAT HOUSING
4. ENGINE CRANKCASE BREATHER HOSE
5. ENGINE OIL LEVEL GAGE
6. COOLANT SURGE TANK
7. TRANSMISSION OIL LEVEL GAGE
8. TURBOCHARGER
9. ELECTRICAL HARNESS RECEPTACLE
10. TRANSMISSION
11. TRANSMISSION OIL PRESSURE SWITCH
12. TRANSMISSION OIL TEMPERATURE SWITCH
13. ENGINE OIL FILTER
14. ENGINE OIL PRESSURE SWITCH
15. ENGINE BREATHER DRAIN COLLECTOR
16. ENGINE/TRANSMISSION OIL COOLER
17. AIR BOX DRAIN HOSE
18. ENGINE COOLANT PUMP
19. RADIATOR COOLANT FAN

117

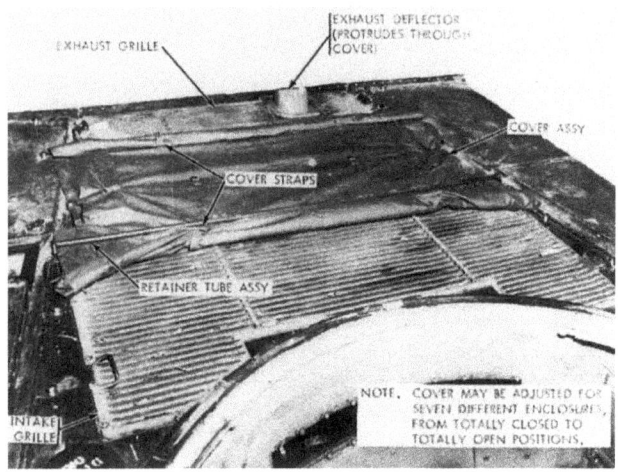

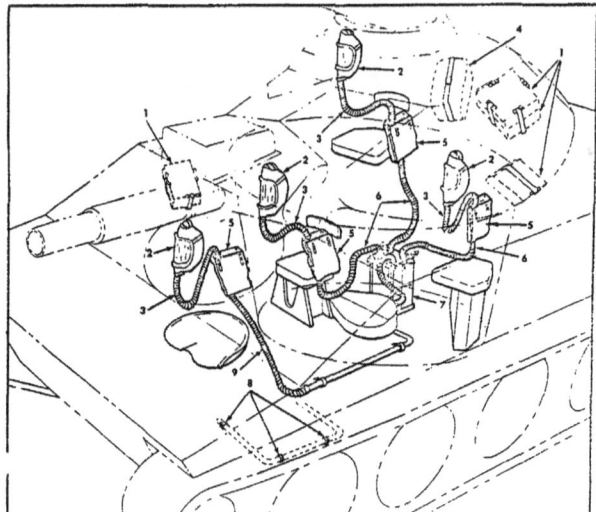

Above, the winterization kit is installed on the rear deck of the M551. At the right, the components of the M8A3 gas-particulate filter system are identified in the sketch.

1. CARRIER WITH PROTECTIVE MASK AND CANISTER (STOWED POSITION).
2. M25 OR M25A1 TANK PROTECTIVE MASK.
3. CANISTER TO MASK HOSE.
4. COMMANDER'S, GUNNER'S AND LOADER'S CANISTER-TO-AIR PURIFIER HOSE STOWAGE BAG.
5. CARRIER WITH CANISTER.
6. CANISTER-TO-AIR PURIFIER HOSE.
7. AIR PURIFIER.
8. DRIVER'S CANISTER-TO-AIR PURIFIER HOSE STOWAGE CLIPS.
9. DRIVER'S CANISTER-TO-AIR PURIFIER HOSE.
10. AIR PURIFIER FOOT GUARD.
11. AIRFLOW CONTROL CAP (4).
12. SPRING CLIP.
13. HOSE, AIR FILTER-TO-CONTACT RING SLIP JOINT.

A winterization kit also was developed for cold weather operations. This included a coolant heater system to prevent the batteries from freezing and to keep the engine warm for easy starting. An M8A3 gas-particulate filter system was provided for the protection of the crew. Separate face masks for each crew member were connected to the single filter system. This unit was effective not only against poison gas, but also in a heavy dust environment.

The vulnerability of the Sheridan to mine damage resulted in the development of a mine protective kit. Initially, this kit consisted of an aluminum spacer plate and a steel armor plate bolted to the bottom front of the hull. Later, two additional steel armor plates were attached, one on each side, above the front road wheels. Needless to say, with the belly armor installed, it was no longer possible to use the driver's floor escape hatch.

The photograph by R.P. Vaughan at the left below shows a Sheridan of the 2nd Armored Cavalry Regiment fitted with the mine protective kit. The front of the belly armor plate can be seen bolted to the bottom front. Details of the side and belly plates are shown in the sketches at the bottom left and right respectively.

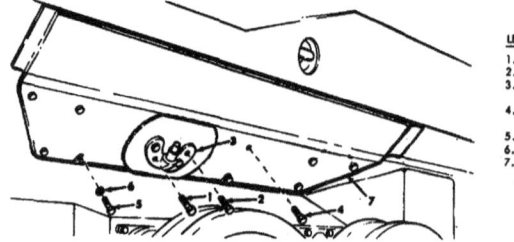

LEGEND
1. SCREW (2)
2. SCREW (2)
3. SHOCK ABSORBER MOUNTING BRACKET
4. SCREW (7 LEFT SIDE, 8 RIGHT)
5. SCREW (2)
6. WASHER (2)
7. PLATE (LEFT SHOWN, RIGHT SIMILAR)

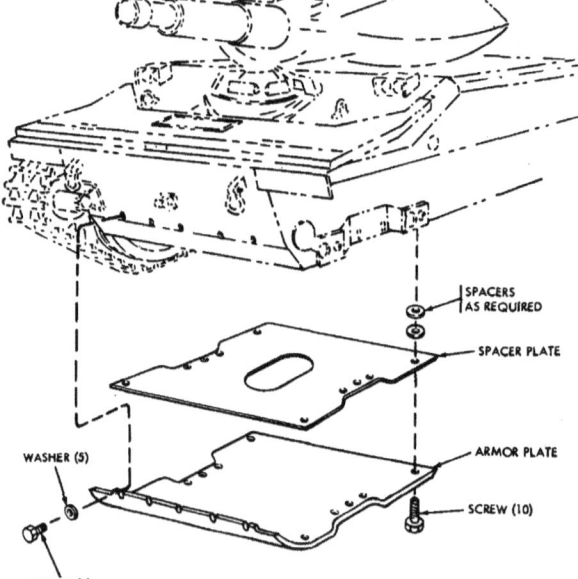

SPACERS AS REQUIRED
SPACER PLATE
ARMOR PLATE
WASHER (5)
SCREW (10)
SCREW (5)

118

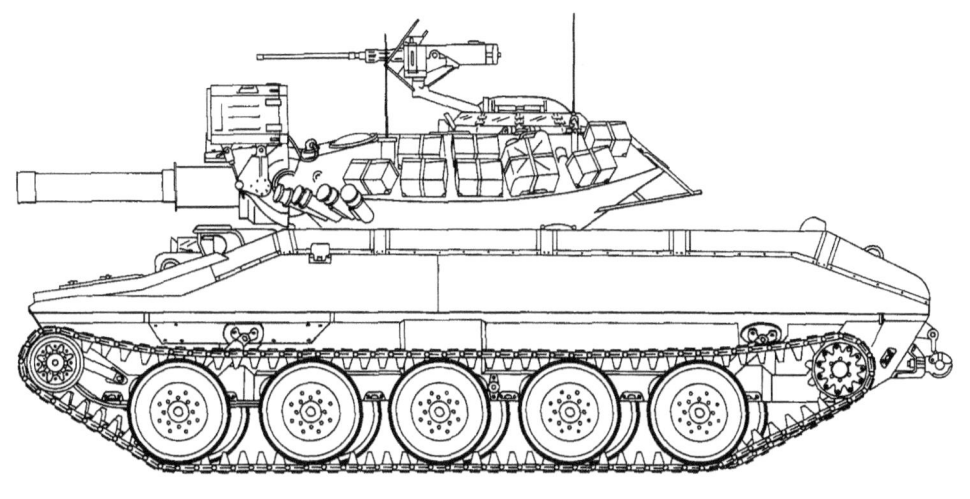

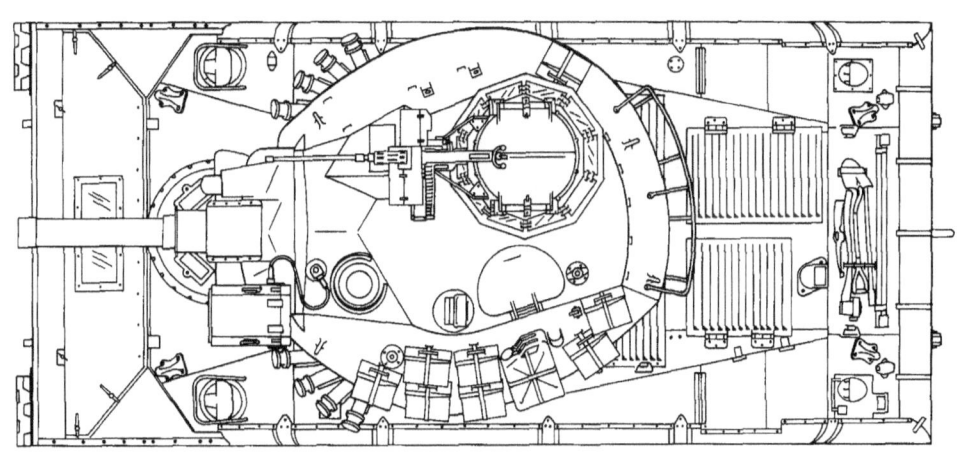

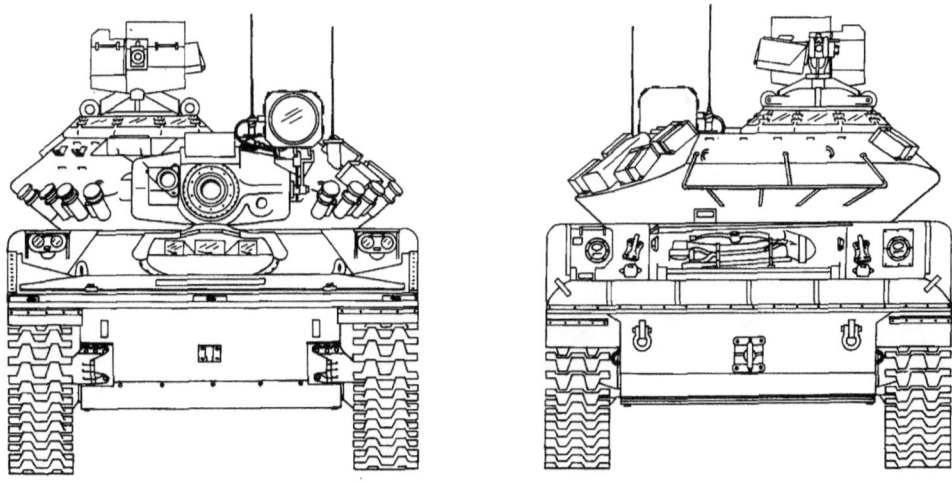

©D.P. Dyer

152mm Gun-Launcher AR/AAV M551

These views show the stowage arrangement on a later production M551. The commander's front armor shields are installed, but the rear enclosure is omitted. Note that the missile transmitter cover is open.

Above, a Sheridan on its platform emerges from a C130 aircraft during a low velocity air drop at Fort Bragg, North Carolina on 11 October 1972. In this case, a single parachute is being used to extract the load from the aircraft.

An important feature of the Sheridan was that it could be delivered directly to the battle area by air. This was achieved by two methods, the first of which was by low velocity air drop (LVAD) using eight 100 foot diameter parachutes. For such an airdrop, the vehicle was rigged on a 24 foot platform with the eight G-11A cargo parachutes. Stacks of crushable aluminum honeycomb on the platform under the vehicle and over the tracks were used to cushion the landing impact. Various components were removed and secured for the airdrop using honeycomb, ½ inch tubular nylon, and tie-down straps. Other movable components were lashed in place. The vehicle itself was lashed to the platform and covered with cotton duck cloth. The eight cargo parachutes were positioned on the front of the vehicle. Two modified 28 foot diameter cargo parachutes were attached to extract the load from the aircraft. After the latter extracted the vehicle and pallet from the aircraft in flight, it was lowered to the ground by the eight cargo parachutes.

Above is a bottom view of the platform as it clears the aircraft and at the right, it is descending on its parachutes.

 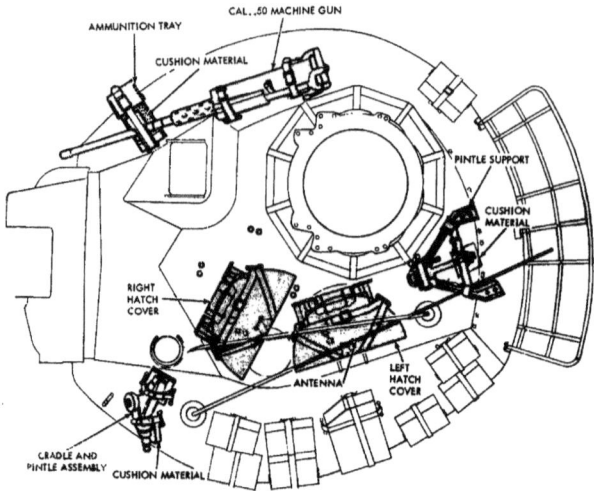

The method of securing components on the Sheridan turret for the low velocity air drop can be seen in the photograph and drawing above. At the left below, the complete vehicle is lashed to the platform prior to loading into the aircraft.

Above at the right, a Sheridan is being loaded into a C130 aircraft and below, the vehicle is on the ground after the air drop.

Above, the first extraction parachute is starting to pull the load out of the C130 aircraft during the low pass over the landing zone.

The second method of air delivery for the Sheridan was by using the low altitude parachute extraction system (LAPES). With this technique, the vehicle was rigged on a 24 foot platform fitted with a low profile rail. Once again, stacks of honeycomb were used to cushion the landing impact. Movable items inside the vehicle were immobilized with nylon webbing. The .50 caliber machine gun, its mount, and the commander's ballistic shield were removed and secured using tubular nylon and honeycomb. The commander's hatch and the driver's hatch were covered with cotton duck cloth and the vehicle was lashed to the pallet. One 15 foot diameter extraction parachute and three 28 foot diameter extraction parachutes were used to pull the load out of the aircraft during a low altitude pass over the landing zone.

Extra Shillelagh missiles also could be delivered by parachute. An eight foot platform was used to airdrop 42 missiles using two 100 foot diameter G-11A cargo parachutes. One 15 foot diameter extraction parachute was used to pull the load out of the aircraft. Stacks of honeycomb were used to cushion the landing impact. Nine Shillelagh missiles could be airdropped in an A-22 cargo bag cushioned by honeycomb on a standard skid. One 64 foot diameter G-12D or G-12C cargo parachute with a pilot chute was used to airdrop the load.

At the right above, the Sheridan has cleared the aircraft, but it has not yet touched the ground. Below, the platform carrying the Sheridan impacts the ground in the landing zone.

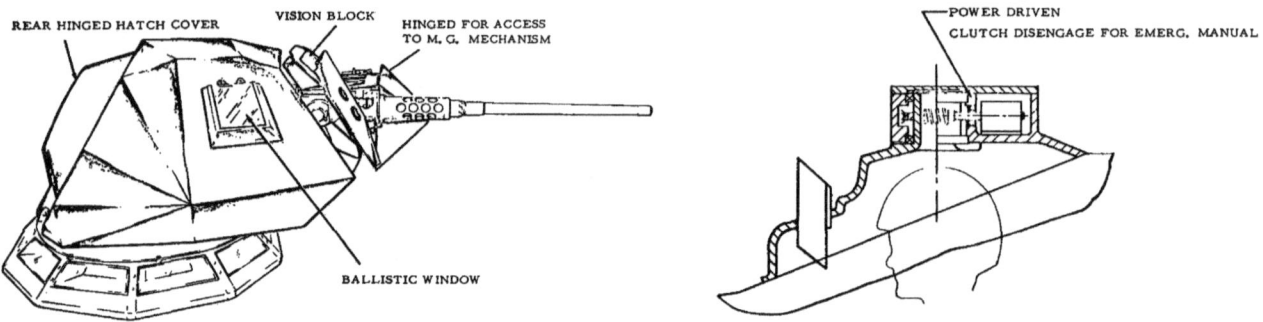

Above are drawings of the proposed commander's armored enclosure (left) and the power operated driver's rotating hatch (right).

MODIFICATION OF THE SHERIDAN

Production of the Sheridan had hardly begun when numerous modifications were under consideration to improve its performance. It also was proposed that the chassis and many components be used as the basis for a whole family of combat vehicles. In November 1966, a meeting at Fort Knox reviewed many of the proposed modifications. They were divided into items considered for three different time periods. The first period covered component improvements available for early introduction into the production vehicle during 1966-1967. These included an armored enclosure to be mounted on top of the commander's cupola which would protect him from .30 caliber armor piercing ammunition. This item had an estimated weight of 250 pounds. Development was proposed of a 152mm incendiary or flame round with a projectile weight of about 49 pounds and a muzzle velocity of 350 feet per second. Power operation was proposed for the driver's rotating hatch to reduce fatigue and permit easy operation on slopes and while moving. It was suggested that the ballistic protection might be improved and the weight reduced in the driver's hatch by replacing the periscopes with a vision block installation. A vehicle weight reduction of about 250 pounds was expected if the foamed hull flotation structure was eliminated and a collapsible fabric flotation barrier was used alone to provide the required volume for flotation.

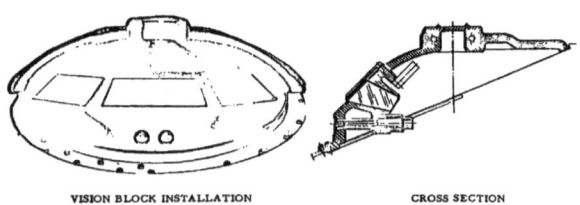

The proposed driver's vision block installation is sketched above and a new hunter-killer fire control system is illustrated below.

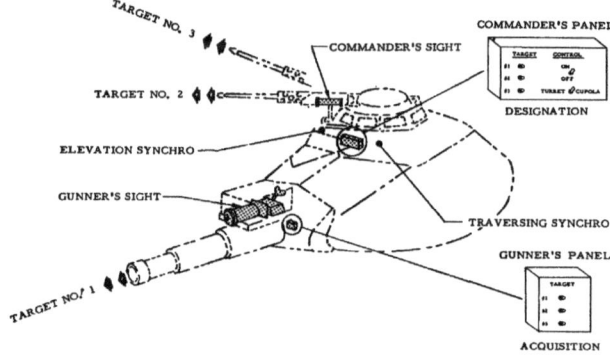

Below, the proposed incendiary projectile is at the left and the new flotation screen design is compared with the original construction at the right.

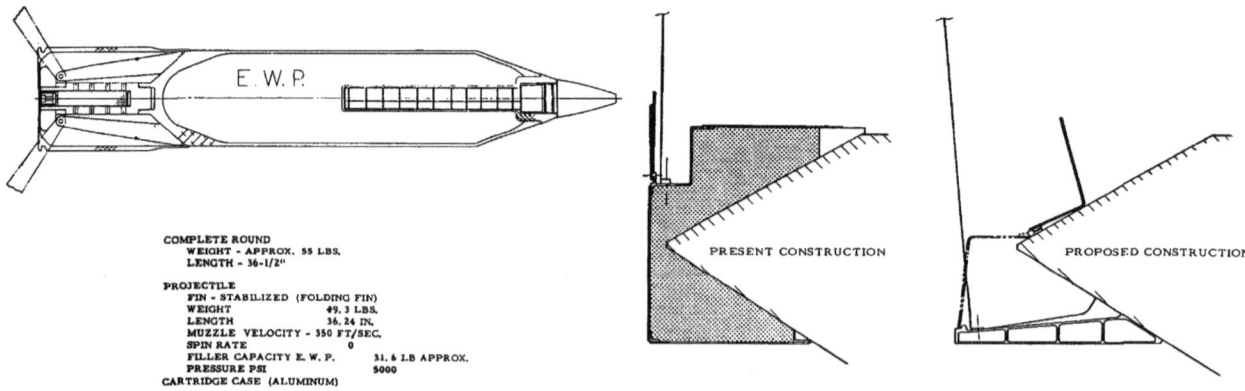

124

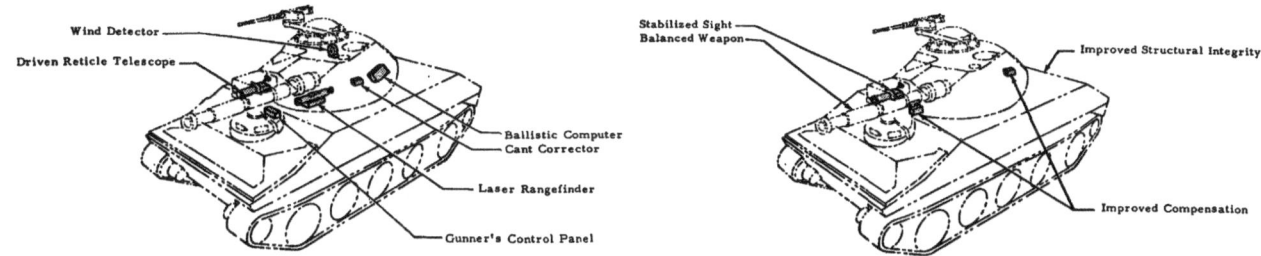

The drawings above show the components of an improved fire control system (left) and a new stabilization system (right). A sketch of the interim laser range finder mounted on the .50 caliber machine gun is at the right below.

Proposals for the 1967-1970 period included an improved fire control system with a laser range finder, a wind velocity detector, a cant corrector, and a ballistic computer. It also included a bore scavenger, an improved version of which was adopted and retrofitted on to the earlier vehicles as the closed breech scavenging system (CBSS). An interim laser range finder was suggested as an attachment on the .50 caliber machine gun. Proposed improvements in protection included the use of applique armor. Many of these items resulted in a heavier weight and the more powerful General Motors 8V53T diesel engine was considered as a new power plant with the X300 transmission. This was expected to increase the vehicle performance, but the new power train itself added about 500 pounds to the weight.

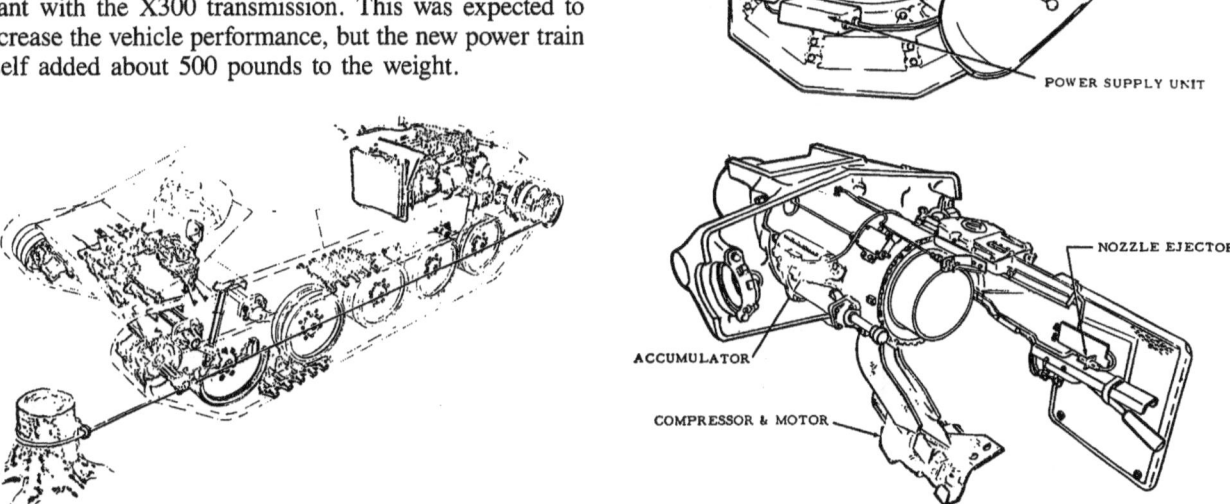

A proposed sprocket mounted winch to permit self recovery of the vehicle is above at the left and an early version of a bore scavenging system is sketched above at the right. Below, the use of applique armor to improve the hull protection is illustrated at the left and the outline of the new 8V53T engine and X300 transmission is shown in the M551 engine compartment at the right.

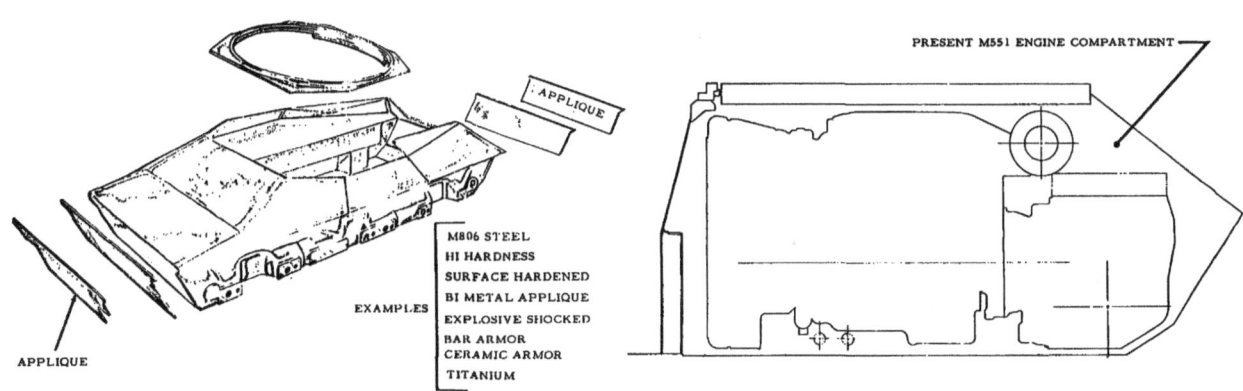

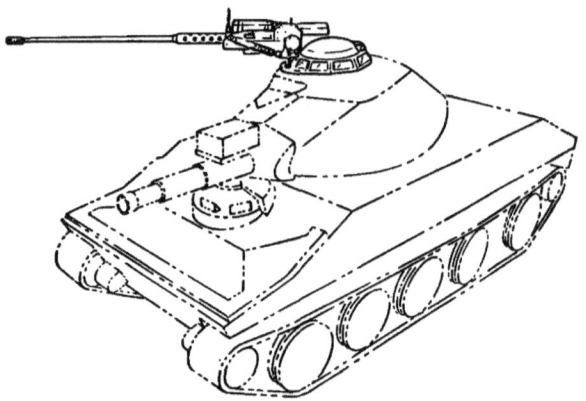

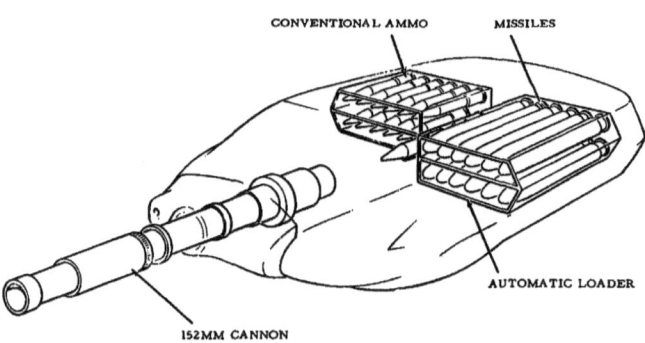

By 1970-1975, it was expected that the .50 caliber machine gun would be replaced by a 20mm or a 25mm weapon. It also was suggested that the XM81E12 gun-launcher be replaced by the long barreled XM150E5 gun-launcher then being developed for the MBT70. This would have added 1607 pounds to the weight and increased the already severe firing shock. Another concept proposed a new turret with an automatic loader and ammunition stowed in the turret bustle. Also, both mechanical and hydropneumatic variable height suspensions were under consideration. Other power trains included the General Motors gas turbine with the X300 transmission and the Continental VCR-465 variable compression ratio diesel engine with the HMPT400 transmission. The reintroduction of water jet propulsion also was proposed to increase the water speed to over five miles per hour.

At the right are sketches of the friction hydropneumatic (upper) and the mechanical tube-over-bar (lower) variable height suspensions. Below is an outline drawing of the VCR-465 engine with the HMPT400 transmission.

At the top left, the commander's .50 caliber machine gun has been replaced by a 20mm or 25mm weapon. Above, the 152mm ammunition is stowed in an automatic loader in the turret bustle. Below, the size of the XM81E12 gun-launcher can be compared with the longer XM150E5.

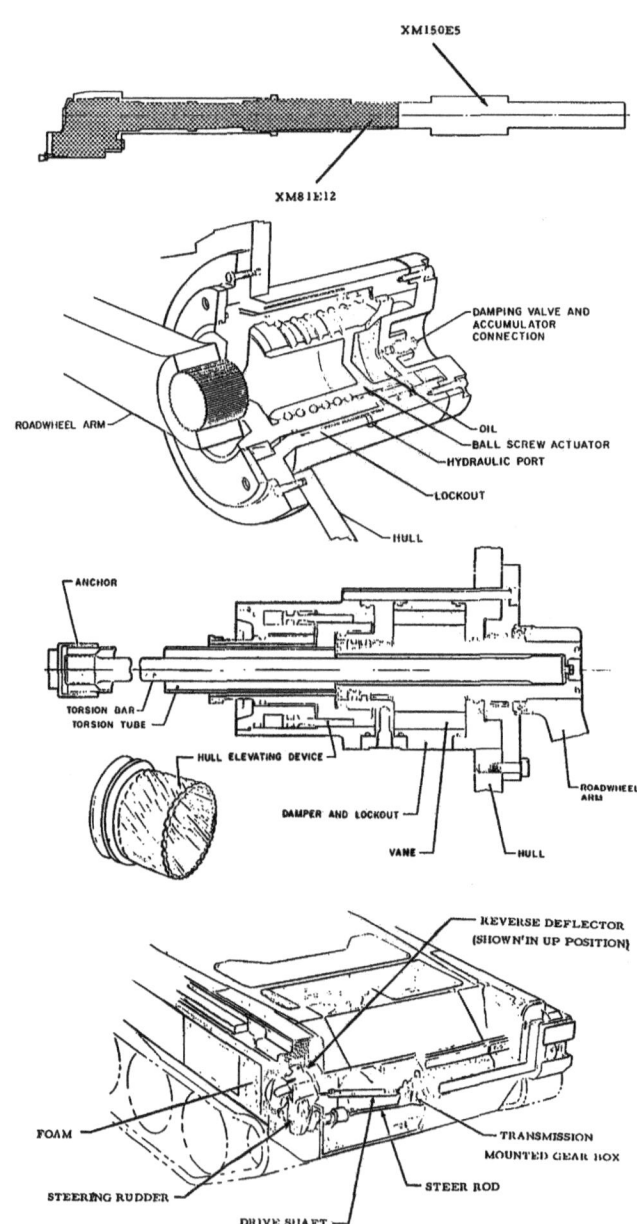

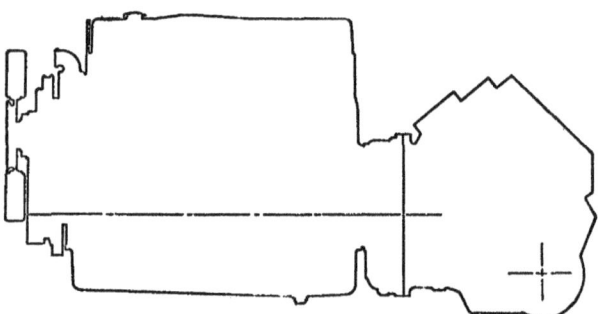

Below is an outline drawing of a proposed power pack consisting of the General Motors gas turbine with the X300 transmission. At the bottom right is a sketch of a dual pump water jet propulsion system.

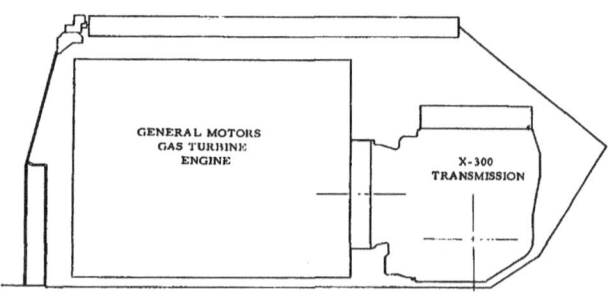

126

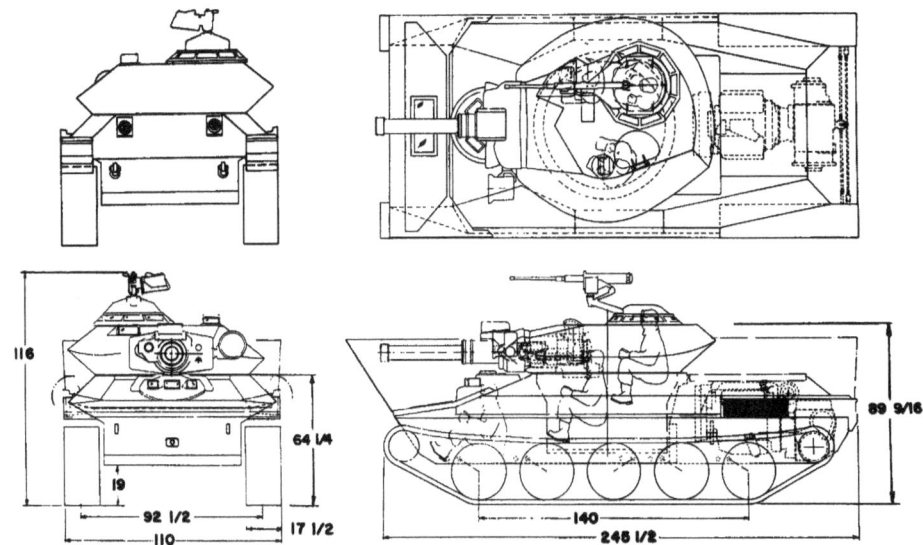

Above is a drawing of the improved Sheridan proposed by General Motors as the M551A1 with the new fire control system and power pack. An artist's concept of this vehicle appears at the bottom of the page.

A concept study presented by General Motors under the unofficial designation AR/AAV M551A1 incorporated many of the proposed new design features. It retained the same armament as the M551, but a laser range finder and a target designation system were added. The original power train was replaced by the 400 gross horsepower 8V53T diesel engine with the X300 transmission. This resulted in an estimated maximum speed of 45 miles per hour and a cruising range of 300 miles. The foam filled flotation structure was eliminated and replaced by a collapsible fabric flotation barrier. The estimated combat weight of the redesigned vehicle was 34,150 pounds. This weight reduced the freeboard of the vehicle to the top of the flotation barrier. When underway, it was expected to be about eight inches in front and 13 inches in the rear. Dual pump water jet propulsion gave an estimated maximum water speed of 5.25 miles per hour.

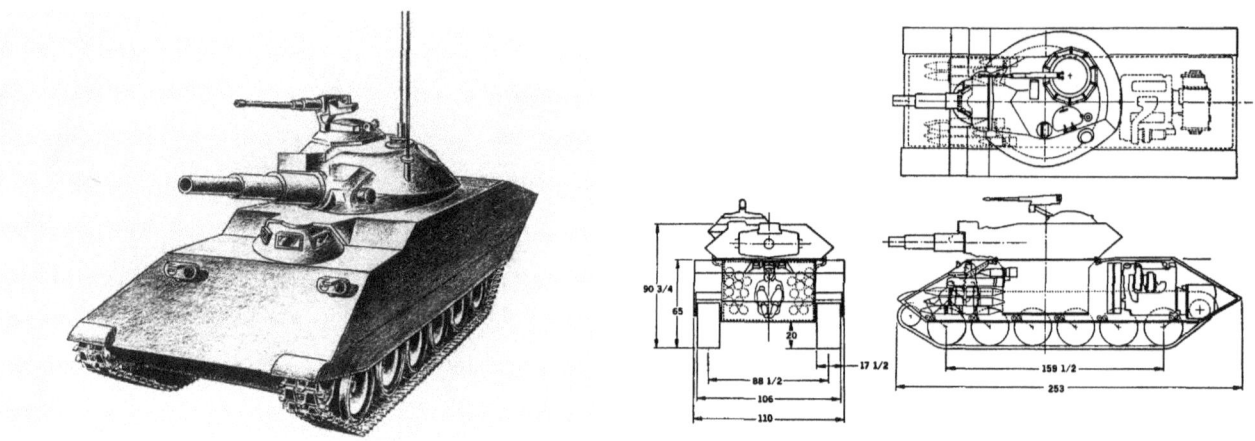

The proposed lightweight main battle tank based upon Sheridan components can be seen in the artist's concept and the three view drawing above.

During this period, numerous studies were in progress for a new main battle tank. One concept that was presented utilized many Sheridan components in the design of a lightweight interim main battle tank. With a new smooth hull front and heavier armor, this vehicle had a lengthened hull with six road wheels per side. It retained the rotating driver's hatch design from the Sheridan.

One proposal from General Motors replaced the main armament in the Sheridan with an M10-8 flame thrower using the standard hull and turret. A shroud was installed to simulate the appearance of the 152mm gun-launcher. A .50 caliber machine gun, a 20mm cannon, or a 25mm cannon was to be mounted coaxially with the flame gun inside the shroud. The normal machine gun armament of the Sheridan was retained. The vehicle carried 150 gallons of flame gun fuel providing a 25 second duration of fire. With a three man crew, the estimated combat weight of the flame thower tank was 32,900 pounds. Its performance was essentially the same as that of the Sheridan.

A concept for a Sheridan based combat engineer vehicle replaced the gun-launcher in the turret with the 165mm XM135 demolition gun. It was manned by a crew of four. Fitted with a steel tipped magnesium alloy bulldozer blade and a 15,000 pound capacity boom and winch, the estimated vehicle combat weight was 39,360 pounds. Although it could be transported by air, it was not air droppable or amphibious. The maximum road speed was estimated to be 35 miles per hour and the cruising range decreased to 250 miles.

The Sheridan chassis also was proposed as a carrier for an armored vehicle launched bridge. With the turret replaced by a flat plate covering the opening, a 20 ton folding bridge was carried on top of the hull. The two man crew consisted of the driver and the bridge operator who rode under a flat hatch in the turret ring cover. The estimated combat weight of the vehicle was 30,000 pounds. It was amphibious and air transportable with the bridge removed. The performance was essentially the same as that of the standard Sheridan.

The proposal for a flame thrower tank based upon the Sheridan is illustrated below. The tanks for the flame gun occupy the former loader's position.

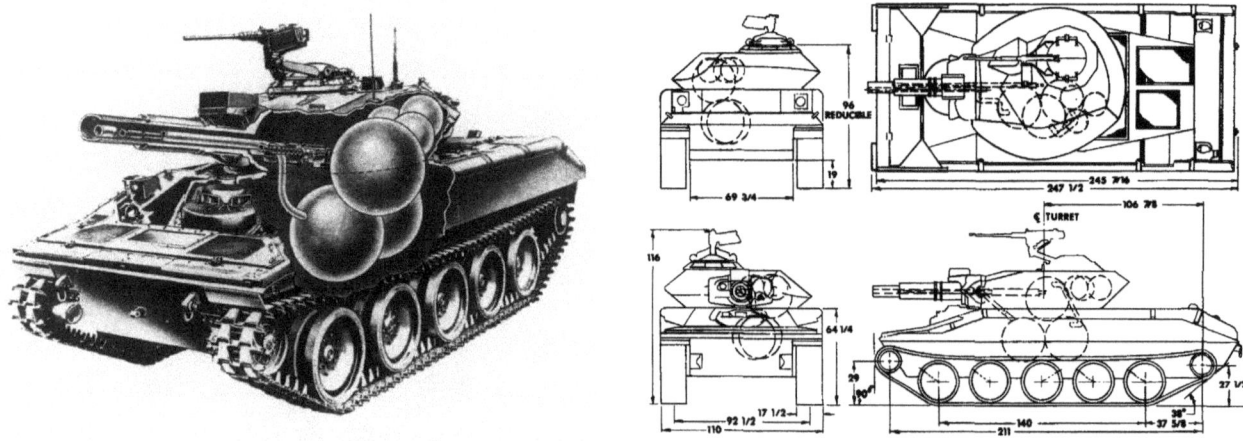

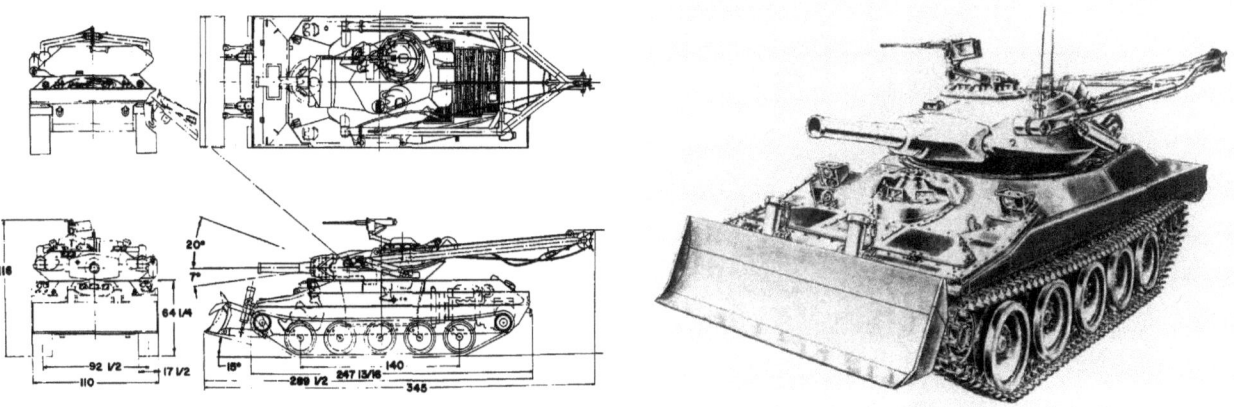

A combat engineer version of the M551 is shown in the drawings above. Below, the Sheridan chassis is fitted with a 20 ton folding bridge.

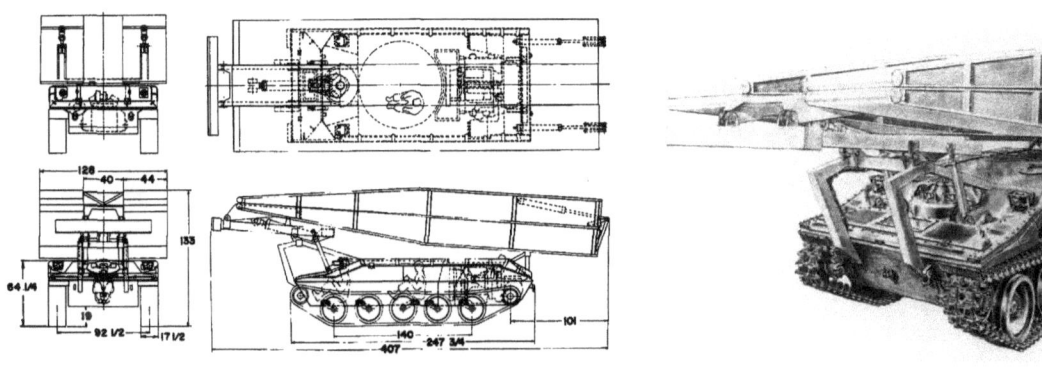

The production Sheridan hull was proposed as the basis of a rapid fire weapon system armed with two 25mm automatic guns. Mounted in a new turret on the Sheridan hull, these weapons would have been effective against both ground and air targets. With stowage space for 1500 rounds of 25mm ammunition, the twin guns had a combined firing rate 520-570 shots per minute. Manned by a crew of three, the rapid fire weapon system had an estimated combat weight of 33,548 pounds. Both air droppable and amphibious, its performance was equal to that of the standard Sheridan.

The rapid fire weapon system armed with twin 25mm automatic guns appears in the artist's concept and four view drawing below.

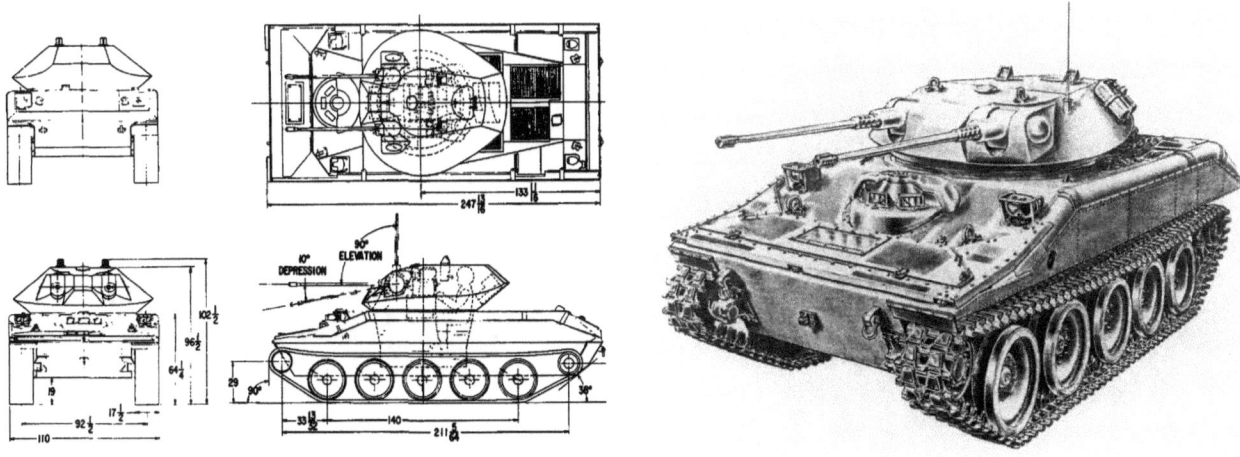

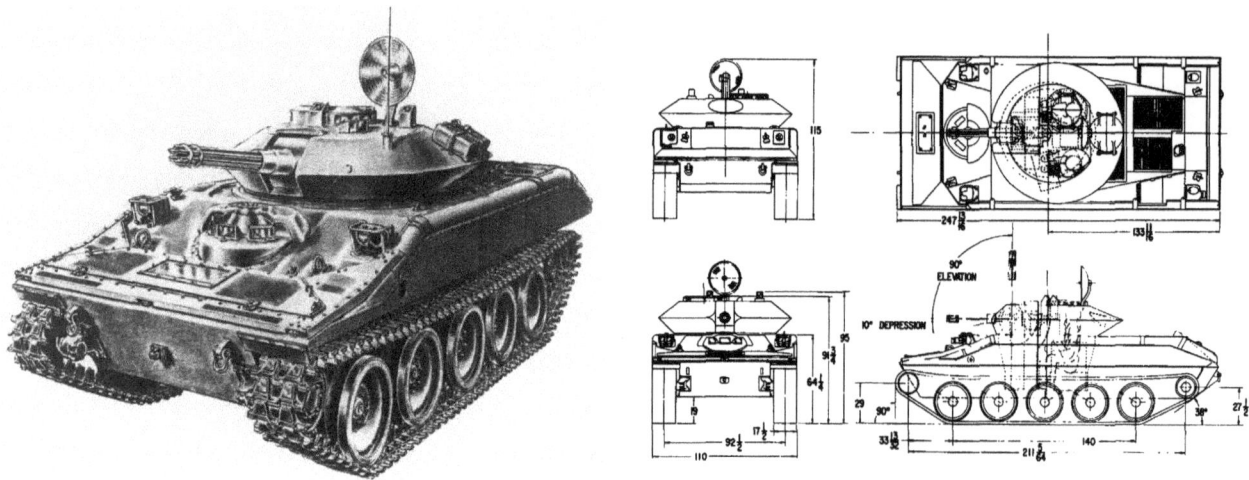

Above is the forward area air defense vehicle utilizing the Sheridan chassis. The turret is armed with a 20mm M61 cannon.

Another proposed antiaircraft weapon was the forward area air defense vehicle. It was armed with the 20mm M61 Vulcan aircraft cannon in a new two man turret on the production Sheridan chassis. Stowage space was provided for 11,000 rounds of 20mm ammunition. The M61 cannon had a firing rate of 6000 rounds per minute. Fire control equipment included an M20 sight and range radar. With an estimated combat weight of 34,338 pounds, the vehicle was both air droppable and amphibious. Its performance was expected to be the same as the standard Sheridan.

A lightweight version of the forward area air defense vehicle was proposed with a new turret on a redesigned chassis. Still using Sheridan components, the power train was relocated placing the engine and transmission in the right front hull. The driver was shifted to the left front hull. The drive sprockets were now in the front and the rear hull was free for the weapon system and the gun crew. The M61 20mm cannon was mounted in a new open top turret manned by a single crew member. The radar operator was located in the right rear hull. Stowage space for 11,000 rounds of 20mm ammunition was provided in the turret and the left rear hull. The combat weight of the vehicle was estimated to be 30,488 pounds and it was air droppable and amphibious. This redesigned chassis also was proposed as a carrier for the Mauler antiaircraft missile system.

A lightweight self-propelled howitzer was proposed using the standard Sheridan hull with a new turret. Armed with the 105mm XM103 howitzer, it had an estimated combat weight of 33,518 pounds. Ammunition stowage in the hull and turret consisted of 75 105mm rounds. A total of 1000 rounds of .50 caliber ammunition was carried for the .50 caliber machine gun mounted on the turret cupola. Manned by a crew of four, the self-propelled howitzer was air transportable, but it was not considered to be air droppable. It was amphibious and, in general, equaled the performance of the standard Sheridan.

The lightweight version of the forward area air defense vehicle appears below with the new turret and redesigned hull.

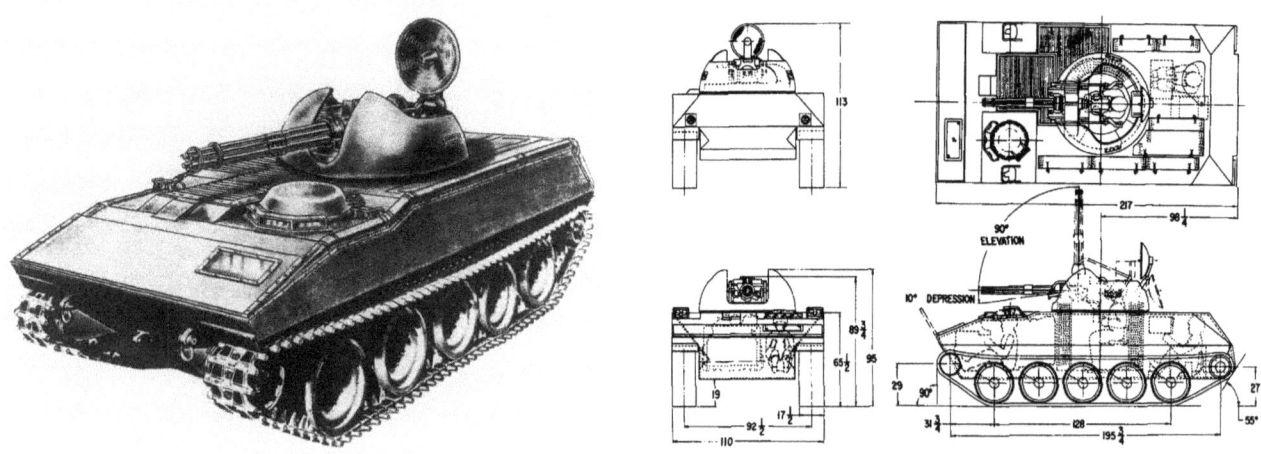

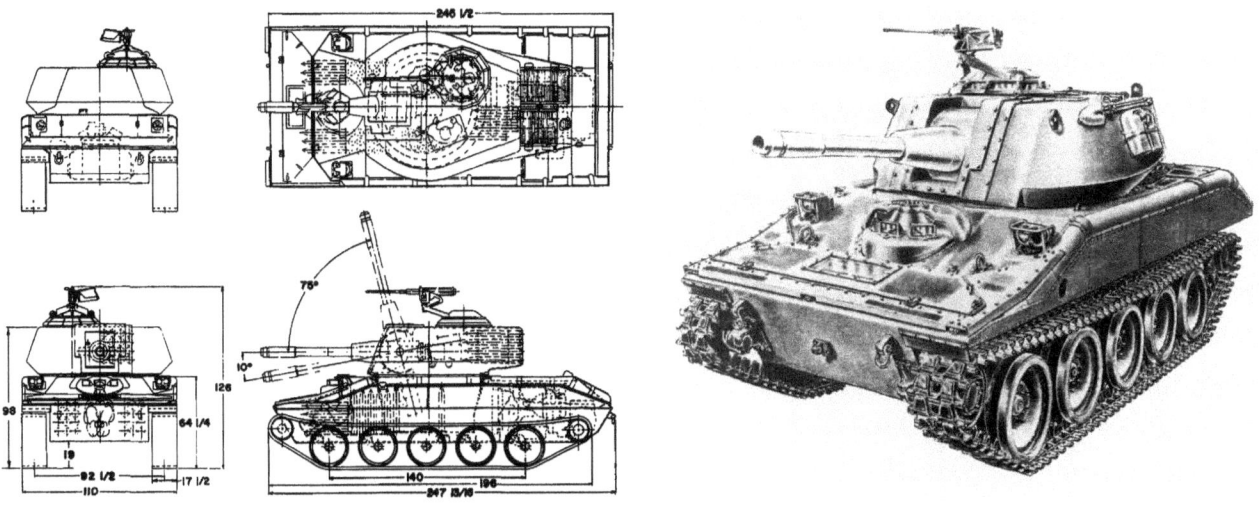

Above, a new turret installed on the M551 chassis is armed with the 105mm howitzer XM103 providing a lightweight self-propelled weapon.

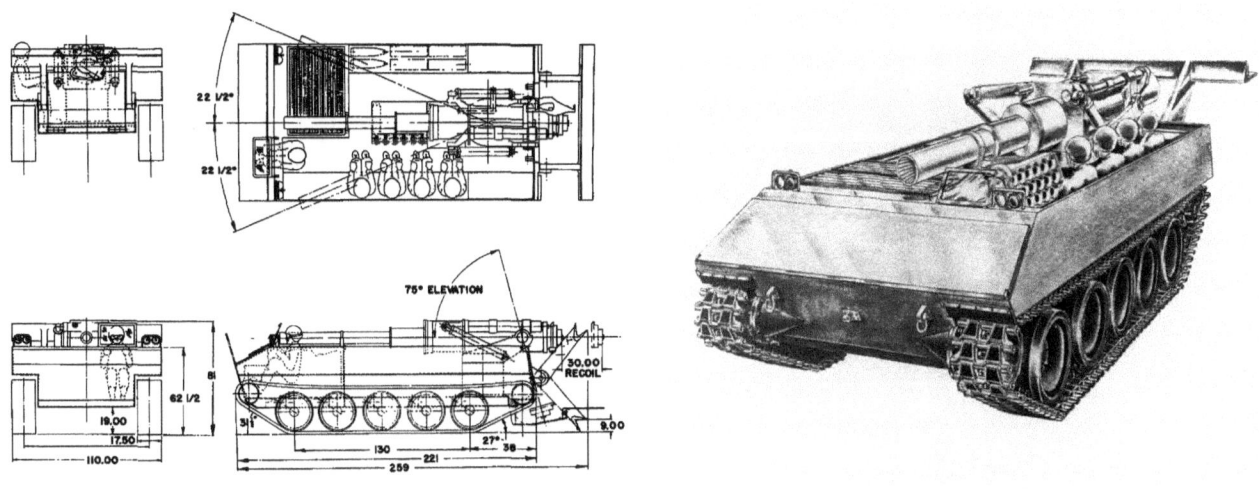

Above, the redesigned chassis using Sheridan components provides the basis for a 155mm self-propelled howitzer.

An even heavier self-propelled weapon was proposed using the redesigned chassis with the power train in front. This self-propelled 155mm howitzer was manned by a crew of five using an open top turretless vehicle with the driver located at the left front. The four cannoneers rode on the left side alongside the howitzer. The vehicle had an estimated combat weight of 27,000 pounds and it was air droppable and amphibious. Twenty rounds of 155mm ammunition were stowed on board. The vehicle performance was about the same as the standard Sheridan, but a fuel capacity of 120 gallons limited the cruising range to about 300 miles.

The front drive chassis also was proposed as the basis for an infantry fighting vehicle, although at that time, it was referred to as an armored ground support vehicle. Carrying 11 men, it had an estimated combat weight of 30,500 pounds. The driver was in the left front alongside the engine. The main armament was a 20mm automatic cannon mounted in a fully powered cupola. Secondary armament was an 7.62mm M73 machine gun. Firing ports were provided for eight infantry weapons, four per side. The mobility of the vehicle was equal to that of the Sheridan and a 204 gallon fuel tank extended the cruising range to about 400 miles. It was both air droppable and amphibious.

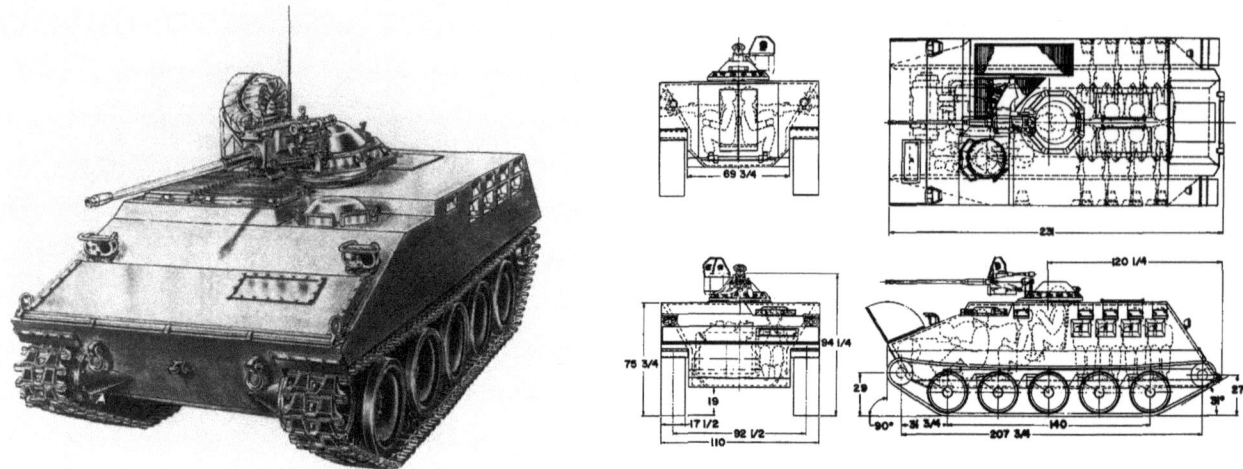

An infantry fighting vehicle (armored ground support vehicle) based upon the M551 appears above. Below is the 107mm mortar carrier also using the redesigned Sheridan chassis.

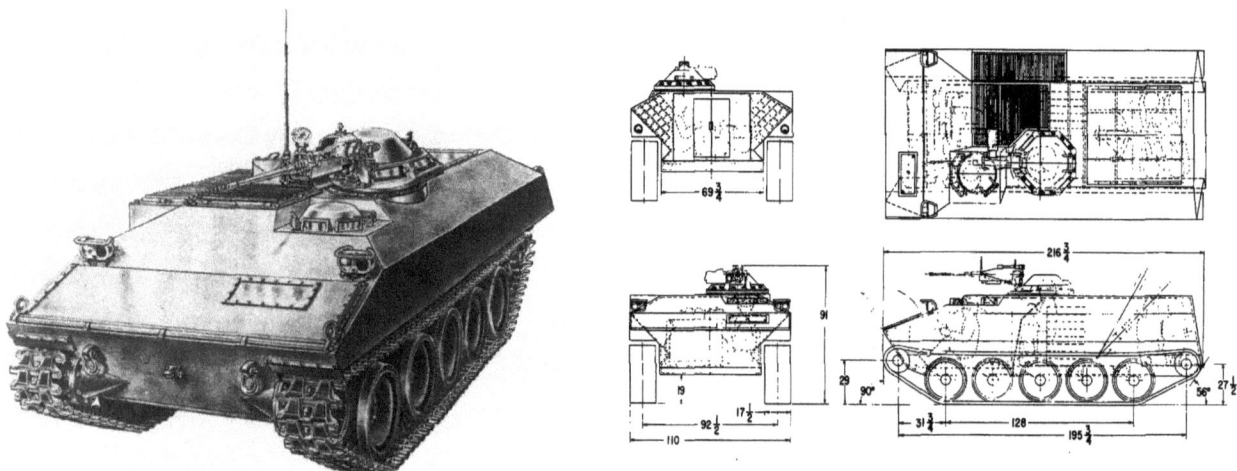

The 107mm (4.2 inch) mortar was mounted in a proposed carrier based on Sheridan components using the front drive chassis. The mortar was installed in the space available at the rear of the hull, firing toward the rear. With a six man crew and a basic load of 188 rounds of mortar ammunition, the estimated combat weight of the vehicle was 28,400 pounds. Its performance was comparable to that of the Sheridan, but the 131 gallon fuel capacity limited the cruising range to about 300 miles. Secondary armament consisted of a cupola mounted .50 caliber machine gun. The vehicle was both air droppable and amphibious.

An attempt to provide a recovery vehicle based upon the Sheridan resulted in the proposed installation of a ten ton crane on the front drive chassis. With a tow winch capacity of 30,000 pounds, its mobility was expected to equal that of the Sheridan. Manned by a crew of three, the vehicle had an estimated combat weight of 34,485 pounds. Its fuel capacity of 250 gallons provided a cruising range of about 450 miles. Armament consisted of a single .50 caliber

machine gun mounted on the cupola. It was not amphibious or air droppable.

The front drive chassis based upon the Sheridan components also was proposed as an armored cargo carrier. With a crew of two, it had an estimated combat weight of 32,247 pounds including a payload capacity of 12,000 pounds. Armament was a single 7.62mm machine gun mounted on the cupola. Its performance was equivalent to that of the Sheridan and it was both air droppable and amphibious.

Although all of the previous vehicles incorporated armor protection similar to that on the Sheridan, the various components also were proposed as the basis of an unarmored cargo carrier. This vehicle, with its crew of four, had an estimated combat weight of 32,647 pounds including a payload of 15,000 pounds. It was both air droppable and amphibious and consisted of an aluminum structure with a fiberglass cab. A fuel capacity of 155 gallons resulted in an estimated cruising range of 306 miles. No armament was specified.

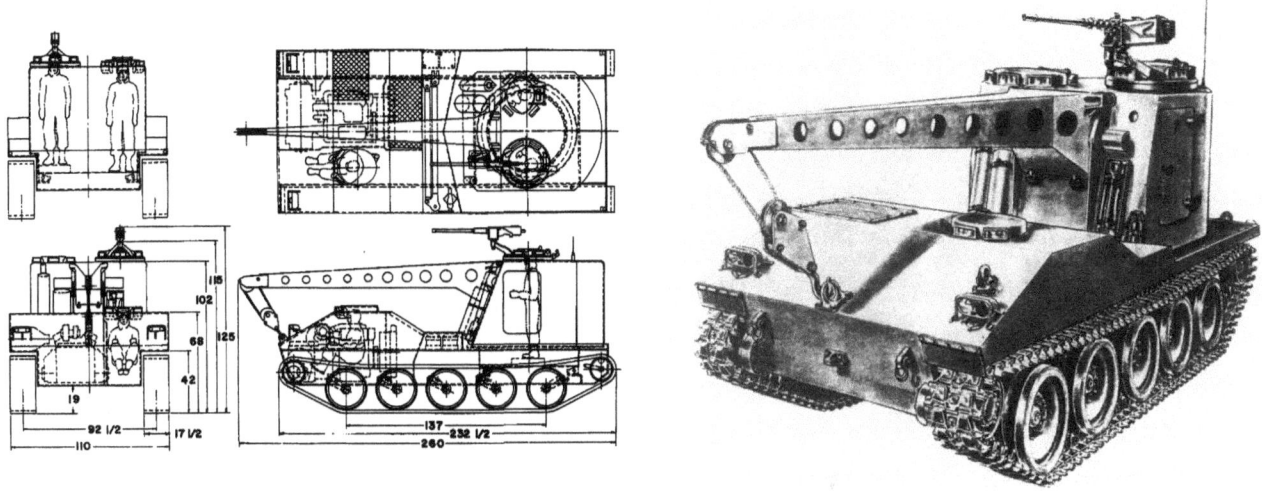

Above is the light recovery vehicle utilizing M551 components. Note the similarity of the cab and crane arrangement to the M578 recovery vehicle.

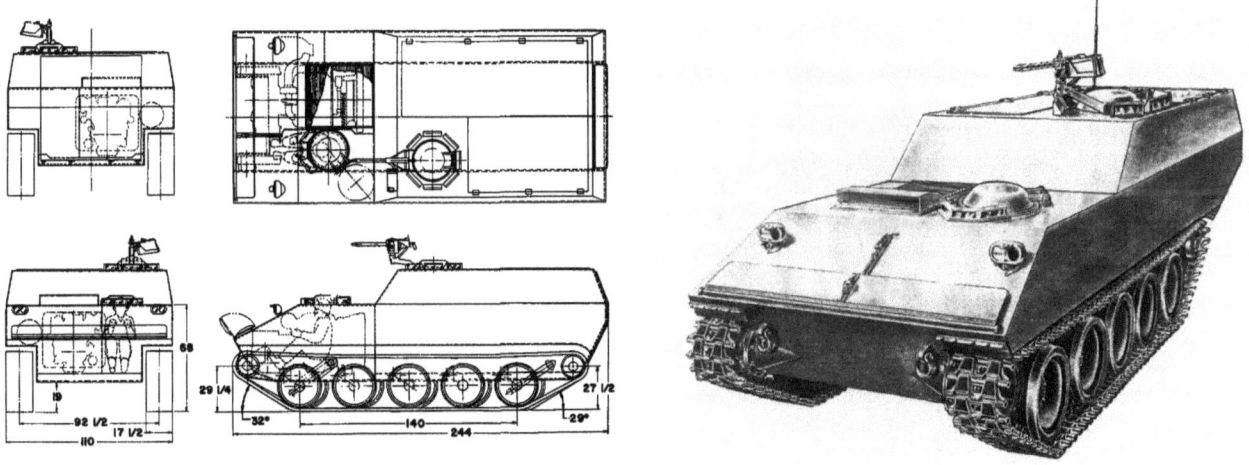

Armored and unarmored cargo carriers based upon the use Sheridan components appear above and below respectively.

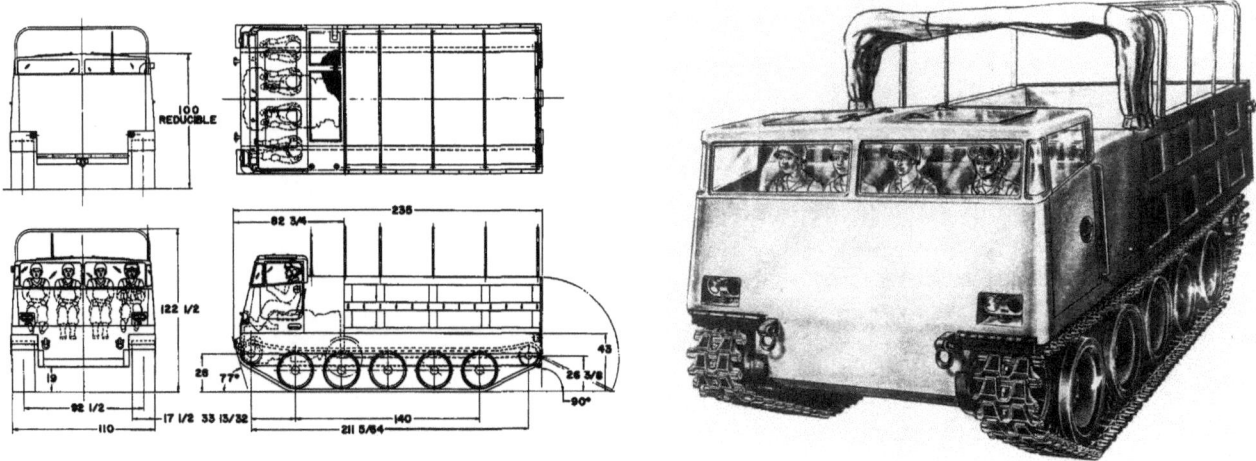

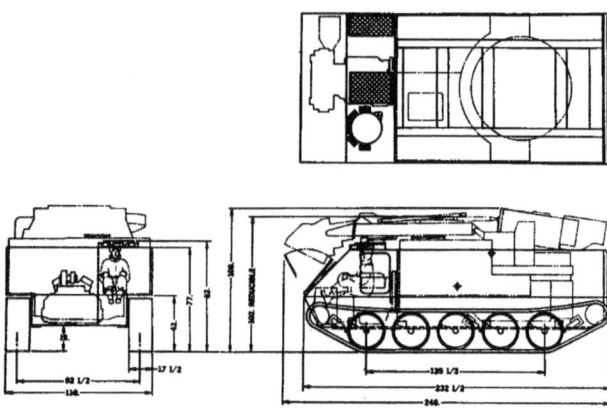

Above, the suspension and power train components from the M551 are utilized in a carrier for the Mauler missile system.

Although a large family of vehicles was proposed based upon the Sheridan chassis, only one was actually constructed. This was the armored tracked recovery vehicle (ATRV). The ATRV utilized the standard Sheridan chassis with the addition of a retractable blade on the rear to stabilize the vehicle during recovery operations. A tow bar assembly was stowed on the rear of the hull below the rack for the pioneer tools. A crane was installed in the front of a turret similar to that on the standard M551. On the turret roof was a ring of vision blocks providing a 360 degree view. This was referred to by General Motors as the "Vista Dome". The only armament on the vehicle was a 7.62mm M60 machine gun mounted at the turret hatch. The ATRV was fitted with the belly armor mine protection kit and had a combat weight of 39,753 pounds. It was manned by a crew of four. The driver remained in the usual position in the center front hull with the standard rotating hatch. The other crew members were located in the turret.

ARMORED TRACKED RECOVERY VEHICLE
(ATRV)

The armored tracked recovery vehicle (ATRV) appears in the photographs above at the right and below. Below, the turret mounted crane is secured for travel at the left and the retracted stabilizer blade can be seen at the right underneath the tow bar stowed on the hull rear.

134

Above, the AN/VVG-1 laser range finder has been installed on the turret of an M551. The cable cover can be seen extending from the transceiver under the machine gun mount back to the armor rear enclosure. All of the photographs on this page were taken by R. P. Vaughan.

Even before production of the Sheridan was complete, the first of what became a long series of product improvement programs was initiated. As early as May 1969, the concept of a cupola mounted laser range finder had been selected. On 22 April 1971, the Department of the Army approved the procurement of 505 laser range finders with deliveries scheduled to begin in early 1972. The transceiver for this AN/VVG-1 laser range finder replaced the cupola vision block directly below the pintle support for the .50 caliber machine gun. Electronic components for this add-on range finder were mounted inside the tank commander's rear armored enclosure. When the range finder was installed, the vehicle was redesignated as the AR/AAV M551A1. It could easily be identified by the laser unit beneath the .50 caliber machine gun mount and the cable cover extending back on the right side of the tank commander's armored enclosure. On the M551A1, the gunner's M127 telescope was replaced by the M127A1 which incorporated a filter for protection against laser light.

Various parts of the laser range finder are shown in these views. Above, the transceiver replaces the cupola vision block under the .50 caliber machine gun mount. Below are two views of the range finder components mounted inside the cupola.

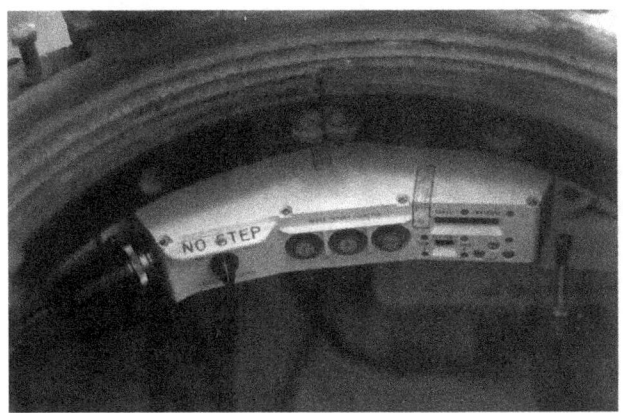

The laser range finder is shown here mounted on a turret trainer at Fort Knox, Kentucky. In the upper photograph, the covers are open on the missile transmitter and the gunner's periscopic sight. Below, the electronic components of the range finder can be seen in the commander's armor rear enclosure.

Additional views of the turret trainer at Fort Knox appear here. Details of the searchlight are clearly visible. The photographs on this and the previous page were taken by R. P. Vaughan.

Service experience continued to reveal the need for numerous modifications. By 1977, many of these requirements were included in a product improvement program (PIP) intended to improve the performance and increase the reliability and durability of the Sheridan. Among the items in this program was an improved design for the power pack mounts, a redesigned generator mounting bracket and drive assembly, a more rugged drive sprocket mount, and the relocation of the electrical receptacle used for slave starting to the outside front hull. If arcing occurred, as it frequently did during the starting operation, the original inside location permitted hot particles to fall on the combustible case ammunition.

An old problem with the Sheridan was the heavy exhaust smoke produced by the diesel engine during rapid acceleration. The was an extremely undesirable feature in a reconnaissance vehicle which did not want to attract attention to itself. The PIP included the installation of a throttle delay mechanism which reduced the level of this exhaust smoke. Another problem concerned the aluminum alloy tow lugs welded to the hull which were frequently torn off during operations. They were replaced by steel lugs bolted to the hull with backing plates on the inside. As mentioned before, the original 6V53T diesel engine had an aluminum alloy block to minimize the weight. By 1977, these engines had been rebuilt many times. The PIP replaced this engine

with a 6V53T engine having a cast iron block. Although the weight increased, the new engine was more rugged and less susceptible to temperature warp. It also had a more efficient turbosupercharger resulting in an increase in net horsepower.

Later modifications to the Sheridan included new smoke grenade launchers. The early arrangement with the four launchers in a row was replaced by a new version with the four launchers clustered in a square pattern. Also, the M73 and M219 7.62mm machine guns had never been fully satisfactory. In May 1979, the M60A3 main battle tank was standardized equipped with the 7.62mm M240 machine gun. This was the U.S. designation for the highly reliable Belgian MAG-58 machine gun that was adopted to replace the previous weapon. Subsequently, the M240 replaced the M219 as the coaxial machine gun in the Sheridan.

In early 1989, the U.S. Army Tank Automotive Command began a program to upgrade the M551A1 for use by the 82nd Airborne Division. The modifications included the installation of a tank thermal sight for the gunner. This was the same sight used in the M60A3TTS main battle tank. The driver also was provided with the night viewing device previously installed in the Bradley fighting vehicle. The conversions were carried out at the Anniston Army Depot and were completed in late 1990. The vehicle was then designated as the M551A1(TTS).

The laser rangefinder and the tank thermal sight are installed on the M551A1(TTS) below. Note the new clustered smoke grenade launchers mounted over the brackets for the earlier type launchers.

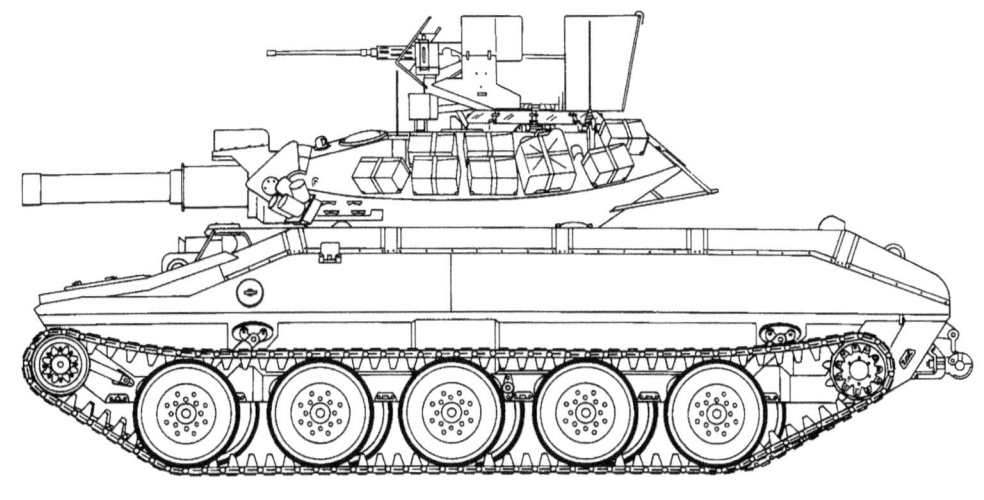

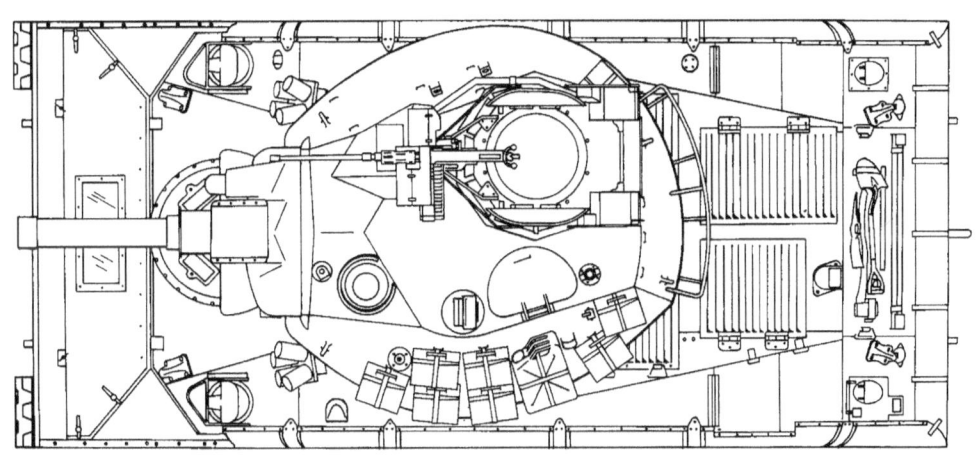

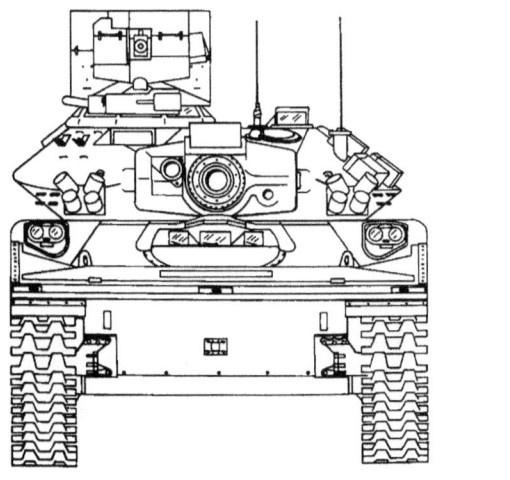

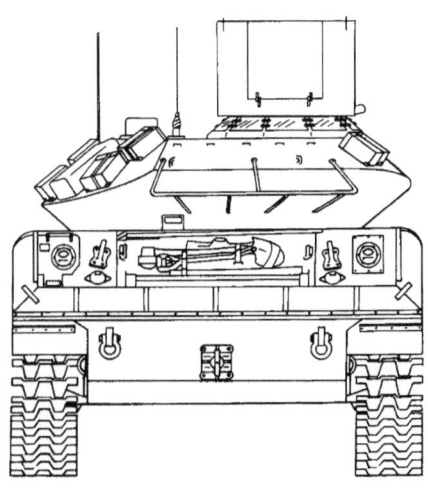

152mm Gun-Launcher AR/AAV M551A1(TTS)

Above, the early X3A100 weapon systems trainer appears at the left and the redesignated XM40 is at the right.

For the instruction of Sheridan crews, two training aids were developed early in the program. The first of these was originally designated as the Sheridan weapon systems trainer X3A100. It provided crew training in the operation and maintenance of the Sheridan turret and was capable of simulating the entire operational mission of the turret portion of the weapon system. Redesignated later as the weapon system trainer XM40, it measured 103 inches high, 196 inches long, 104 inches wide, and weighed approximately 8300 pounds.

The second training aid was originally designated as the Sheridan weapon system conduct of fire trainer X3A101. It was intended for use in the field to develop and maintain the gunner's proficiency in the employment of the Shillelagh missile. Two vehicles were required as part of this training device. A visual effects simulator and an instructor's control unit were installed on a training vehicle and an infrared transmitter in a lighthouse assembly was mounted on a target vehicle. This equipment provided a visual simulation in the gunner's telescope of the launch and flight of

the missile. It indicated the gunner's tracking response and recorded the hits or misses. The training vehicle was later designated as the launcher, conduct of fire trainer XM41 and the target vehicle became the target, conduct of fire trainer XM42. The complete system was then designated as the conduct of fire trainer XM35. The original assembly of this conduct of fire trainer utilized two of the second generation XM551 pilots for the launcher and target vehicles.

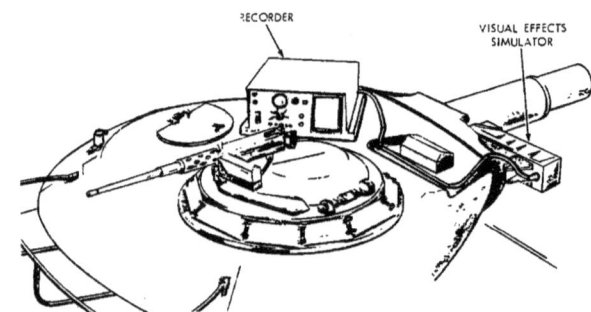

Below the XM42 target vehicle is at the left and the XM41 launcher vehicle is at the right. Both are based upon second generation XM551 pilots. The sketch above shows the equipment installed on the launcher vehicle turret.

The conduct of fire trainer target vehicle above is based upon a production model Sheridan.

As its official designation indicated, the Sheridan was designed to fill a dual role as both an armored reconnaissance vehicle and an airborne assault vehicle. Thus it was required to have a wider range of capabilities than any previous light tank. This included increased firepower and mobility combined with armor protection at a weight level suitable for airborne operations. To meet such specifications required the use of new concepts in armament, armor protection, and design, some of which were still highly experimental. As should have been expected, this required a long development cycle. The decision to prematurely standardize the Sheridan and place it in production resulted in a long and expensive modification program. It was then necessary to retrofit the modifications to the earlier production vehicles. Eventually, the major problems were solved, but the Sheridan remained a complex weapon system that required an excellent maintenance program and a high state of training and motivation for the troops that used it. As a result, it was not popular with soldiers that did not require its wide range of capabilities.

In 1978, the Sheridan was withdrawn from all units except one and the vehicles were placed in storage. The exception was the only airborne armor unit in the Army. In 1969, the 4th Battalion (Airborne), 68th Armor had been equipped with the Sheridan as part of the 82nd Airborne Division. Subsequently, this unit was redesignated as the 3rd Battalion (Airborne), 73rd Armor. As the only vehicle available to meet the requirement for airborne armor, the Sheridan was destined to serve with the 82nd Airborne Division into the 1990s.

Over 300 Sheridans were allocated to the National Training Center at Fort Irwin, California for use as opposing force vehicles. Intended to represent Soviet equipment, they were employed in training exercises against U.S. Army units. The Sheridans were converted for their new role using visual modification (VISMOD) kits to make them resemble Soviet combat vehicles. These VISMOD conversions included the T72 main battle tank, the BMP infantry fighting vehicle, the M1974 122mm self-propelled howitzer, and the ZSU-23-4 self-propelled antiaircraft gun.

The Sheridans on this page have been converted using VISMOD kits to simulate the Soviet T72 tank (above) and the BMP with the Sagger missile (below). The photographs on this and the two following pages were taken at the National Training Center, Fort Irwin, California during October 1982.

VISMOD kits have converted these Sheridans to simulate the Soviet 122mm self-propelled howitzer M1974 (above) and the Soviet ZSU-23-4 self-propelled antiaircraft weapon (below).

The gun-launchers have been removed from the Sheridans above and they have been converted for use by the umpires at the National Training Center. Below is a closeup view of a Sheridan converted to simulate the Soviet T72 tank.

144

PART III

A REPLACEMENT FOR THE SHERIDAN

PART III

A REPLACEMENT FOR THE SHERIDAN

Above is a model of the HIMAG-A. The sketches at the bottom of the page show the chassis test vehicle at the left and the complete HIMAG-A at the right.

NEW RESEARCH AND DEVELOPMENT PROGRAMS

By 1970, the development of the shaped charge chemical energy warhead as well as the kinetic energy long rod penetrator had made even the most heavily armored main battle tank vulnerable to destruction. During the development of the MBT70, the great mobility of the tank with its high power to weight ratio and hydropneumatic suspension had been considered as a major factor in its ability to survive on the battlefield. Such mobility and agility had always been the survival mode for the light tanks since the beginning of World War II. In 1972, the Defense Advanced Research Projects Agency (DARPA) initiated a study to determine the effect of new technologies on the performance of various combat vehicles. These technologies included new automotive components as well as new weapon systems. This program eventually resulted in the development of several test vehicles to evaluate a variety of design features and experimental components.

In the Spring of 1975, the decision was made to build a high performance variable test bed vehicle to determine the effect of mobility and agility on the survivability of a medium weight (25 to 40 tons) vehicle. It also was to evaluate a hydropneumatic suspension system and its effect upon weapon accuracy and crew performance during high speed operations. In addition, the program was to determine the benefits of a hypervelocity, high firing rate, automatic cannon and the advantages of precision fire control, gun control, and stabilization systems during rapid cross-country maneuvers.

In April, proposals were solicited by DARPA for a feasibility study to define a system meeting the program objectives. A review of these proposals resulted in the selection of the National Water Lift Company to design the proposed system. Work began in March 1976 and it was designated as the high mobility agility (HIMAG) vehicle.

The HIMAG program was divided into two major phases. The first phase was to evaluate the mobility and agility of the chassis while phase 2 was to test the complete vehicle with the armament, fire control, and gun control systems. The main weapon to be installed was the

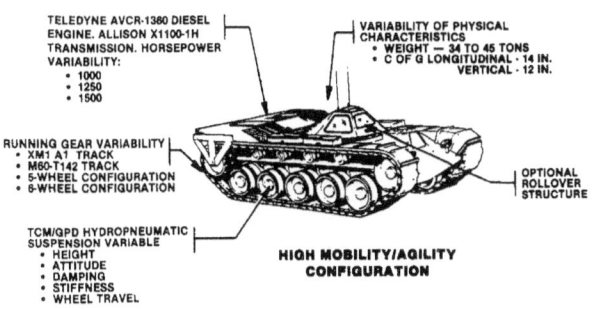

TELEDYNE AVCR-1360 DIESEL
ENGINE. ALLISON X1100-1H
TRANSMISSION. HORSEPOWER
VARIABILITY:
• 1000
• 1250
• 1500

VARIABILITY OF PHYSICAL
CHARACTERISTICS
• WEIGHT — 34 TO 45 TONS
• C OF G LONGITUDINAL - 14 IN.
VERTICAL - 12 IN.

RUNNING GEAR VARIABILITY
• XM1 A1 TRACK
• M60-T142 TRACK
• 5-WHEEL CONFIGURATION
• 6-WHEEL CONFIGURATION

OPTIONAL
ROLLOVER
STRUCTURE

TCM/GPD HYDROPNEUMATIC
SUSPENSION VARIABLE
• HEIGHT
• ATTITUDE
• DAMPING
• STIFFNESS
• WHEEL TRAVEL

**HIGH MOBILITY/AGILITY
CONFIGURATION**

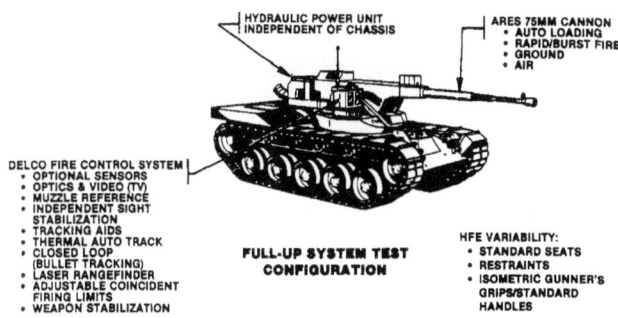

HYDRAULIC POWER UNIT
INDEPENDENT OF CHASSIS

ARES 75MM CANNON
• AUTO LOADING
• RAPID/BURST FIRE
• GROUND
• AIR

DELCO FIRE CONTROL SYSTEM
• OPTIONAL SENSORS
• OPTICS & VIDEO (TV)
• MUZZLE REFERENCE
• INDEPENDENT SIGHT
  STABILIZATION
• TRACKING AIDS
• THERMAL AUTO TRACK
• CLOSED LOOP
  (BULLET TRACKING)
• LASER RANGEFINDER
• ADJUSTABLE COINCIDENT
  FIRING LIMITS
• WEAPON STABILIZATION

**FULL-UP SYSTEM TEST
CONFIGURATION**

HFE VARIABILITY:
• STANDARD SEATS
• RESTRAINTS
• ISOMETRIC GUNNER'S
  GRIPS/STANDARD
  HANDLES

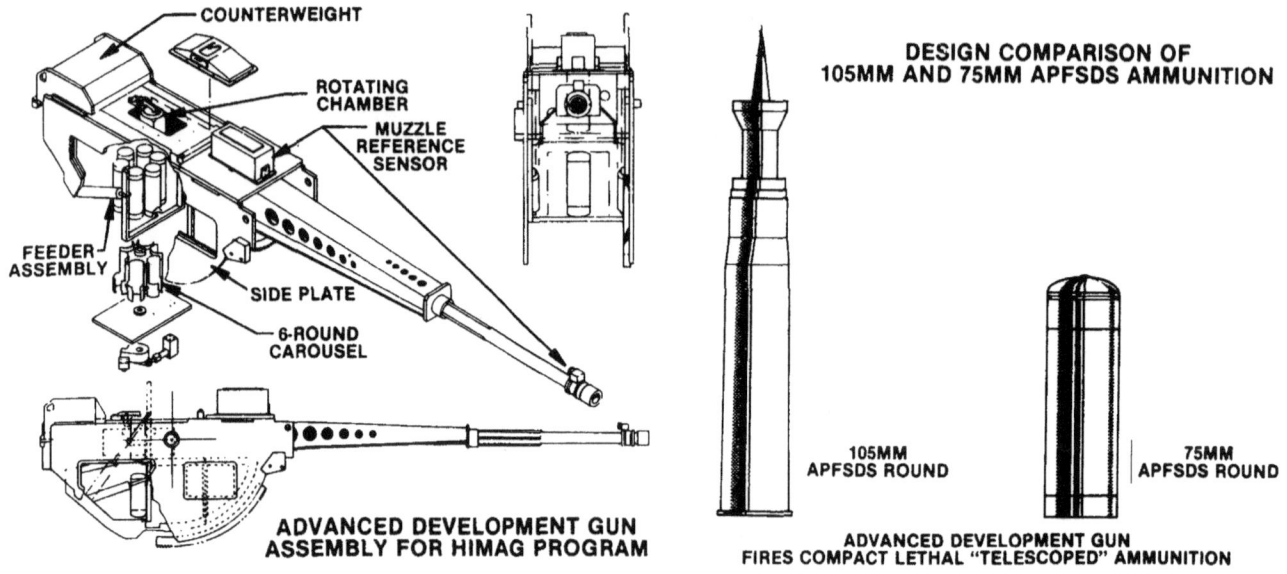

COUNTERWEIGHT

ROTATING
CHAMBER

MUZZLE
REFERENCE
SENSOR

FEEDER
ASSEMBLY

SIDE PLATE

6-ROUND
CAROUSEL

**ADVANCED DEVELOPMENT GUN
ASSEMBLY FOR HIMAG PROGRAM**

**DESIGN COMPARISON OF
105MM AND 75MM APFSDS AMMUNITION**

105MM
APFSDS ROUND

75MM
APFSDS ROUND

**ADVANCED DEVELOPMENT GUN
FIRES COMPACT LETHAL "TELESCOPED" AMMUNITION**

The 75mm medium caliber, antiarmor, automatic cannon is illustrated above at the left and the configuration of the 75mm telescoped ammunition is compared to the standard 105mm APFSDS round at the right.

medium caliber, antiarmor, automatic cannon (MC-AAAC). This 75mm weapon, developed under a DARPA contract by Ares, Incorporated, was a hypervelocity, smooth bore, gun firing telescoped, solid propellant, ammunition. Although only 75mm in caliber, its penetration performance was greatly enhanced through the use of newly developed armor piercing, fin stabilized, discarding sabot (APFSDS) projectiles. Originally, it had been intended to include a liquid propellant gun in the program, but that weapon was diverted to a separate development project.

The HIMAG was assembled with a lightweight rigid hull and the test weight of the vehicle could be varied using removable lead weights. It was powered by the Teledyne Continental Motors AVCR-1360 diesel engine. As its designation indicated, this was variable compression ratio, air-cooled, engine with a displacement of approximately 1360 cubic inches. The maximum power output could be established for the various test programs at 1000, 1250, or 1500 gross horsepower. This engine drove the HIMAG through the X-1100-1H transmission produced by the Detroit Diesel

Allison Division of General Motors Corporation. The driver rode in the center front hull.

The hydropneumatic suspension system could vary the height of the vehicle and tilt it forward, backward, or to either side. Although the early tests were performed with five dual road wheels per side, the vehicle was designed to accommodate six wheels per side.

The chassis test bed was delivered to Fort Knox for tests beginning in February 1978 and continuing into September. During this time, work was in progress on the turret assembly and it was mounted on a General Motors XM1 tank chassis for preliminary tests in June 1978. However, the original sliding sleeve type automatic 75mm gun failed during dispersion tests at Ares in January 1979 and it was replaced by the later rotating breech gun.

The HIMAG turret was a self-contained modular design with a 100 horsepower gas turbine auxiliary power unit installed at the rear of the gun pod. The cannon was fed from a six round carousel at the bottom of the gun pod. With its rotating breech vertical, the telescoped round was

The HIMAG chassis test vehicle without the turret is shown below during the test program.

The complete HIMAG-A vehicle armed with the 75mm automatic gun appears above.

inserted into the chamber. The breech then rotated into the horizontal position in line with the gun barrel. When fired, the cartridge case extended to form a seal between the case and the rear of the barrel. The case then contracted to permit the breech to rotate to the vertical position. A new round was rammed into the chamber from the bottom ejecting the empty case upward out of the gun pod. The rate of automatic fire was about one round per second. The two man turret crew were seated in the basket low within the hull and were protected by the armor steel top from any explosion in the external gun pod. The gun turret drive and stabilization systems were installed in the turret along with the various fire control system components.

After tests by the contractor at the Yuma Proving Ground, the complete HIMAG system was transferred to Aberdeen Proving Ground for evaluation by the government between March and September 1980. The vehicle was then shipped to Fort Knox where the testing was completed by the end of March 1981. After the end of the test program, the HIMAG was refurbished and converted to the six road wheel configuration. It was then assigned to the Tank Automotive Command for use in other research programs.

The test results had indicated that the hydropneumatic suspension improved the mobility and increased the life of the suspension, road wheels, and tracks. The single shot dispersion of the 75mm automatic cannon was similar to that of the standard large caliber tank guns. Two round bursts were found to be the most effective method of operation when firing on the move. The muzzle reference system was a necessary requirement to compensate for the barrel bend of the 75mm cannon.

The fire control system of the HIMAG-A is sketched below and a front view of the complete vehicle is at the right.

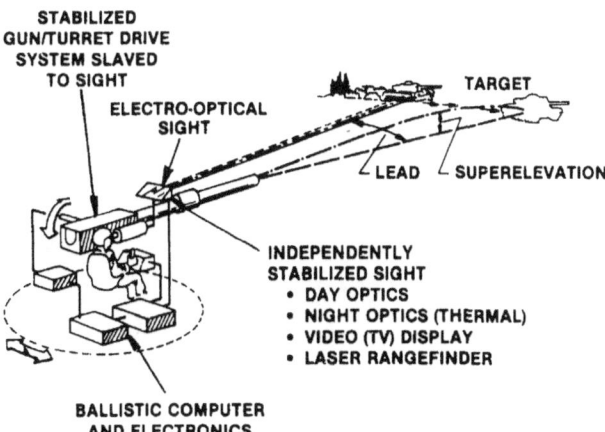

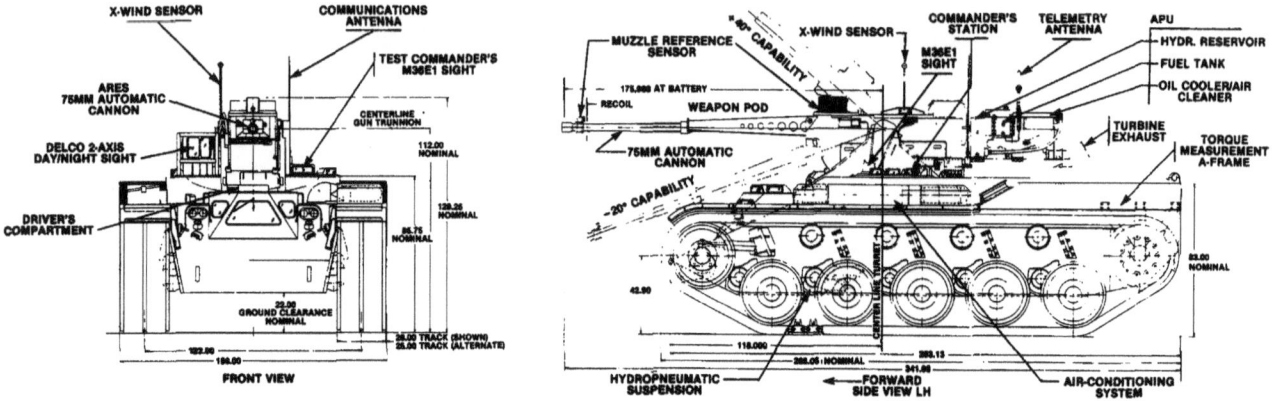

The dimensions of the HIMAG-A are shown in the drawings above and the internal arrangement of the turret assembly is sketched below.

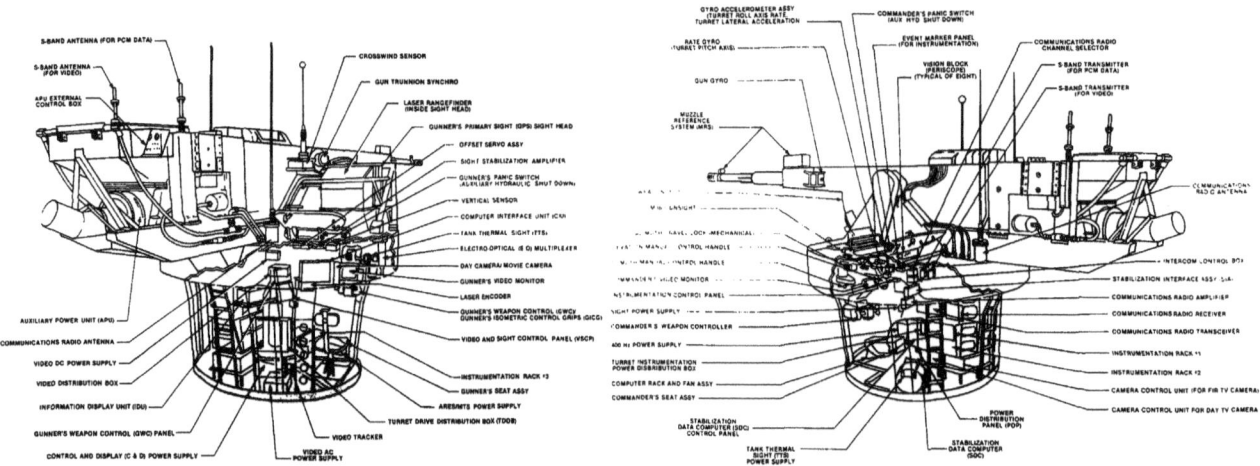

Although the HIMAG program originated under the DARPA contract, in 1976 it became the joint responsibility of DARPA, the U.S. Army Tank Automotive Command (TACOM), and the U.S. Marine Corps under the Armored Combat Vehicle Technology (ACVT) program. In 1977, TACOM assumed responsibility for the DARPA portion of this program. The vehicle described above actually was designated as the HIMAG-A. The HIMAG-B was a crew position test vehicle evaluated at Fort Knox.

The complete HIMAG-B appears at the right and two views of the vehicle without the turret are below.

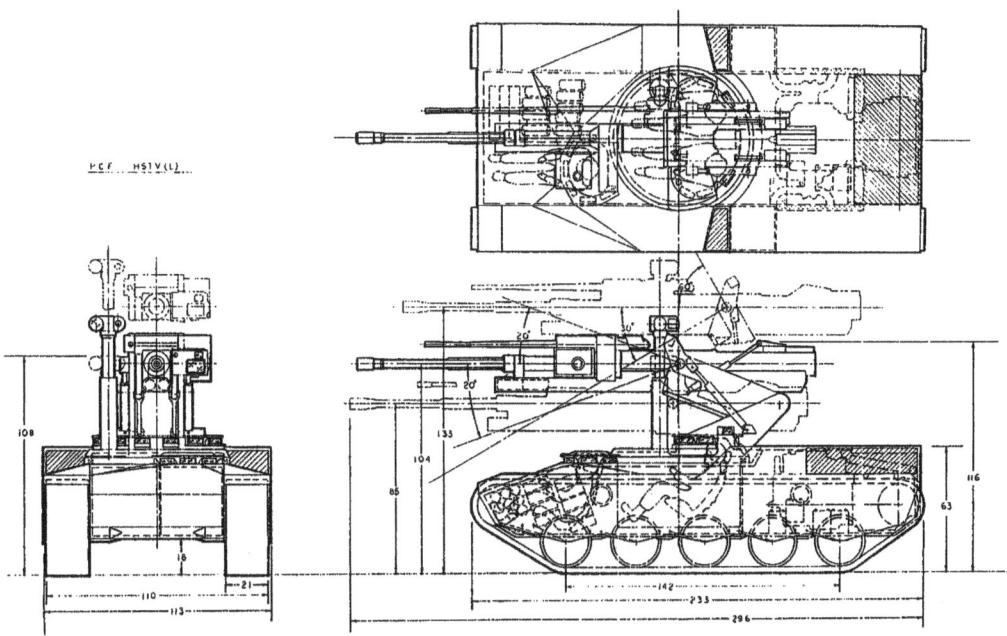

PCF HSTV(L)

The HSTV-L configuration proposed by the Pacific Car and Foundry Company is shown in the drawing above. Note the range of height possible with the elevating trunnion system.

Another vehicle developed under the ACVT program was the high survivability test vehicle - lightweight (HSTV-L). This project had its origin in some of the early proposals for the HIMAG program. Although not selected for the HIMAG, they included features that were considered worthy of further development. Studies were carried out at AAI Corporation and Pacific Car and Foundry Company during the period 7 February to 7 July 1977. Their competing proposals were received by the Tank Automotive Research and Development Command on 11 July 1977.

Both HSTV-L proposals were for vehicles armed with the Ares 75mm automatic cannon, but the methods of mounting the weapon were entirely different. The AAI proposal installed the 75mm gun in a conventional turret mount while the Pacific Car and Foundry vehicle carried the cannon in an external elevating trunnion system. This arrangement allowed the vehicle to fire from behind terrain features with only the cannon and sighting equipment exposed. The tank commander and gunner rode in the turret basket inside the hull with an elevating periscope for the commander. Secondary armament of the Pacific Car and Foundry proposal consisted of a 25mm Bushmaster cannon and a 7.62mm machine gun. The Bushmaster was mounted coaxially with the 75mm gun and the machine gun was installed on the commander's hatch. Originally, a 40mm grenade launcher also was included, but this was eliminated during review of the proposal. The ready round magazine for the automatic loader was located below the gun tube and it carried 22 75mm rounds (11 APFSDS, 11 HE). An additional 33 rounds of 75mm ammunition were stowed in the left front hull alongside the driver. The proposed Pacific Car and Foundry HSTV-L was powered by a General Motors 8V71T diesel engine through an HMPT-500 transmission. The advanced torsion bar suspension was fitted with five road wheels per side.

The mock-up of the Pacific Car and Foundry HSTV-L appears in these photographs. The bottom view shows the cannon raised to the maximum height.

151

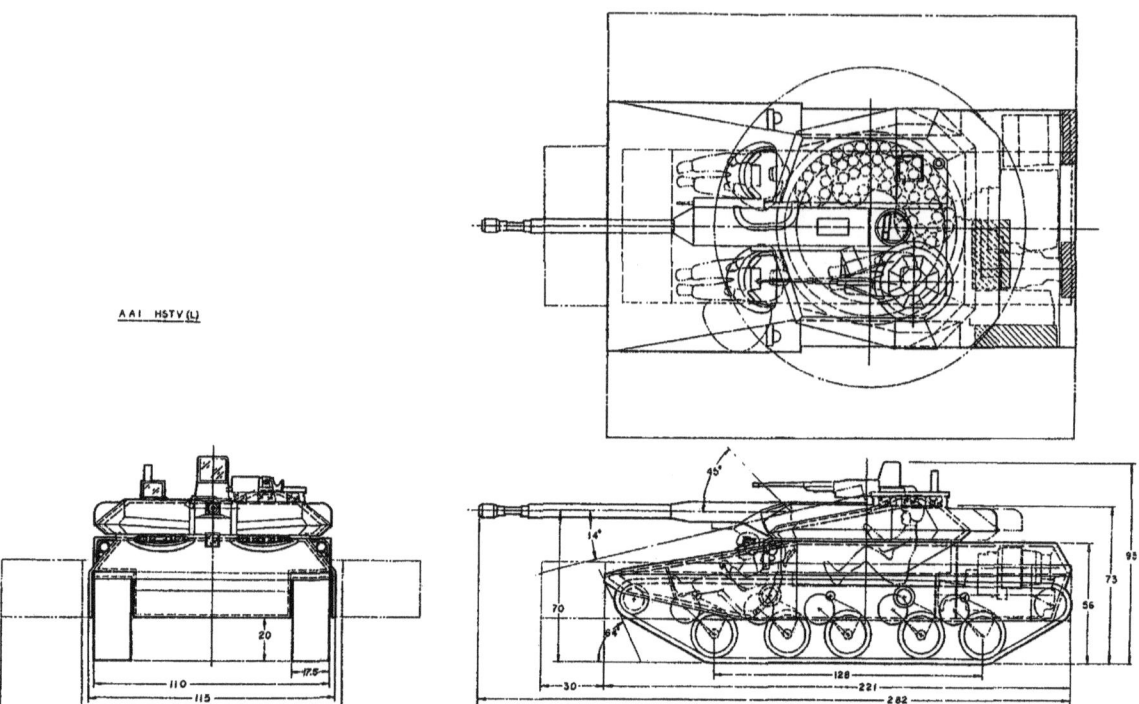

AAI HSTV (L)

The drawing above shows the HSTV-L design proposed by AAI Corporation.

The three man crew arrangement in the AAI HSTV-L concept differed from that in the Pacific Car and Foundry vehicle. Only the tank commander was located in the turret. The gunner and driver were seated in the front hull with the driver on the left and the gunner on the right. Either man could drive the vehicle. Any one of the three crew members could fire the armament. Thus in an emergency, either the gunner or the driver could operate the vehicle. Another advantage of the AAI design was that the entire 60 round basic load of 75mm ammunition was available to the automatic loader. Thus no shifting of ammunition would have been required in combat. The proposed secondary armament consisted of a 7.62mm coaxial machine gun and a .50 caliber machine gun on a skate mount at the commander's hatch. The AAI concept was to be driven by a Lycoming 800 series gas turbine with the Detroit Diesel Allison X-300-4A transmission. The vehicle was supported at five road wheel stations per side by the National Water Lift hydropneumatic suspension.

After the proposals were evaluated in late 1977, the contract for the construction of the HSTV-L was awarded to AAI Corporation. The final vehicle differed in some respects from the original proposal. It was slightly longer and the .50 caliber machine gun at the commander's hatch was replaced by an M240 7.62mm machine gun. The weight of the vehicle with test instrumentation and partial applique armor was about 20 tons. As originally constructed, the 75mm ammunition supply was limited to 26 rounds for the test program. However, the turret design was such that this could be increased to the 60 rounds specified in the original proposal. The test program for the HSTV-L provided valuable data for future light combat vehicle design.

The mock-up of the AAI Corporation HSTV-L appears below.

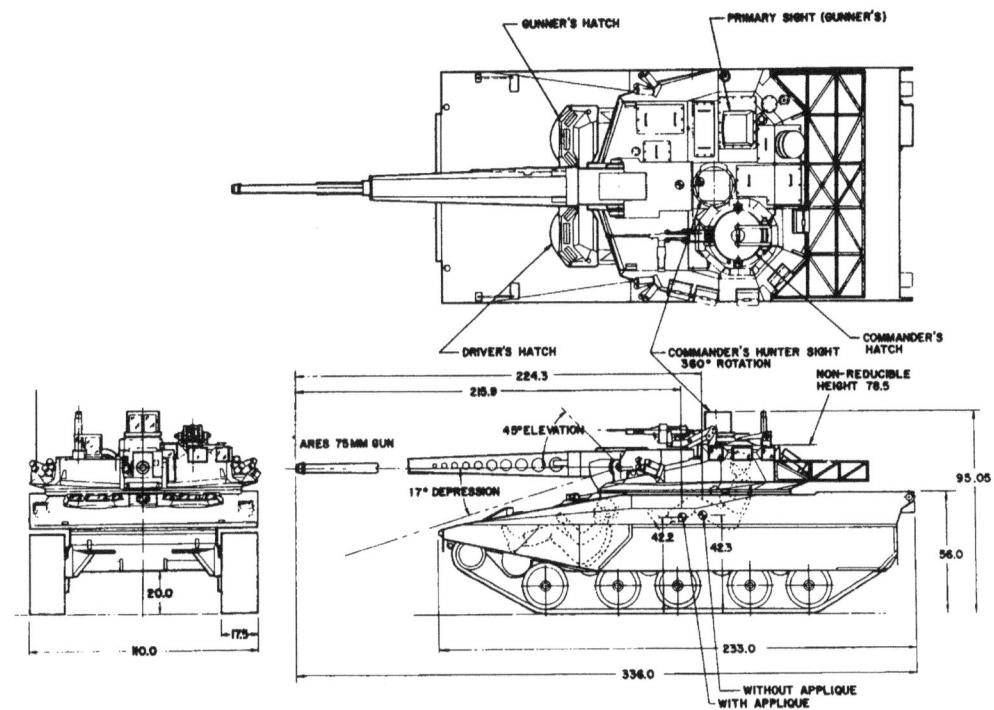

The drawing above reflects the final design of the HSTV-L as built by the AAI Corporation.

These two photographs show the HSTV-L after completion at AAI Corporation.

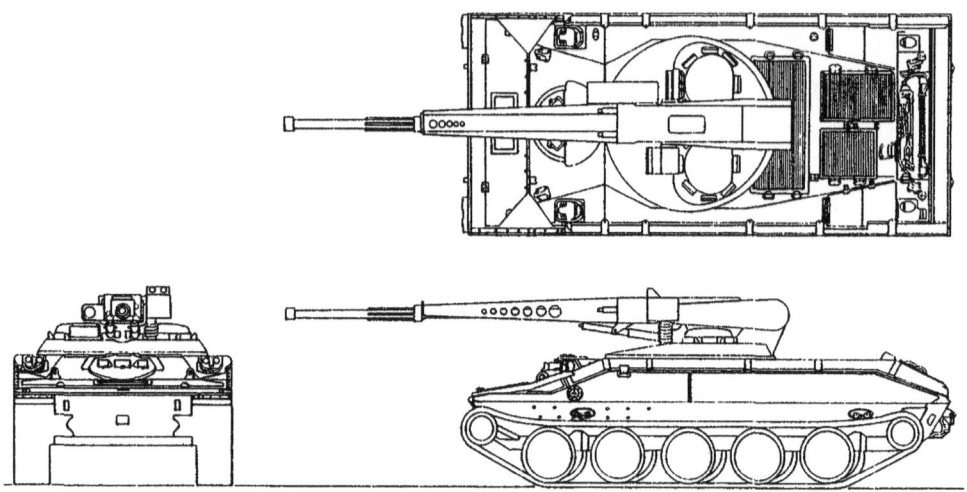

The ELKE test bed can be seen in the drawing above and in the photograph below at the right.

Although the Pacific Car and Foundry proposal for the HSTV-L was not accepted, the elevated trunnion gun mount concept was considered to be worth further evaluation. As a result, Pacific Car and Foundry received a contract to build the elevated kinetic energy (ELKE) technology demonstrator. This consisted of a special turret armed with the Ares 75mm automatic cannon in an elevated trunnion mount. It was installed on the chassis of the M551 Sheridan.

The assembly of the ELKE test bed was completed in the third quarter of fiscal year 1982 and it was shipped to the Yuma Proving Ground for preliminary tests. In addition to evaluating the technology of the elevating trunnion gun mount, the ELKE was used to develop tactics using defilade positions which could be effectively employed by a vehicle equipped with such a gun mount.

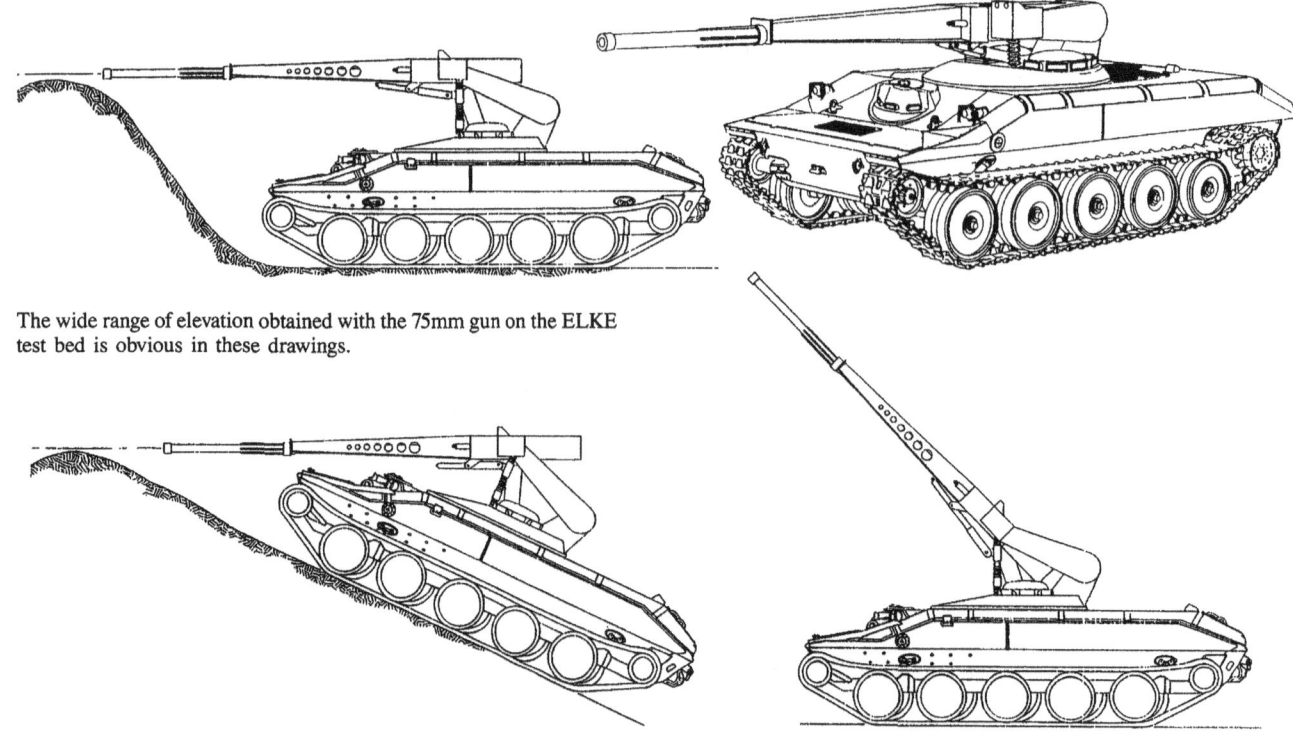

The wide range of elevation obtained with the 75mm gun on the ELKE test bed is obvious in these drawings.

154

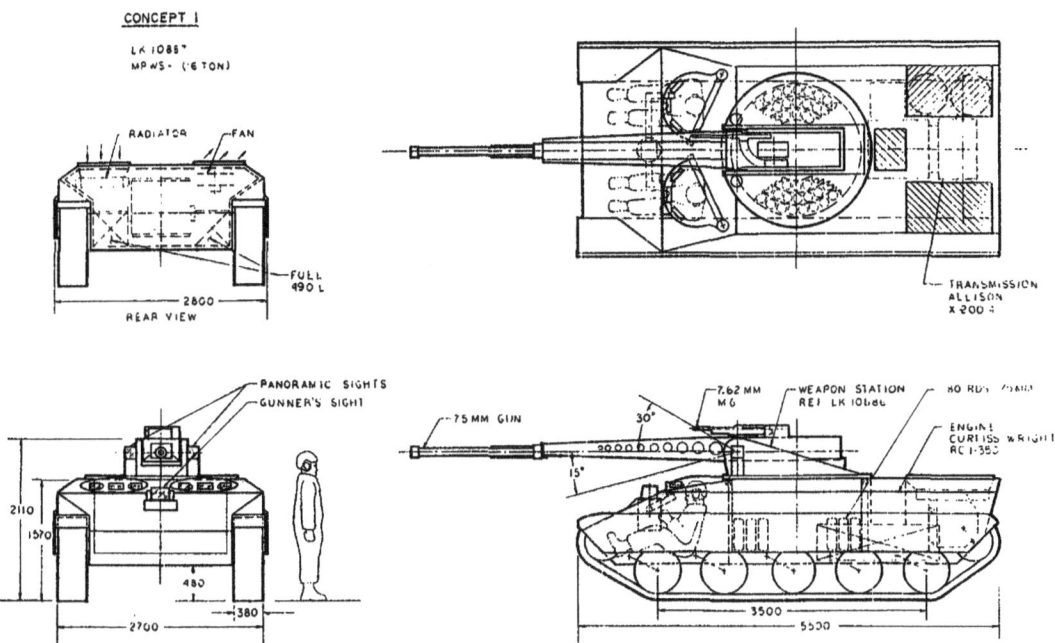

Above is a drawing of the concept 1 vehicle proposed under the ACVT program. Concept 9 appears at the bottom of the page.

As part of the ACVT program, the Tank Automotive Research and Development Command produced 25 vehicle design concepts divided into three categories. These were dedicated antiarmor, scout, and infantry fighting vehicles. Among these were 11 concepts that were intended to meet the 16 ton weight limit required by the U.S. Marine Corps for their mobile protected weapon system (MPWS). This limit was to permit the vehicle to be lifted by the CH53 helicopter. Three of the 11 designs also were proposed to meets the Army requirement for a light tank in the light division. However, it is suspected that the level of protection was insufficient for this application. The weight of the other concepts in the study ranged up to over 60 tons. The latter was an XM1 tank rearmed with an experimental 90mm Ares automatic cannon and manned by a crew of three.

Among the 11 lightweight designs, concept number 5 was a wheeled vehicle. Designs 1 through 4, 6, 7, and 9 through 11 were tracked vehicles with mobility equivalent to the XM1 tank, The chassis were almost identical and the proposed power plant was the Curtiss Wright RC1-350, single rotor, rotary engine developing 375 horsepower. It was coupled to an Allison X-200-4 transmission. An advanced torsion bar suspension supported the vehicle on five road wheels per side. The major difference between these nine vehicles was in the armament installation and the crew arrangement. Concepts 1 and 9 were manned by

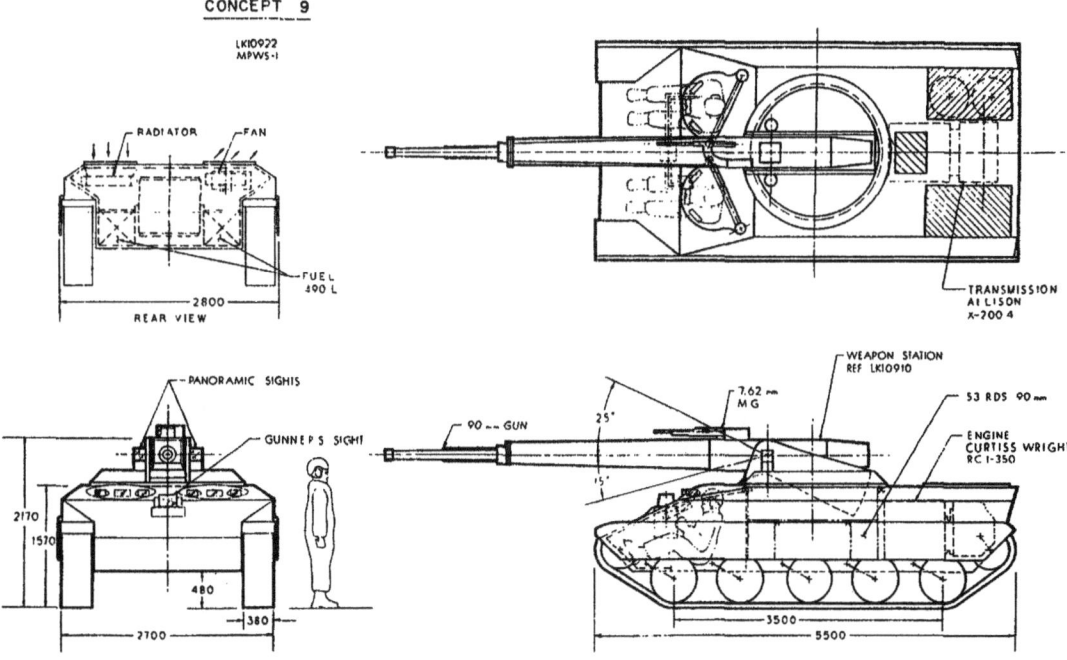

155

CONCEPT 2

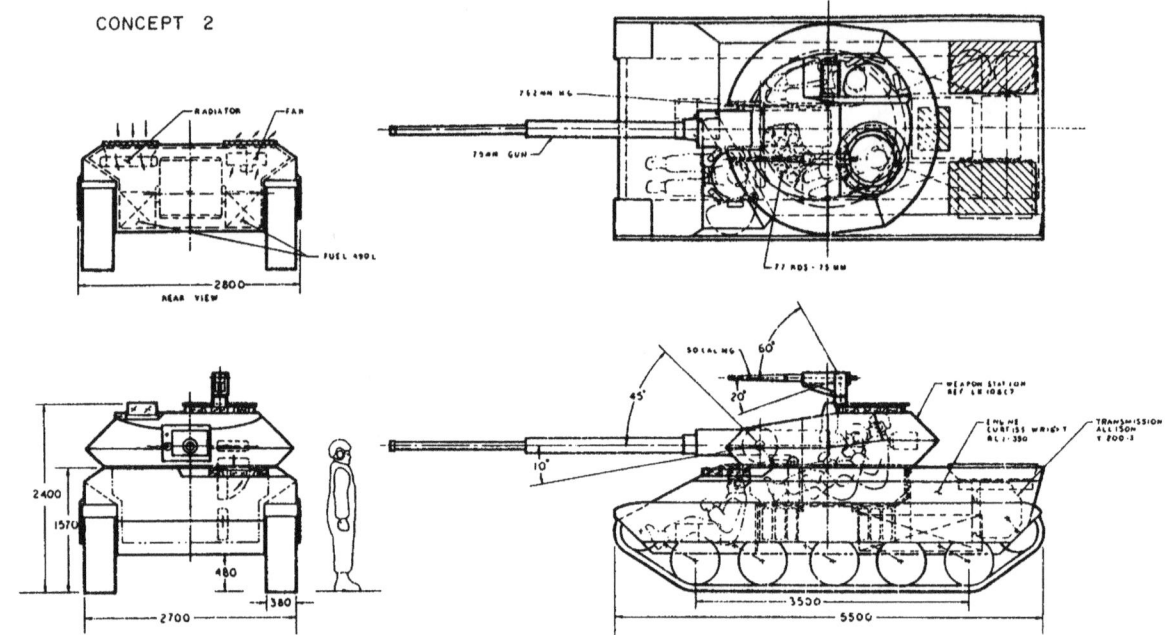

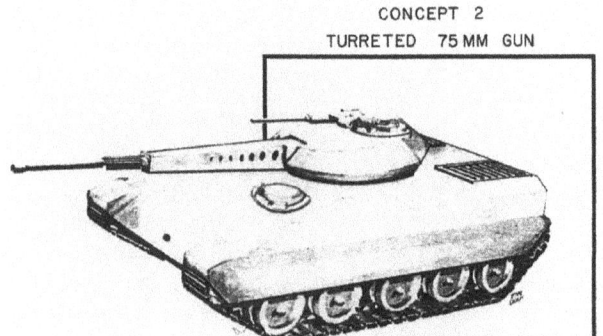

CONCEPT 2
TURRETED 75 MM GUN

Concept 2 under the ACVT program is shown above in the drawing and at the left.

a crew of two with both men seated in the front hull. Each man could drive the vehicle or operate the armament. Design number 1 was armed with a remote controlled 75mm Ares automatic cannon and number 9 was fitted with a 90mm version of the same weapon. A coaxial 7.62mm machine gun was provided in both cases.

Proposals 2, 7, and 10 each had a crew of three with the Ares automatic cannon mounted inside the turret. The

Concept 7 appears in the drawing below.

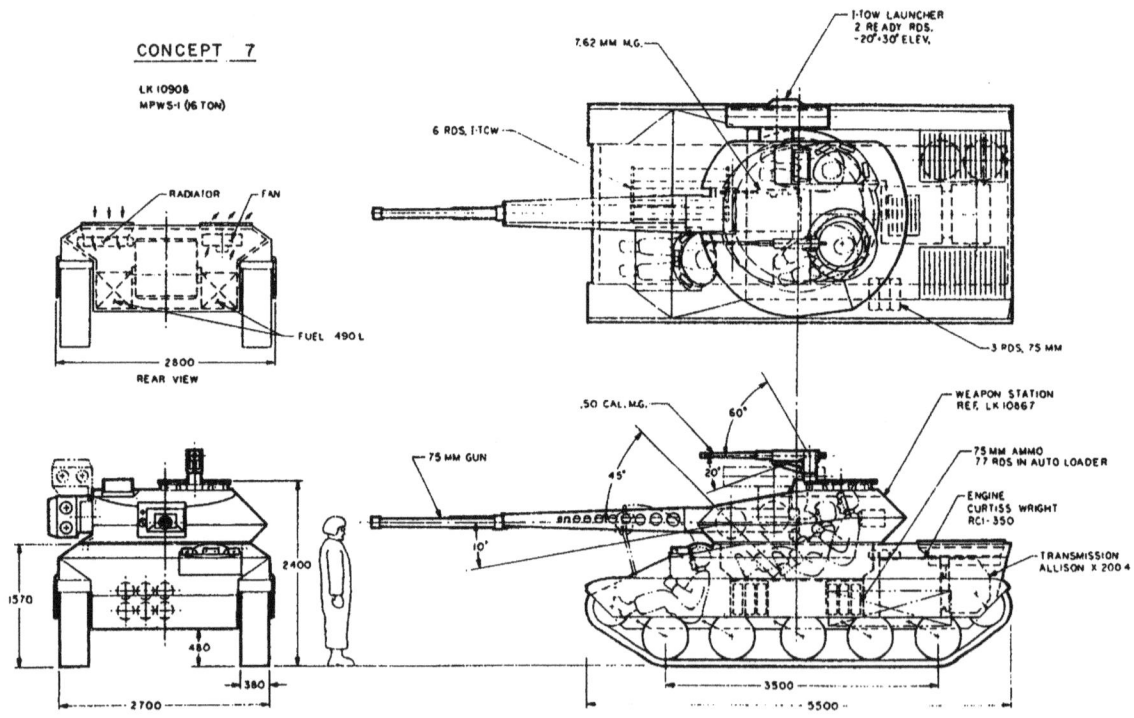

156

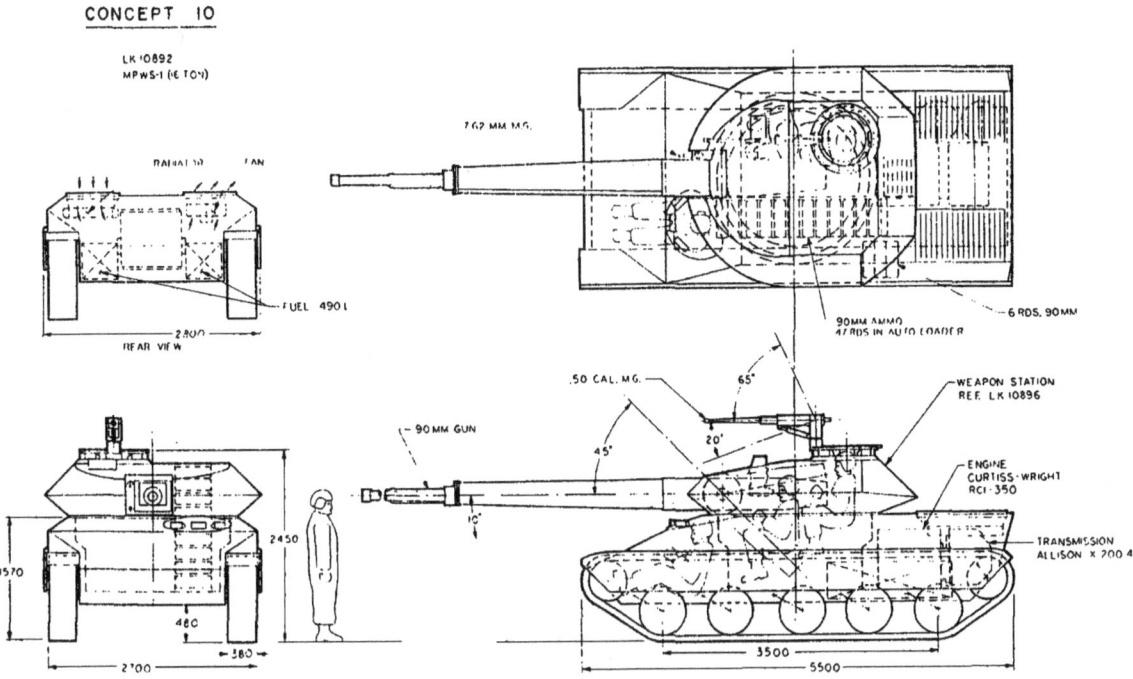

Above is a drawing of concept 10 under the ACVT program.

75mm weapon was installed in numbers 2 and 7 while number 10 carried the 90mm gun. All three had a 7.62mm coaxial machine gun and a .50 caliber machine gun on the commander's hatch. Number 7 also was fitted with a TOW missile launcher.

Concept number 6 also was armed with a turret mounted Ares 75mm gun, but the automatic loader was omitted and the crew was increased to four. It also had a 7.62mm coaxial machine gun and a .50 caliber machine gun for the tank commander.

The design of concept 6 can be seen in the drawing below.

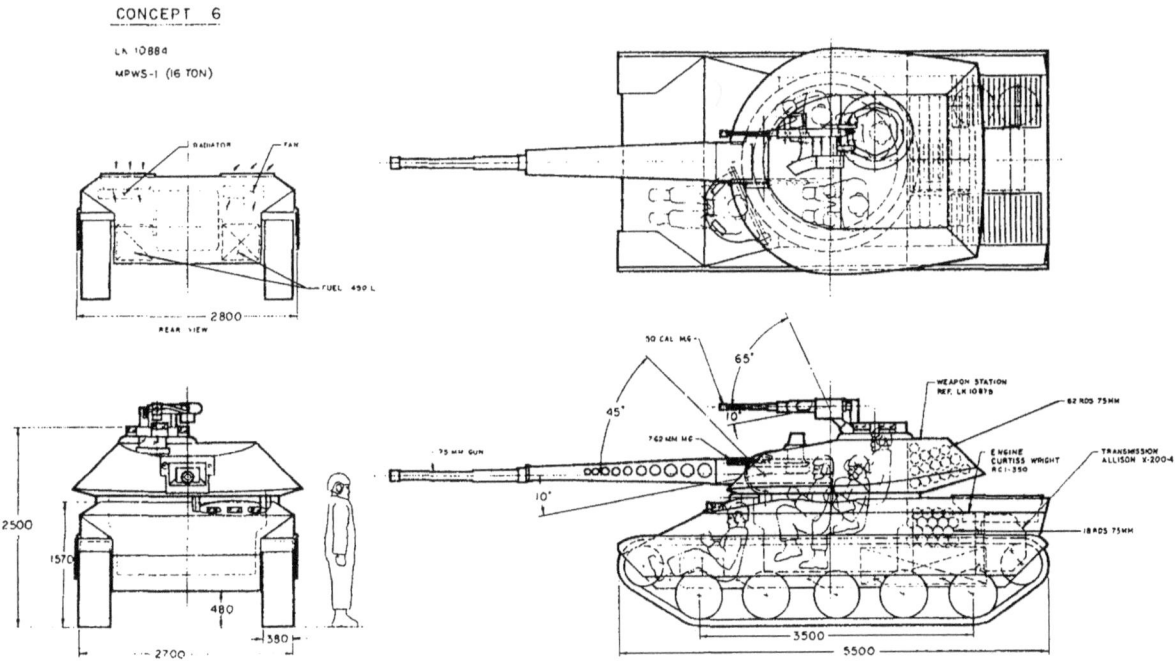

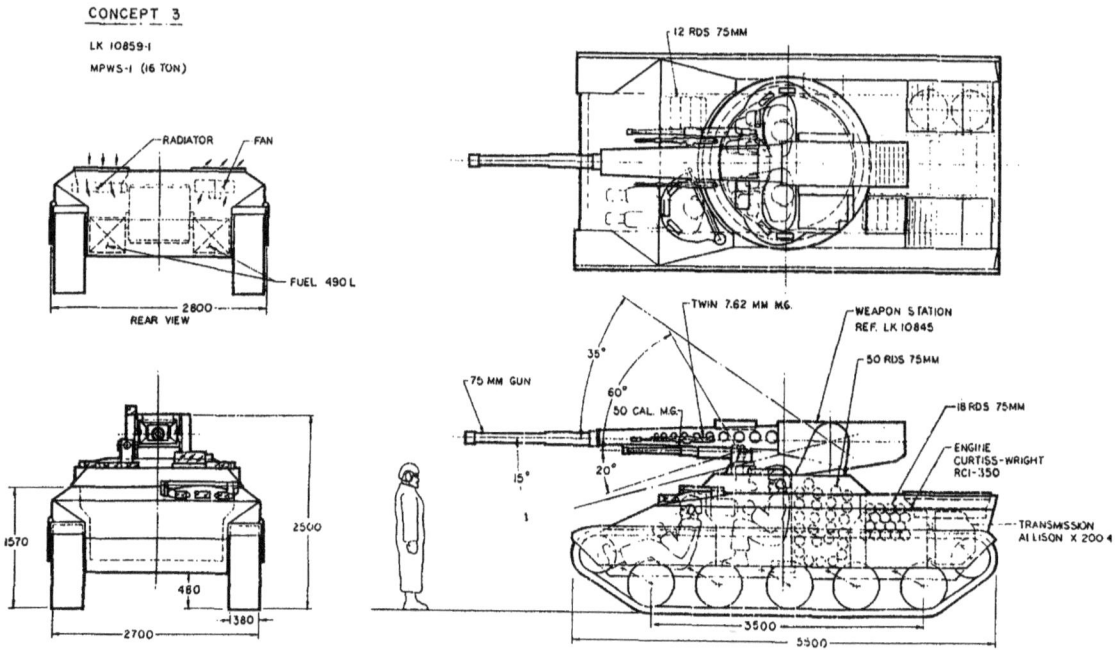

CONCEPT 3

LK 10859-1
MPWS-1 (16 TON)

RADIATOR   FAN

FUEL 490 L

2800
REAR VIEW

1570   2500

480
380
2700

12 RDS 75MM

TWIN 7.62 MM M.G.

WEAPON STATION
REF. LK 10845

50 RDS 75MM

75 MM GUN

35°
60°
50 CAL. M.G.
15°
20°
18 RDS 75MM

ENGINE
CURTISS-WRIGHT
RCI-350

TRANSMISSION
ALLISON X 200 4

3500
5500

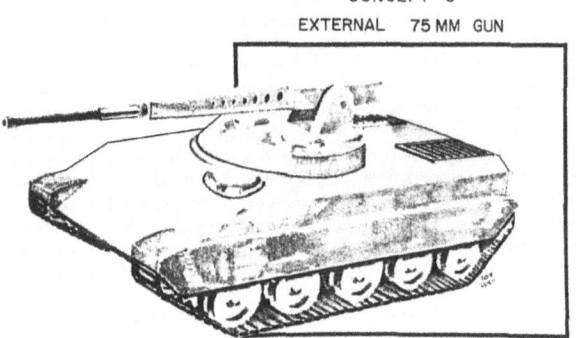

CONCEPT 3

EXTERNAL 75 MM GUN

ACVT concept 3 appears at the left and in the drawing above.

Proposals 3 and 11 each had a crew of three and were armed with the Ares automatic gun in an external mount. This was the 75mm gun on number 3 and the 90mm gun on number 11. Both vehicles had twin 7.62mm coaxial machine guns and a .50 caliber machine gun for the commander.

Concept number 4 carried the 75mm Ares automatic gun on an external variable height trunnion mount as on the Pacific Car and Foundry HSTV-L proposal. A 25mm

Below is the drawing of concept 11.

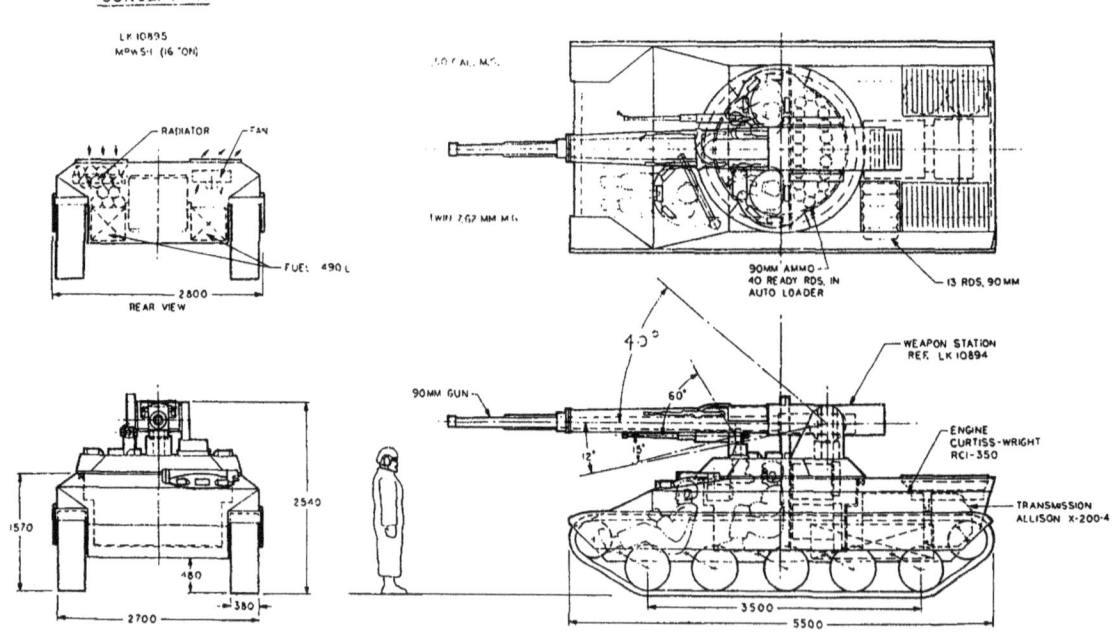

CONCEPT 11

LK 10895
MPWS-1 (16 TON)

RADIATOR   FAN

FUEL 490 L

2800
REAR VIEW

1570   2540

480
380
2700

.50 CAL. M.G.

TWIN 7.62 MM M.G.

90MM AMMO-
40 READY RDS. IN
AUTO LOADER

13 RDS. 90 MM

40°

WEAPON STATION
REF. LK 10894

90MM GUN
60°
12°   15°

ENGINE
CURTISS-WRIGHT
RCI-350

TRANSMISSION
ALLISON X-200-4

3500
5500

158

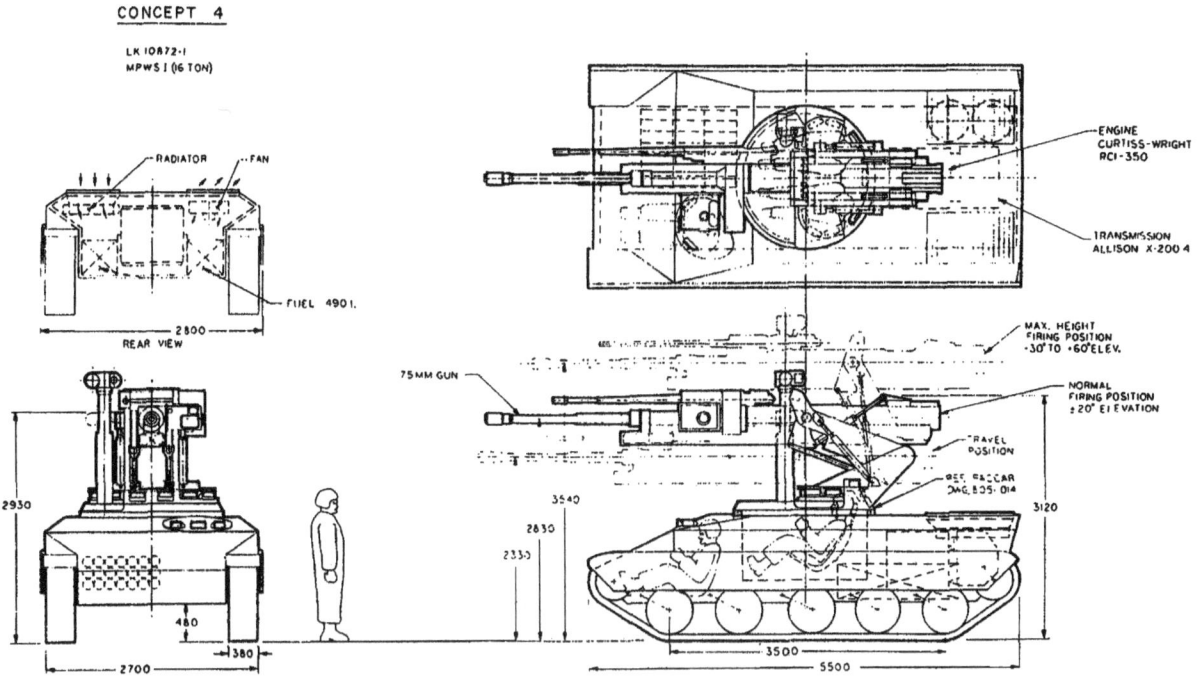

CONCEPT 4

LK 10872-1
MPWS I (16 TON)

Concept 4 used the variable height trunnion gun mount as shown in the drawing above.

Bushmaster cannon and a 7.62mm machine gun were coaxial weapons. It was manned by a crew of three.

Design proposal number 8 had a lengthened chassis with six road wheels per side for the advanced torsion bar suspension. A Curtiss Wright RC2-350, two rotor, rotary engine developing 600 horsepower was proposed as the power plant with the Allison X-300 transmission. The mobility was expected to be equal to that of the HIMAG. The vehicle was manned by a crew of two in the front hull. Each man could act as gunner or driver. Main armament was the remote controlled 75mm Ares automatic cannon with a 7.62mm coaxial machine gun.

Below is the longer concept 8 vehicle powered by the Curtiss Wright RC2-350 rotary engine.

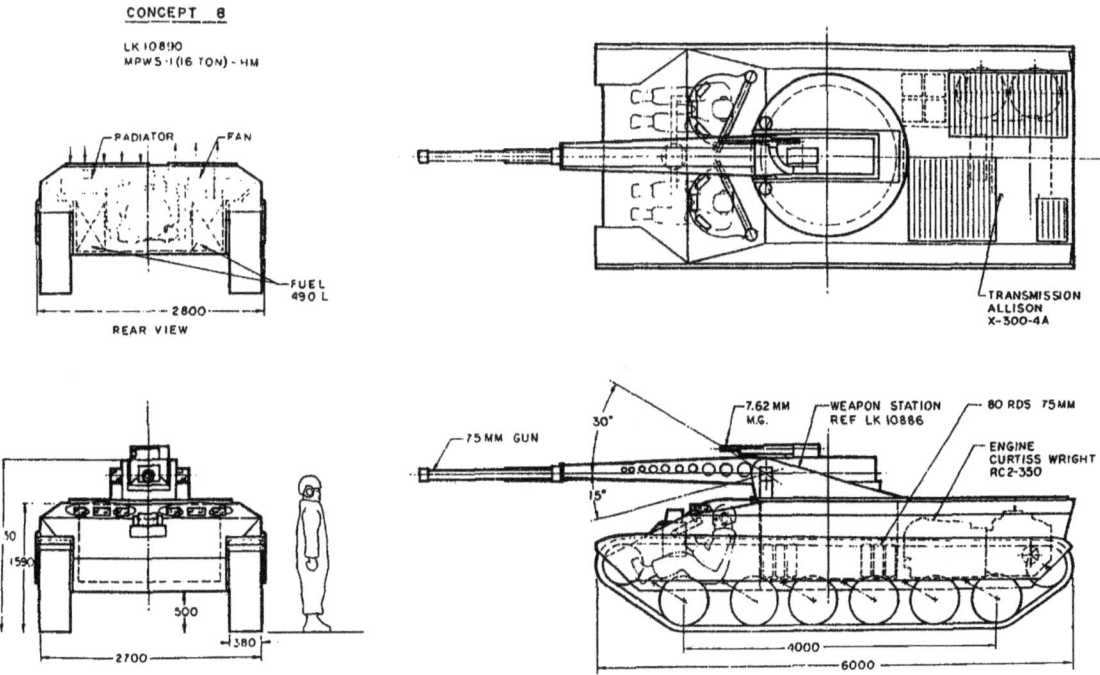

CONCEPT 8

LK 10890
MPWS-1 (16 TON)-HM

One of the Sheridans converted under the ARMVAL program appears in the photographs above and below. Note that the driver's rotating hatch, the armament, and the flotation screen have all been removed.

The Advanced Antiarmor Vehicle Evaluation (ARMVAL) program was a joint effort of the U.S. Army and the U.S. Marine Corps. The Tank Automotive Research and Development Command rebuilt ten M551 Sheridans by removing the armament and some armor. The original engine was replaced by an uprated General Motors 8V53T diesel engine to increase the power to weight ratio. The suspension also was modified to improve the cross-country mobility. The ARMVAL program ran from September 1978 to December 1980. In addition to the vehicle performance, the project evaluated the fully stabilized Staget sighting system which included a laser gun simulator and daylight television with zoom optics.

Below, the ARMVAL vehicle is operating during the test program.

The AAI Corporation rapid deployment force light tank (RDF/LT) is shown in the photographs above and at the bottom of the page. This version located the tank commander in the turret with the gunner and the driver in the front hull.

THE REQUIREMENT FOR A NEW LIGHT TANK

Based on the development of the HSTV-L, AAI Corporation in 1980 produced a prototype designated by the Company as the rapid deployment force light tank (RDF/LT). At that time, it was hoped that it would meet the requirements of both the Army and the Marine Corps. Like the HSTV-L, the RDF/LT was armed with the Ares 75mm automatic cannon now designated as the XM274. The fully stabilized gun had a high tracking rate and a maximum elevation of about 40 degrees providing both antiarmor and antiaircraft capability. For the latter application, proximity fuzed high explosive ammunition was required. A total of 60 rounds of 75mm ammunition were directly available to the automatic loader. Secondary armament consisted of a 7.62mm M240 coaxial machine gun. Constructed of aluminum armor, the RDF/LT had a combat weight of 14.8 tons.

Steel applique armor could be added increasing the weight to about 16 tons. Thus it met the Marine Corps weight limit for helicopter transportation. The crew arrangement was similar to that on the HSTV-L with the commander in the turret and the driver and gunner in the front hull on the left and right respectively. However, the commander was shifted from the left side of the turret to the center behind the cannon. The gas turbine power plant in the HSTV-L was replaced in the RDF/LT by a General Motors 6V53T diesel engine developing 350 gross horsepower. It was coupled to the Allison X-200 transmission. The complete power pack could slide out of the rear hull on extendable rails for easy access during maintenance. A torsion bar suspension was substituted for the hydropneumatic suspension on the earlier vehicle.

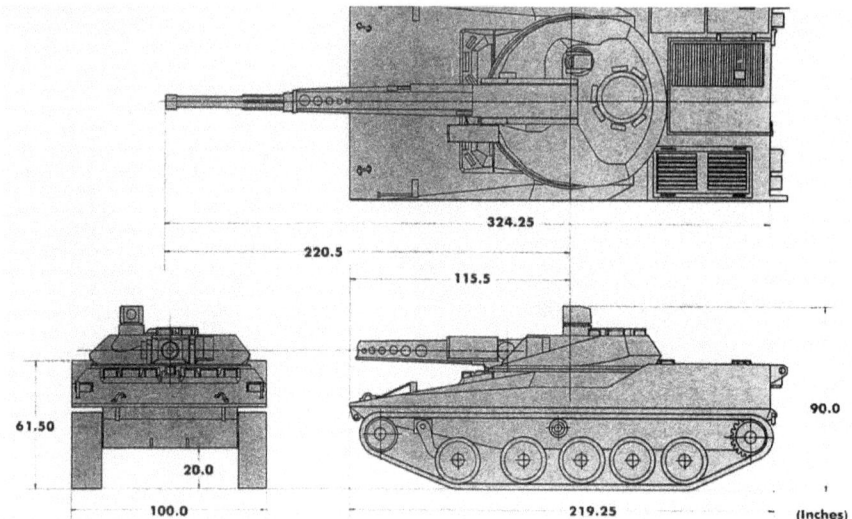

The dimensions of the AAI RDF/LT can be seen in the drawing above. Below, the number of these vehicles that could be carried on various aircraft is shown at the left and the turret and gun mount installed on the RDF/LT appears at the right.

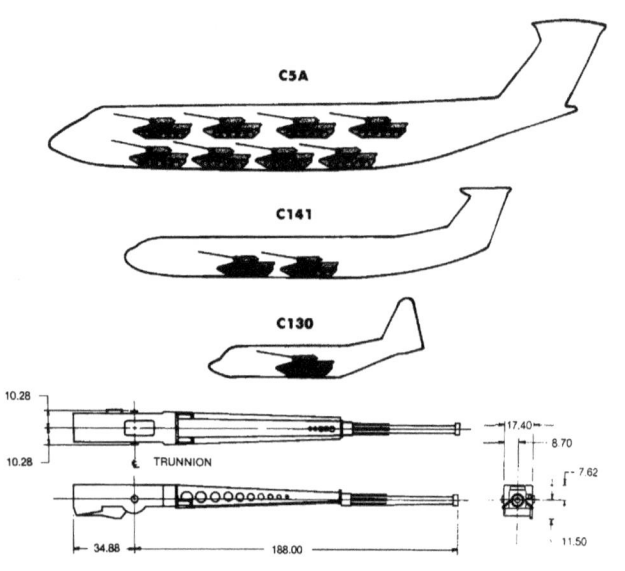

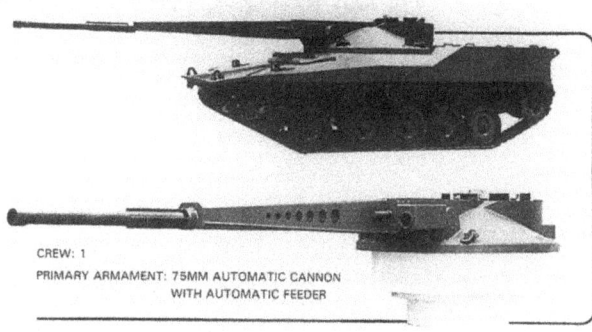

CREW: 1

PRIMARY ARMAMENT: 75MM AUTOMATIC CANNON
WITH AUTOMATIC FEEDER

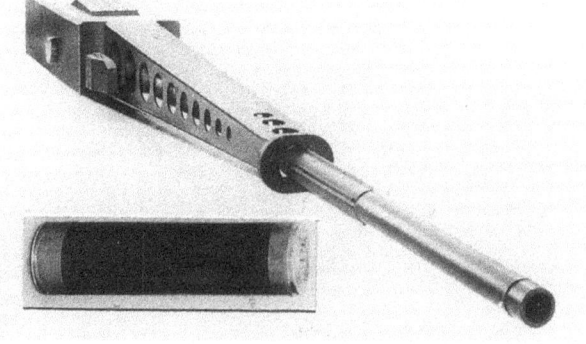

The Ares XM274 75mm gun is above and at the right. Below are two views of the RDF/LT. Note the maximum elevation obtainable in the left photograph.

162

The RDF/LT appears above and at the right during its test program.

The turret of the RDF/LT could be replaced by an unmanned turret carrying an externally mounted 75mm XM274 cannon thus reducing the crew to two. This turret also could be equipped with two pods carrying a total of eight antiaircraft missiles.

The two man version of the RDF/LT appears at the left with the Ares 75mm automatic cannon fully elevated. Below the same vehicle has been armed with eight ground to air missiles.

The RDF/LT appears in the photographs above and at the bottom of the page armed with the 76mm gun M32. Note that this vehicle, modified from the original pilot, retains the hatch in the right front hull. The low silhouette of the RDF/LT and its stowage on board a C130 aircraft are illustrated in the drawings below.

Since the Ares automatic cannon was not released for foreign sales, AAI Corporation developed a version of the RDF/LT with a new turret armed with the M32 76mm gun previously used in the M41 light tank series. This required some rearrangement of the crew placing the gunner in the turret with the tank commander. The driver remained in the left front hull. It was intended to stow 30 rounds of 76mm ammunition in the right front hull with an additional 20 rounds in the turret basket. AAI developed a new APFSDS round for the 76mm gun which greatly improved its penetration performance. The new 76mm gun turret was installed on the chassis of the original RDF/LT for demonstration purposes. The latter retained the crew station in the right front hull.

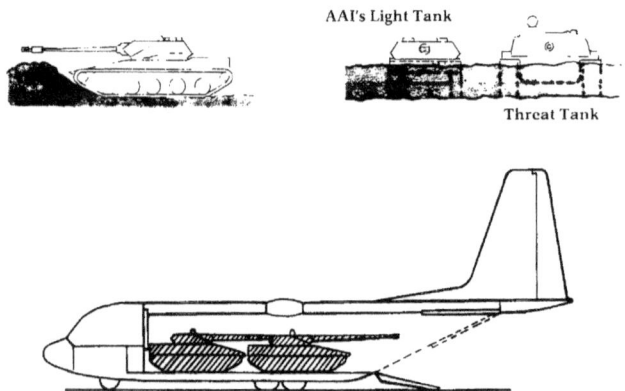

AAI's Light Tank

Threat Tank

164

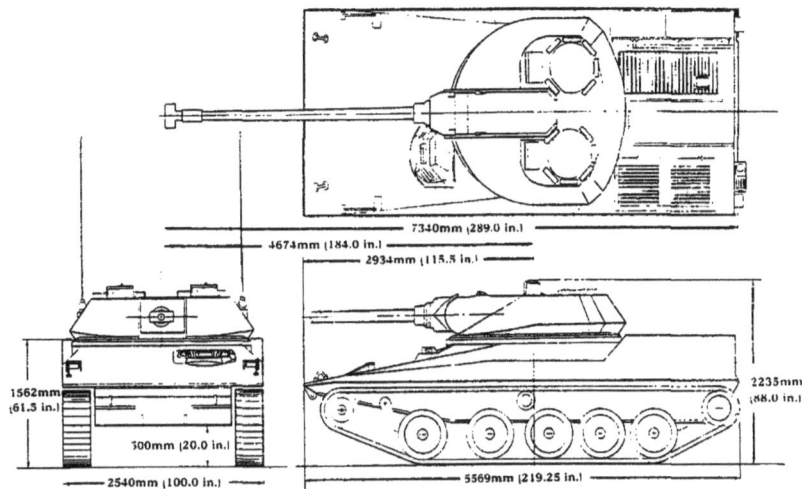

The dimensions of the RDF/LT with the 76mm gun M32 are shown above. The new 76mm APFSDS-T round is compared at the left below with the standard 105mm armor piercing rounds. The internal arrangement of the RDF/LT with the 76mm gun can be seen in the cutaway drawing below at the right.

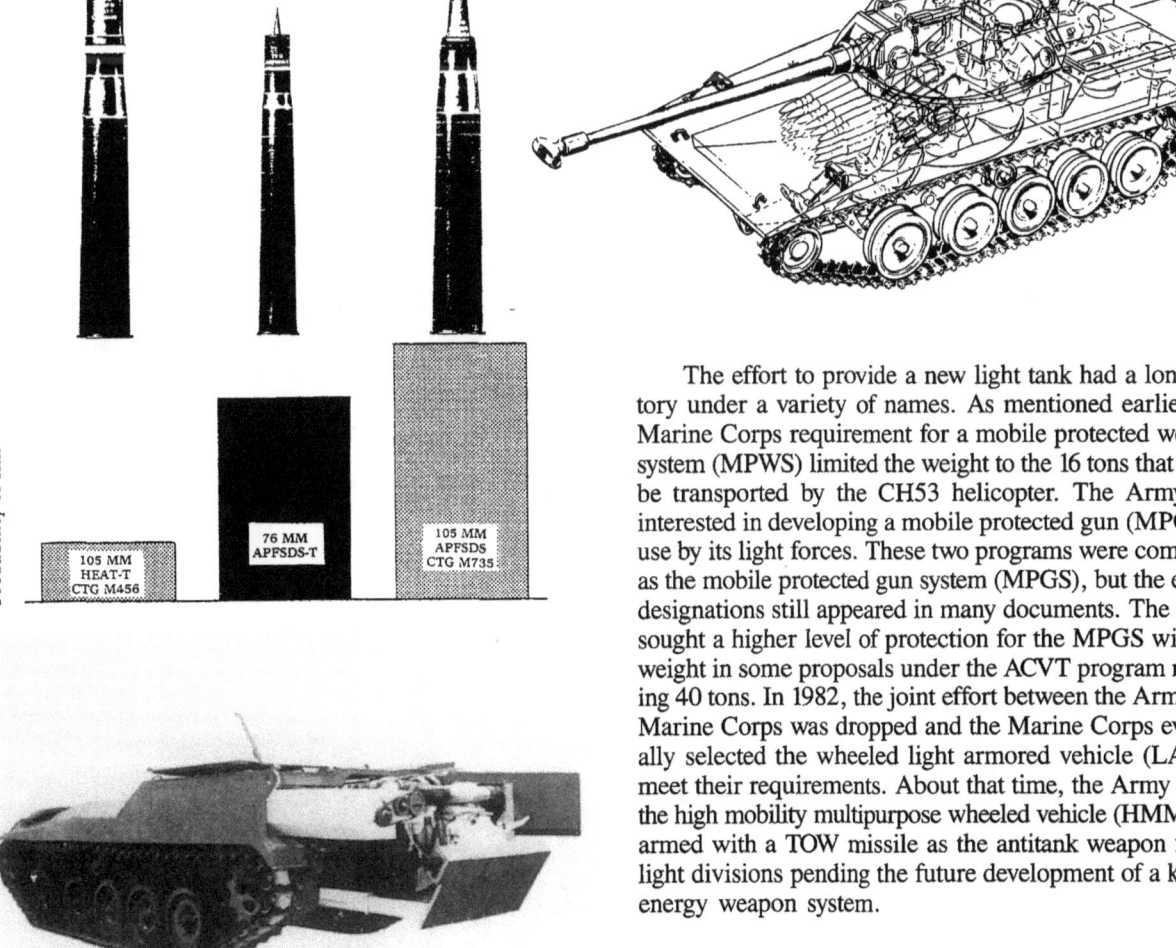

The effort to provide a new light tank had a long history under a variety of names. As mentioned earlier, the Marine Corps requirement for a mobile protected weapon system (MPWS) limited the weight to the 16 tons that could be transported by the CH53 helicopter. The Army was interested in developing a mobile protected gun (MPG) for use by its light forces. These two programs were combined as the mobile protected gun system (MPGS), but the earlier designations still appeared in many documents. The Army sought a higher level of protection for the MPGS with the weight in some proposals under the ACVT program reaching 40 tons. In 1982, the joint effort between the Army and Marine Corps was dropped and the Marine Corps eventually selected the wheeled light armored vehicle (LAV) to meet their requirements. About that time, the Army chose the high mobility multipurpose wheeled vehicle (HMMWV) armed with a TOW missile as the antitank weapon for its light divisions pending the future development of a kinetic energy weapon system.

The photograph at the left shows the power pack of the RDF/LT rolled out of the engine compartment for easy maintenance.

165

Above is the Sheridan armed with the 105mm gun M68 installed by the Naval Surface Weapons Center. The view at the bottom of the page shows the weapon fitted with a muzzle brake.

THE DEVELOPMENT OF THE ARMORED GUN SYSTEM

By the early 1980s, the Army concluded that a more powerful weapon than the 75mm gun would be required for use by the light forces. Characteristics were outlined for a new MPGS, but the designation was revised once again. It now became the armored gun system (AGS).

The search for new weapon continued with the 105mm gun considered as the most likely candidate. The Naval Surface Weapons Center installed the 105mm gun M68 in the turret of an M551 Sheridan using the recoil system for the 152mm gun-launcher. The cannon was test fired with and without a muzzle brake. Although this installation would

have required considerable further development before it would have been satisfactory for service use, the tests indicated that it would be practical to mount the 105mm gun on such a light vehicle. Other work also showed that a soft recoil system could solve the problem of mounting the 105mm gun on a lightweight chassis. Such systems were under development by Rheinmetall in Germany and Royal Ordnance in England as well as by others in the United States and Europe. In addition to mounting the 105mm gun, the new AGS would have to be capable of parachute delivery from the C130 transport aircraft. As the study continued,

Above, the Commando Stingray turret is mounted on the Sheridan chassis for test firing. Below are views of the complete Commando Stingray.

it was considered to be highly desirable to select a non-development item for the new AGS now designated as the XM4. Several potential candidates existed that might meet the requirements or could be modified to do so.

The Cadillac Gage Company had developed the Stingray light tank aimed primarily toward the export market. However, with its 21 ton combat weight, it was a suitable prospect for the AGS role. The Stingray was armed with a turret mounted low recoil force (LRF) version of the L-7 105mm gun developed by Royal Ordnance in England. The reduction of the recoil force was achieved by increasing the length of recoil and the addition of a muzzle brake. The

turret and gun were successfully test fired after installation on the chassis of an M551 Sheridan. The four men in the Stingray crew were located in a conventional arrangement with the driver in the center front hull and the commander, gunner, and loader in the turret. A total of 36 rounds for the 105mm gun were carried including eight ready rounds. A 7.62mm coaxial machine gun was mounted alongside the cannon and either a 7.62mm or a .50 caliber machine gun could be installed at the commander's hatch. Two axis stabilization was available as an option as well as a laser range finder and a thermal sight. The turret and hull were assembled by welding the Cadloy high hardness steel armor.

The internal layout of the Stingray appears below in the cutaway drawing.

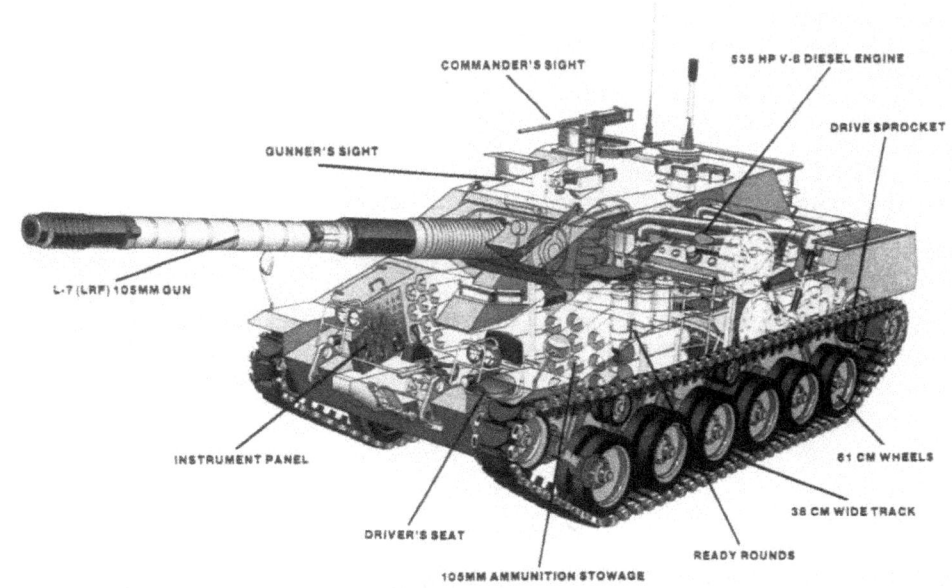

Additional views of the Stingray are shown above. At the top right, the vehicle is moving at high speed.

Flat sections were used to permit the easy addition of applique armor if increased protection was required. The Stingray was powered by the Detroit Diesel Allison 8V92TA diesel engine developing 535 horsepower at 2300 revolutions per minute. It was coupled to the Detroit Diesel Allison XTG-411-2A transmission. The vehicle was supported by a torsion bar suspension with six dual road wheels per side. The 15 inch wide steel, double pin, tracks had detachable rubber pads. The Stingray had a maximum level road speed of 43 miles per hour and a cruising range of about 300 miles.

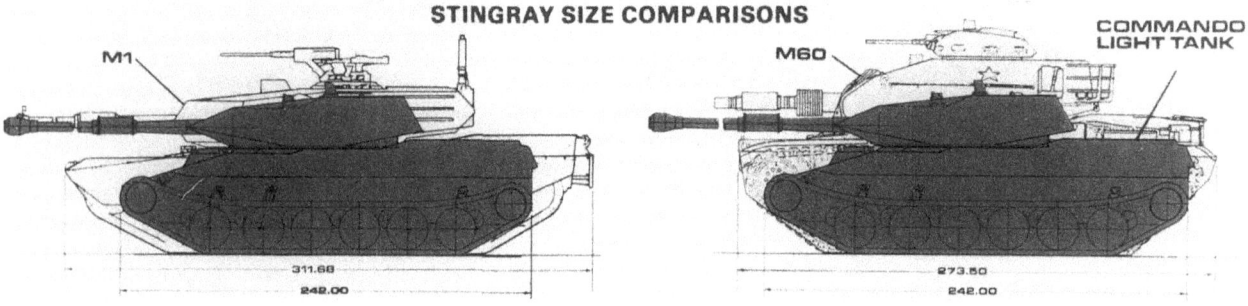

Above, the dimensions of the Stingray are compared with those of the M1 and M60 main battle tanks which also were armed with the 105mm gun. Below, three Stingrays are operating in wooded terrain.

168

Above is the original mock-up of the Teledyne Continental Motors armored gun system (AGS) armed with the 75mm Ares automatic cannon. At the bottom of the page, the early pilot AGS is now fitted with the 105mm gun.

A more radical design was presented by Teledyne Continental Motors with the engine and transmission installed in the front to increase the protection for the crew. This vehicle featured an external mount for the M68A1 105mm gun with a long recoil system and a muzzle brake. The weapon was fed from a nine round magazine in the turret. Another 20 rounds were available to the automatic loader from two ten round replenisher drums in the hull. An additional 13 rounds in the hull brought the total 105mm ammunition stowage to 42. An M240 7.62mm coaxial machine gun was mounted in the gun pod on the left side of the cannon. Either a 7.62mm or a .50 caliber machine gun could be installed for use by the tank commander. The commander and gunner rode in the turret basket below the external gun

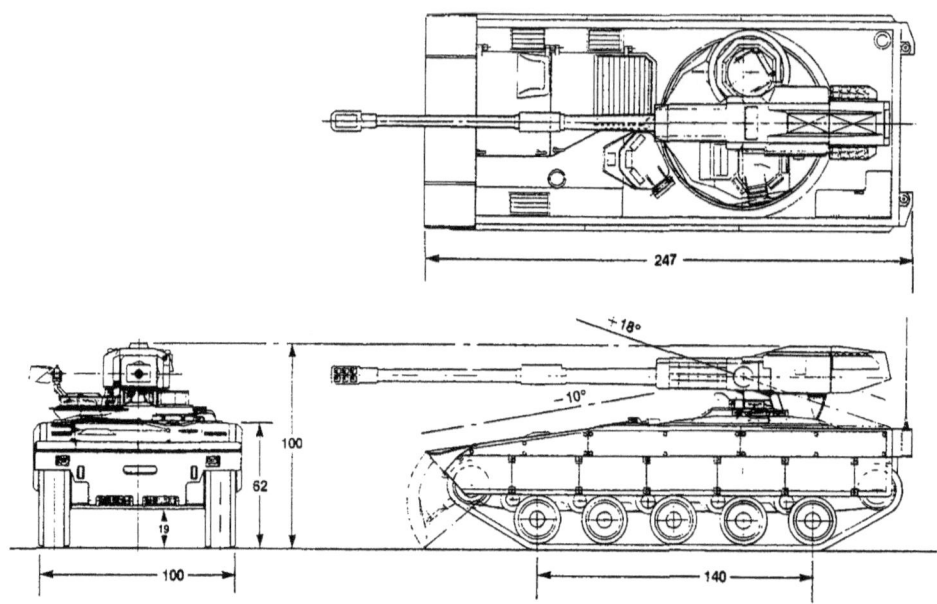

The three view drawing above shows the layout and dimensions of the Teledyne Continental Motors AGS candidate.

mount with the commander on the right and the gunner on the left. The driver was located in the left front hull. With a combat weight of 21 tons, the vehicle was driven by a Cummins VTA-903 diesel engine through a General Electric HMPT-500 transmission. This engine developed 500 gross horsepower. A model 2884 hydropneumatic suspension supported the vehicle on five dual road wheels per side. The maximum road speed was about 45 miles per hour with a cruising range of approximately 300 miles. A modification proposed for the Teledyne AGS was the installation of antitank missiles on wing-like extensions on each side of the gun pod. The external gun also could be replaced by a more powerful weapon such as the 120mm gun. The increased size would have required transportation in a C141 aircraft instead of the C130. Other proposed variants on the basic chassis included an infantry fighting vehicle and a self-propelled air defense system.

Details of the Teledyne AGS can be seen in these photographs. Note the low silhouette of the vehicle compared to the crew member.

The machine gun provided for the tank commander is obvious in the view of the Teledyne AGS above. Below, the crew stations and the internal layout of the vehicle are shown in these two drawings.

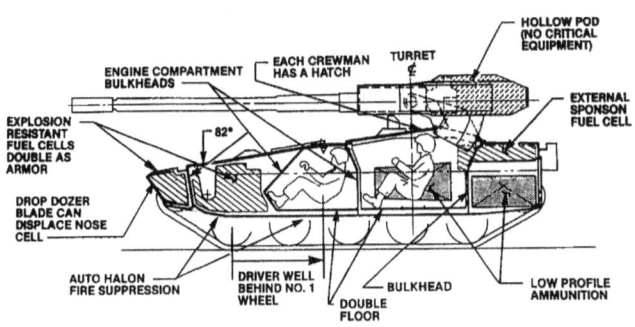

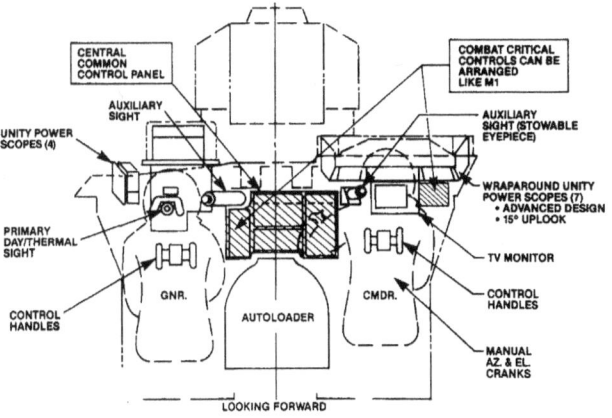

Below, the proposed installation of antitank missiles can be seen at the left and the possible future 120mm gun armament appears at the right. The latter vehicle could only be transported in a C141 or larger aircraft.

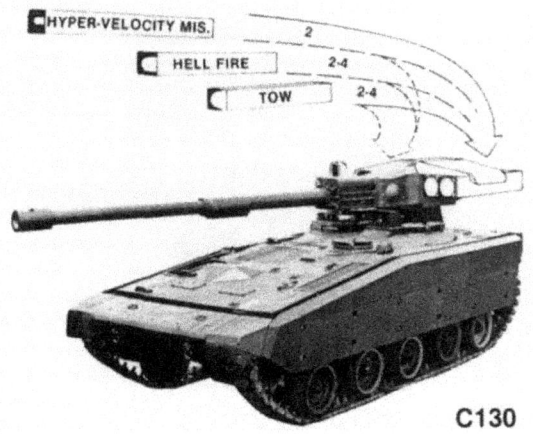

C130

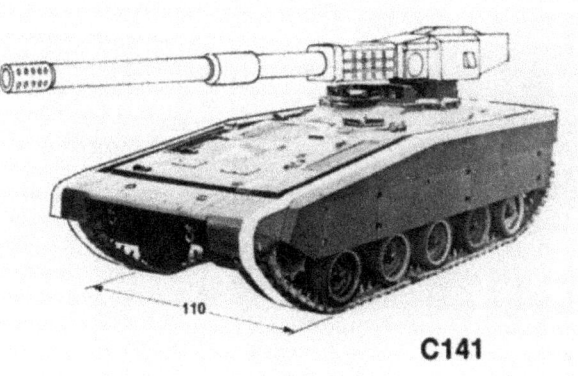

C141

Above is the mock-up of the FMC Corporation close combat vehicle light (CCVL) which was a candidate in the AGS competition. The pilot CCVL itself appears at the bottom of the page.

The FMC Corporation produced a more conventional configuration as their candidate. Designated as the close combat vehicle light (CCVL), it was armed with the Rheinmetall modified 105mm gun M68A1 in a soft recoil turret mount. The gun was provided with an automatic loader reducing the crew to three. The gunner and tank commander were located in the right side of the turret with the automatic loader on the left. A bulkhead separated the turret crew from the gun and the automatic loader. There were 19 ready rounds in the magazine of the automatic loader. An additional 24 rounds brought the total 105mm ammunition stowage to 43. Secondary armament consisted of a 7.62mm M240 coaxial machine gun. The gunner was provided with a Hughes stabilized day/night thermal sight and an independent thermal viewer was available as an option for the tank commander. A Cadillac Gage system stabilized the gun and turret. The structure of the CCVL was a welded assembly of aluminum armor with the driver seated in the center front hull. It had a combat weight of 21.5 tons and was powered by a Detroit Diesel Allison 6V92TA diesel engine with a General Electric HMPT-500-3 transmission. The engine developed 575 gross horsepower at 2400 revolutions per minute. The maximum speed on a level road was about 43 miles per hour. A 150 gallon fuel capacity gave the CCVL a cruising range of approximately 300 miles at 25 miles per hour.

Details of the FMC CCVL can be seen in the photograph above. Note the commander's independent thermal viewer in front of the cupola.

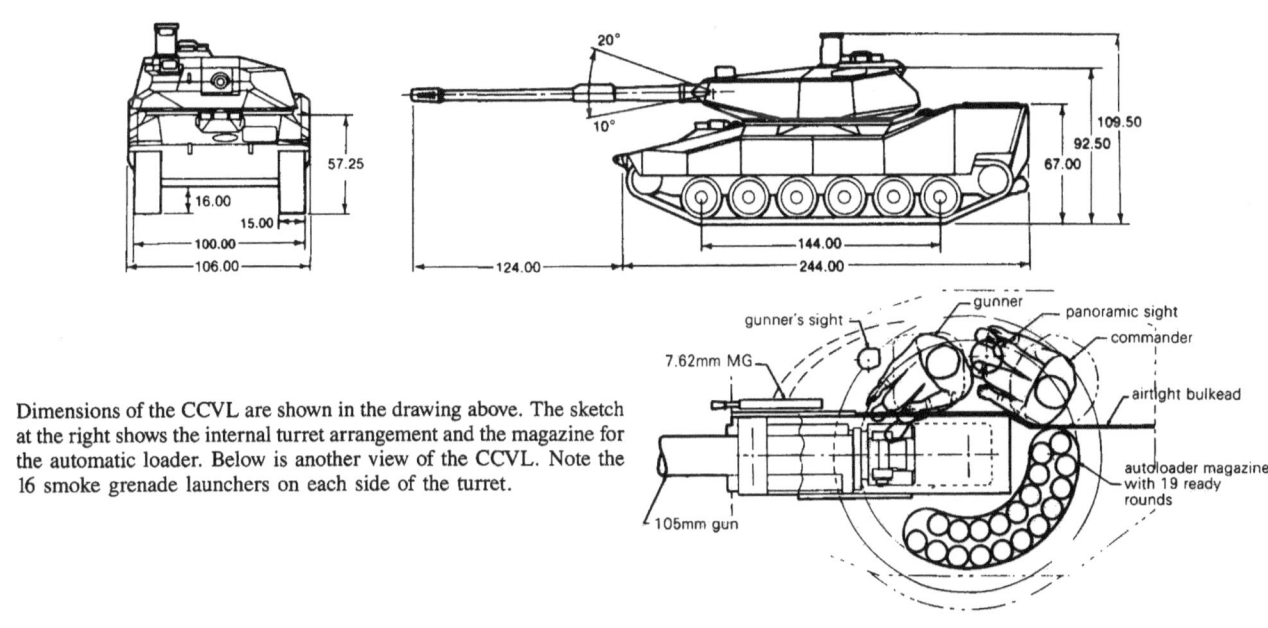

Dimensions of the CCVL are shown in the drawing above. The sketch at the right shows the internal turret arrangement and the magazine for the automatic loader. Below is another view of the CCVL. Note the 16 smoke grenade launchers on each side of the turret.

The CCVL above has been modified. Note the absence of the commander's independent thermal viewer. Below, the CCVL is being transported in the commercial equivalent of the Air Force C130 aircraft.

Above is the Hägglunds IKV-91 tank armed with the 90mm gun. The conventional crew arrangement is shown in the sketch below at the right. The various crew members are indicated as follows: (1) tank commander, (2) gunner, (3) loader, and (4) driver.

Another possible candidate for the XM4 requirement was the Hägglunds IKV-91 built in Sweden. However, it would need to be rearmed with the 105mm gun specified for the armored gun system.

Although the XM4 AGS had been supported by the Army Vice Chief of Staff, the limitations imposed by the Gramm-Rudman Act eliminated funding for the project. Subsequently, the Armored Family of Vehicles Task Force (AFVTF) was directed to solve the light forces requirement for a kinetic energy tank killing system.

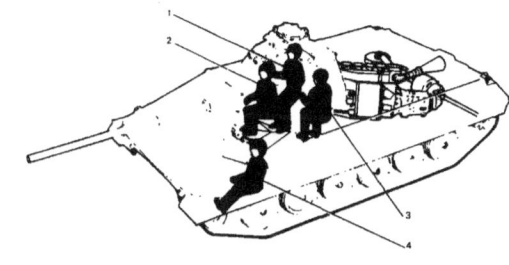

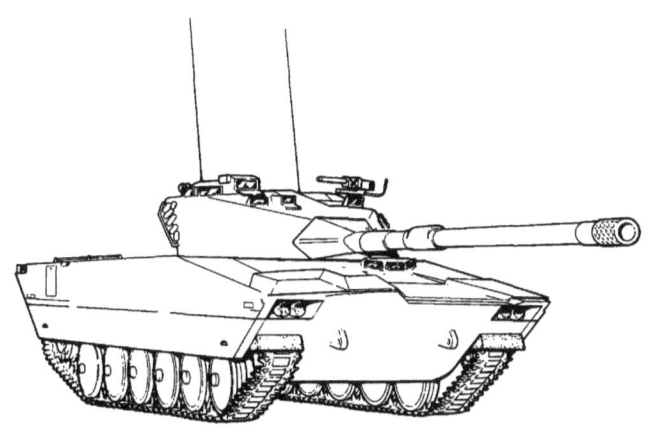

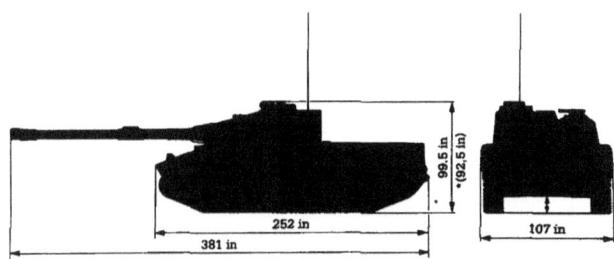

The configuration of the IKV-105 is sketched at the left and its dimensions are shown above. Below, the 105mm gun is firing.

175

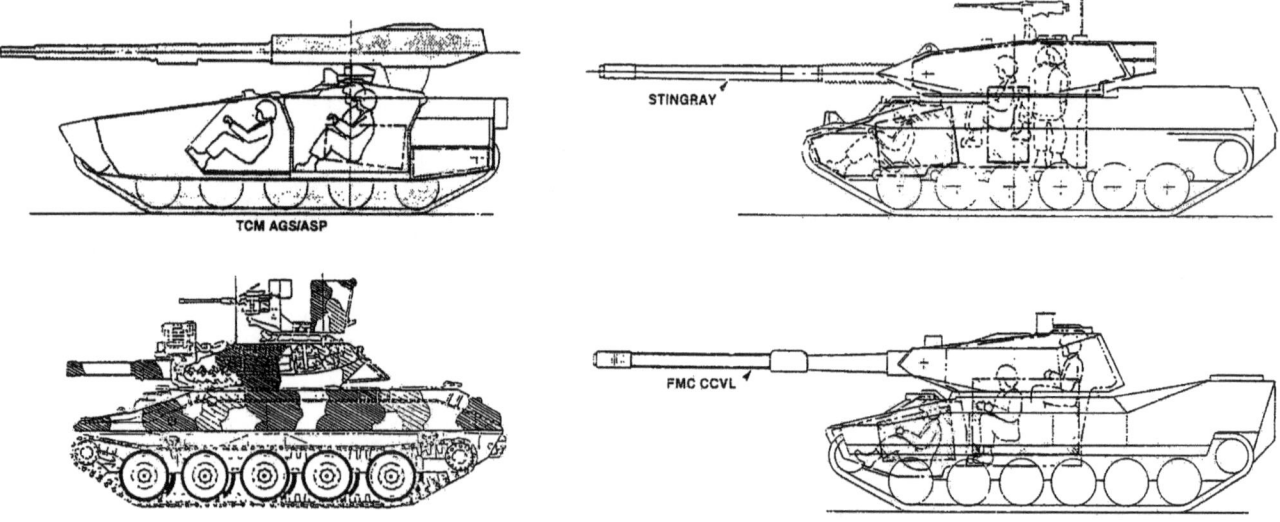

Above, the side view of the Sheridan (lower left) is compared with three of the AGS candidates. These are the Teledyne Continental Motors ASP (upper left), The Cadillac Gage Stingray (upper right), and the FMC CCVL (lower right).

The AFVTF under Major General Robert J. Sunell reviewed the requirements of the light forces and the history of the previous development. The Task Force also developed the concept of an all purpose fire support platoon (APFSP) using two types of vehicles. These were the armored support platform (ASP) and the rearm platform (RAP). The ASP was, of course, another name for the armored gun system (AGS). The AFVTF studied a long list of possible candidates to find one suitable for use as the AGS. This list was reduced to four for final consideration. These were the FMC Corporation CCVL, the Cadillac Gage Stingray, the General Motors light armored vehicle with a 105mm gun (LAV-105), and two versions of the Teledyne Continental Motors vehicle with the 105mm gun in an external mount. The two versions of the latter differed in the ammunition stowage arrangement for the automatic loader.

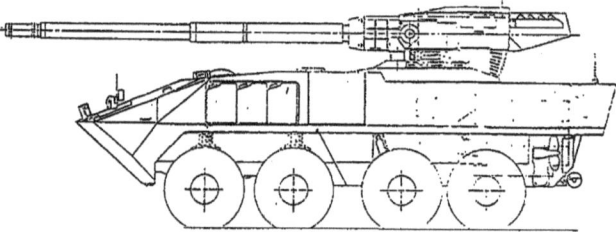

At the right is a side view of the General Motors LAV-105. Below is a four view drawing of the Teledyne Continental Motors ASP armed with the lightweight EX-35 105mm gun. This version stowed the 105mm ammunition in the hull drum feed.

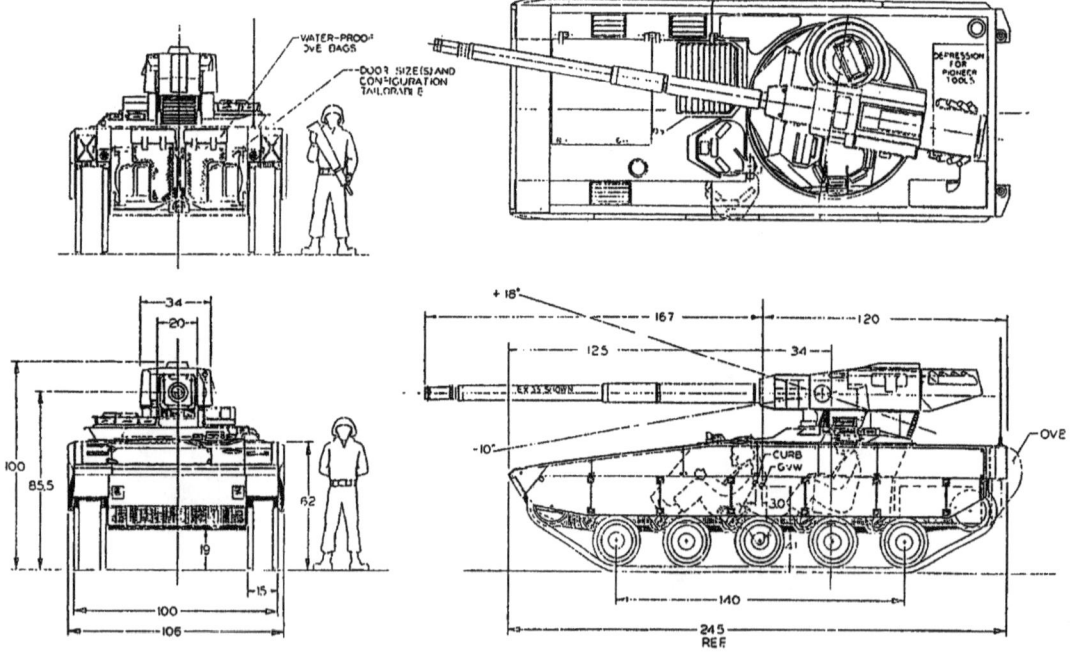

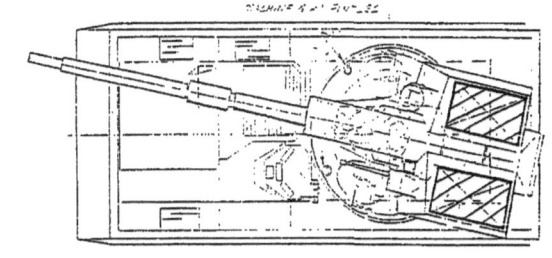

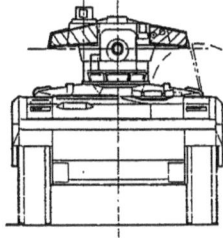

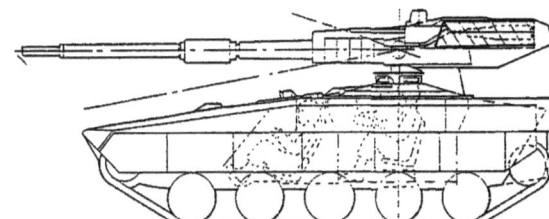

The drawing above shows the version of the Teledyne Continental Motors ASP with the 105mm ammunition stowed in side extensions to the external gun pod.

One used a drum feed in the hull while the other carried ammunition in the gun pod itself. At the conclusion of its study, the AFVTF recommended that one version of the Teledyne Continental Motors vehicle be selected as the new AGS or ASP. The version selected would depend upon the survivability after live fire tests. The RAP was to be based upon the ASP or upon the M113 personnel carrier.

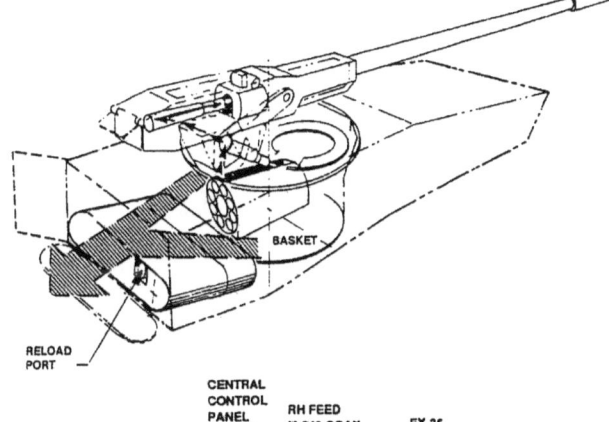

Emergency exit paths for the ASP crew members are shown in the two sketches at the right. A 30 ton version of the ASP appears in the drawing below using applique armor for increased protection. The rearm platform can be seen supplying the ASP at the bottom right.

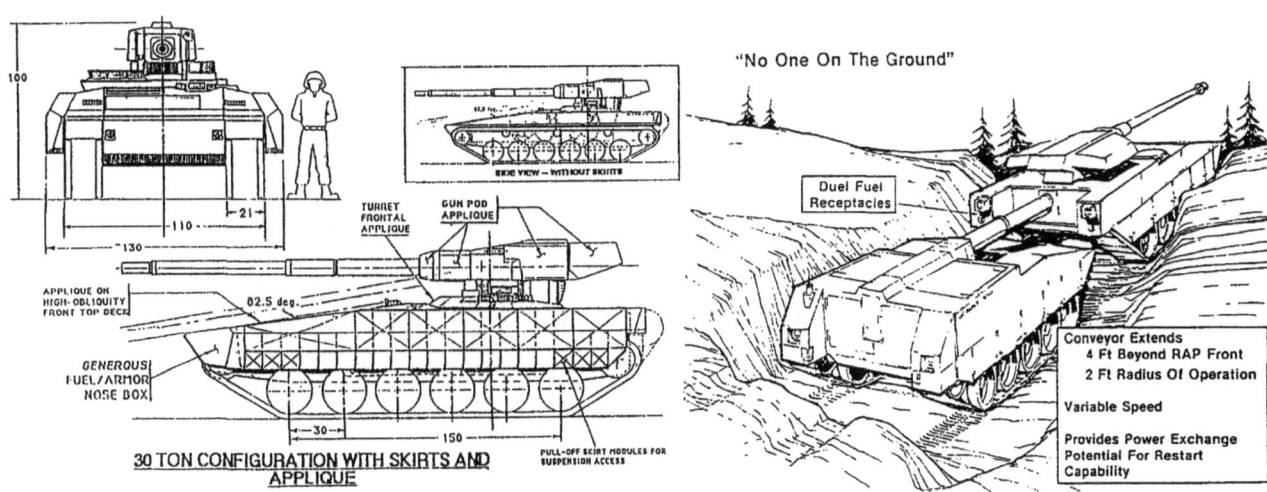

177

In June 1987, the new Army Chief of Staff, General Carl E. Vuono, announced the decision to field a new kinetic energy gun system for use by the light forces. However, several questions remained that required answers before this could be achieved. It was still considered to be highly desirable to purchase a non-developmental item for the AGS application. Unfortunately, none of the candidates available met all of the specified requirements. In fact, there was considerable controversy within the Army as to what those requirements should be. The 82nd Airborne Division and the XVIII Airborne Corps were adamant in their demand that the AGS be suitable for parachute delivery from the C130 aircraft. This required a weight limit of a little under 18 tons and restrictions on the height and width of the vehicle. Other factions in the Army wanted a higher level of protection than could be obtained at that weight and wanted to increase the limit to anywhere from 21 to 23 tons. Their concept envisaged an AGS able to roll off of a transport aircraft ready for immediate action. This would, of course, require the prior seizure of a suitable landing strip.

In March 1991, the Army reduced the air droppable requirement to a desired, but not essential capability. This was over the strong objections of the airborne troops. In June 1991, Congress cut 37.9 million dollars in AGS funding from the 1992 budget and directed the Army to use the LAV-105 to meet the AGS requirement. This was the wheeled light armored vehicle being developed for the Marine Corps armed with the low recoil force 105mm gun. However, this approach was killed when the Army objected and pointed out that the AGS mission differed widely from that intended for the Marine Corps vehicle. Also, about that time, the Marine Corps determined that they could not afford the LAV-105 and the project was cancelled.

Originally, the initial bids for the new AGS contract were due to be received by 13 November 1991. However, at the end of October, the date was extended to 13 December 1991 to permit the Army to further review the requirements. Four contenders submitted bids for the AGS contract prior to the deadline. They were FMC Corporation, a team formed by General Dynamics Land Systems and Teledyne Continental Motors, Cadillac Gage Textron, and Hägglunds USA.

Although the AGS proposals from the various sources were based on vehicles produced earlier, they also were modified to meet the latest specification requirements. The entry from FMC Corporation was their CCVL with certain changes. For example, the number of ready rounds available to the automatic loader in the turret was increased from 19 to 21. An additional nine rounds in the hull brought the total 105mm ammunition stowage to 30. All of this ammunition was in separate compartments with blowout panels to vent any explosion away from the crew. Like all of the AGS contenders, the main armament was the lightweight XM35 105mm gun. This low recoil cannon developed by the Benet Laboratory at Watervliet Arsenal was originally designated as the EX35. The conflicting weight and armor requirements were solved on the FMC vehicle by providing three levels of protection. For the low velocity air drop (LVAD), the weight could be reduced to 35,500 pounds. This would be increased to an estimated combat weight of 38,300 pounds by additional stowage and crew after landing. The installation of applique armor increased the roll-on/roll-off (RO/RO) level 2 combat weight to about 42,300 pounds. Additional applique armor brought the level 3 combat weight up to an estimated 49,500 pounds. The original CCVL turret was modified relocating the 7.62mm M240 coaxial machine gun and the hunter-killer fire control system with the commander's independent thermal viewer was eliminated. The proposal included a weapon at the commander's hatch which could be either an M2 .50 caliber machine gun, a 40mm Mark 19 grenade launcher, or an M240 7.62mm machine gun. A total of 16 smoke grenade launchers were to be installed on the turret.

An artist's concept drawing of the FMC AGS candidate is shown below. Note the changes in the turret configuration compared to the CCVL.

The General Dynamics Land Systems-Teledyne Continental Motors AGS candidate was based upon the pilot shown above. However, it was to be rearmed with the lightweight 105mm gun now designated as the XM35.

The General Dynamics Land Systems-Teledyne Continental Motors proposal was based upon the earlier Teledyne AGS with some modifications. For maximum survivability, the total number of 105mm rounds was reduced to 30. One proposed arrangement stowed eight rounds in the drum of the automatic loader between the gunner and the tank commander with the remaining 22 rounds in a low profile replenisher. The rate of fire with the automatic loader was 10-12 rounds per minute. The drum of the automatic loader could be replenished either manually or automatically. In addition to the M240 7.62mm coaxial machine gun, either a .50 caliber or a 7.62mm machine gun was provided for the tank commander. The full solution digital fire control system utilized a roof mounted day/thermal sight with a laser range finder. A Kollmorgen periscope was used as an auxiliary sight with eyepieces for both the gunner and the tank commander. The vehicle was to be fitted with a collapsible screen kit for amphibious operations. Referred to in some sources as the Hammer, the proposed AGS retained essentially the same power train and suspension as the earlier Teledyne vehicle. Its weight could be reduced to 35,500 pounds for parachute delivery. For this operation, it carried only ten rounds of 105mm ammunition and 100 gallons of fuel. The addition of the three man crew, 20 105mm rounds, and 70 gallons of fuel raised the combat weight to about 37,000 pounds. Applique armor could further increase the weight to 43,000-49,000 pounds. Fully loaded, the vehicle had a maximum level road speed of 45-50 miles per hour and a cruising range of approximately 300 miles.

The advantage of the external gun mount when firing from a defilade position is obvious below at the left. At the right below, the vehicle is loaded aboard a C130 aircraft.

The Cadillac Gage AGS candidate appears above. Note the improved protection with the additional armor compared to the original Stingray.

The Cadillac Gage Textron entry in the AGS competition was a modified version of their Stingray. The new vehicle was armed with the EX35 105mm gun in the low recoil force mount. It was fitted with an automatic loader developed by Fairey Hydraulics, Ltd. reducing the crew to three men. The total 105mm ammunition stowage was 30 rounds including 16 ready rounds. The fire control system included a two axis stabilized day/night sight with an integral laser range finder. This was combined with an electromechanical gun/turret drive and stabilization system. Secondary armament consisted of a 7.62mm M240 coaxial machine gun and another M240 on the turret roof for use by the tank commander. Sixteen smoke grenade launchers were installed with eight on each side of the turret. Spaced, high hardness, steel armor plate provided level 2 protection with a combat weight of 44,000 pounds. Additional armor could be added to achieve level 3 protection with a combat weight of 49,500 pounds. Compared to the original Stingray, the hull was lengthened and the road wheels on the torsion bar suspension were reduced in diameter from 24 to 23 inches. The track width was increased from 15 to 16 inches. The new vehicle was powered by a Detroit Diesel 6V92 DDEC TA engine with a General Electric HMPT-500-3 transmission. This power plant developed 550 gross horsepower at 2300 rpm. The maximum speed on a level road was estimated to be 42 miles/hour with a cruising range of 300 miles at 30 miles/hour. The vehicle had a fording depth of 40 inches.

The configuration and dimensions of the Cadillac Gage AGS are shown in the four view drawing below.

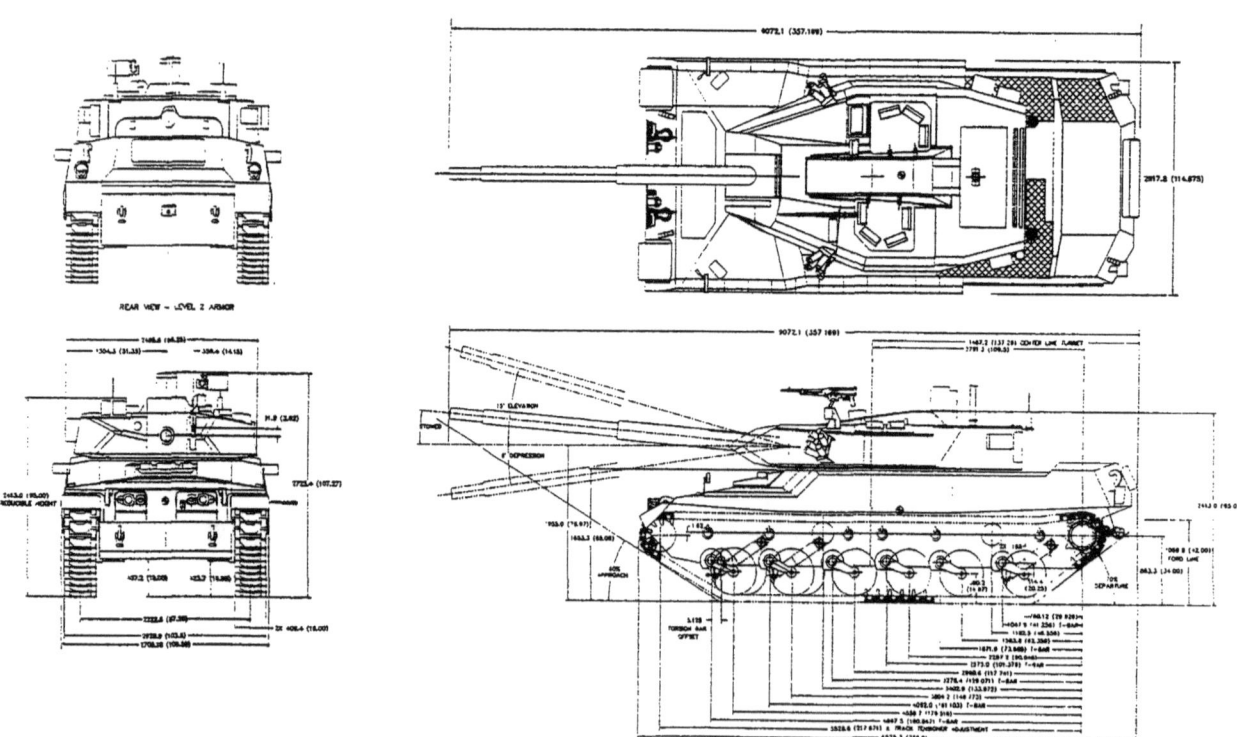

227 in.
98 in.
235 in.
89 in.
58 in.
16 in.
107 in.

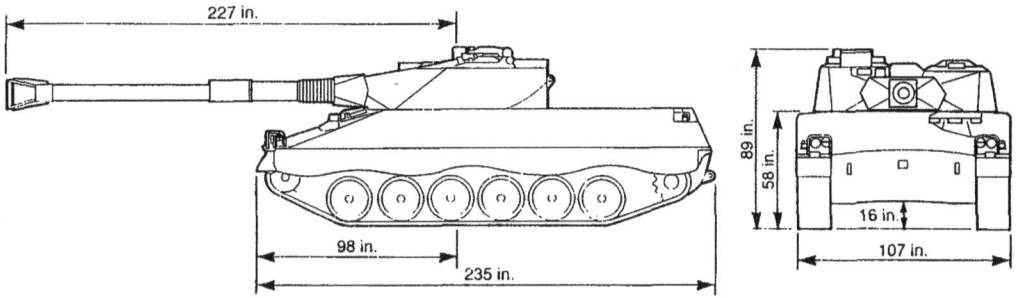

The size of the Hägglunds AGS is shown in the dimensional sketch above. Below, the cutaway view reveals the internal arrangement of the vehicle.

Team Hägglunds USA presented an AGS candidate based upon the proven chassis of their CV90 and IKV91 developed for the Swedish Army. The turret, a product of GIAT Industries, was to be armed with the U.S. XM35 cannon like the other entries, but it did not include an automatic loader. Thus the crew was increased to four men. The weight could be reduced to about 35,000 pounds for air transportation and the combat weight was estimated to be 37,500 pounds. Applique armor could increase the combat weight to about 44,000 pounds. Powered by the General Motors 565 gross horsepower 6V92TA diesel engine with the X-300 transmission, the maximum level road speed was estimated to be 43 miles per hour. The cruising range was about 300 miles.

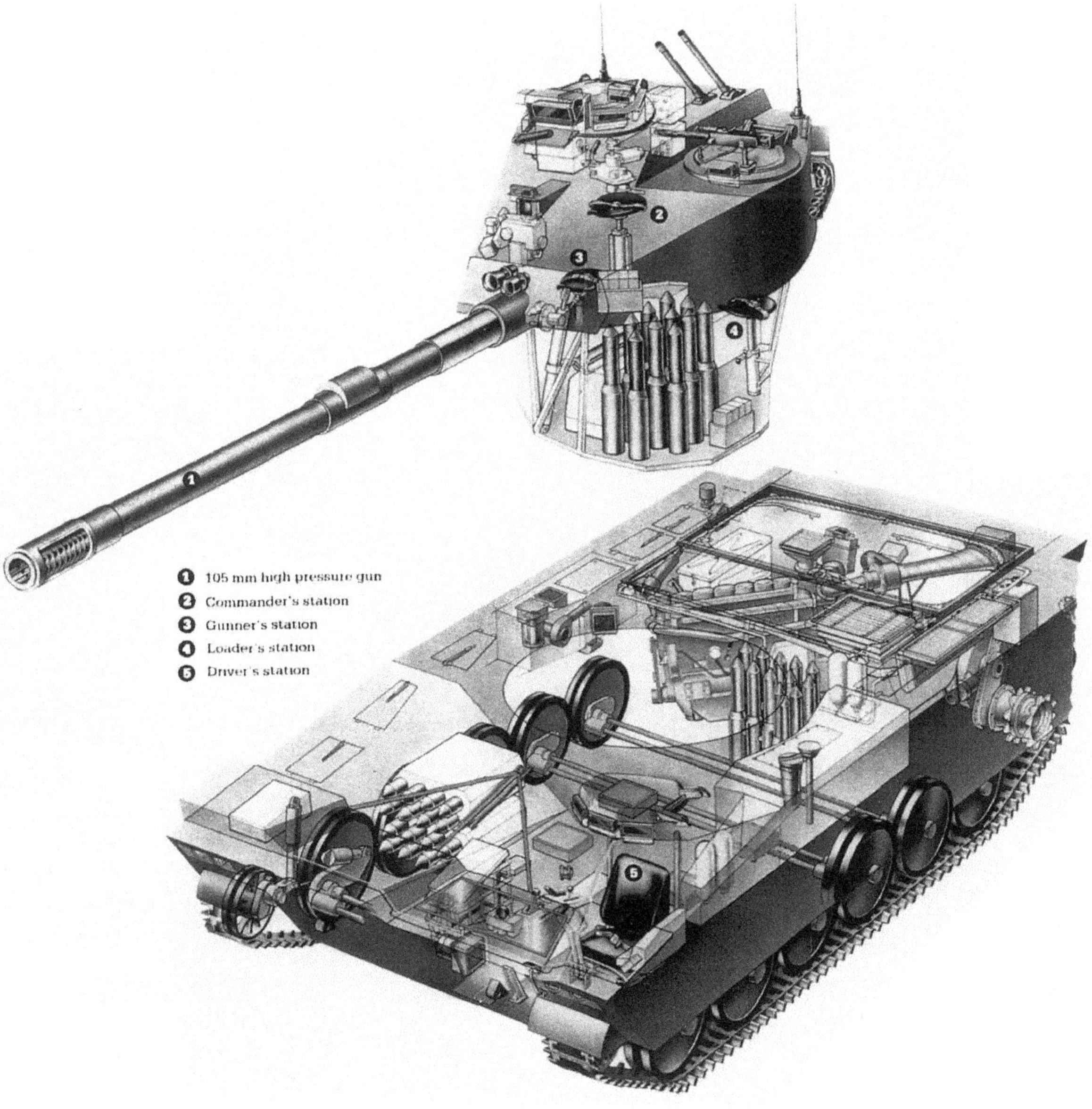

1 105 mm high pressure gun
2 Commander's station
3 Gunner's station
4 Loader's station
5 Driver's station

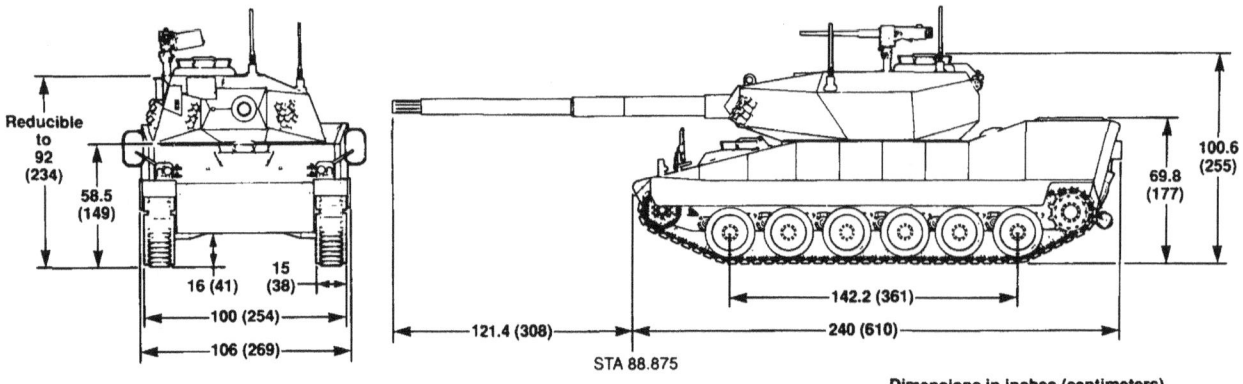

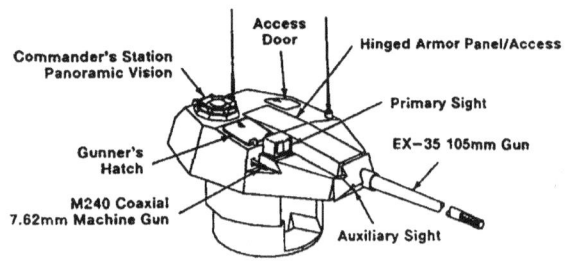

Dimensions in inches (centimeters)

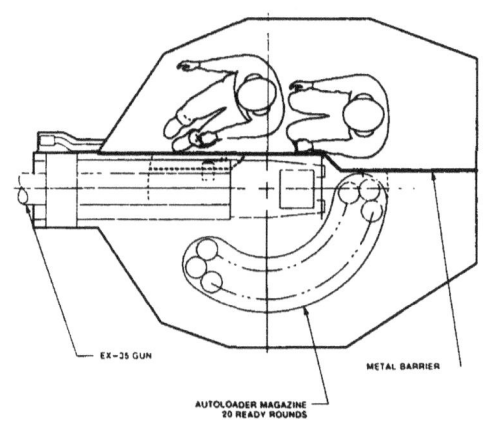

The dimensions of the XM8 armored gun system and details of the turret can be seen in these drawings.

On 14 June 1992, FMC Corporation was awarded the contract for the AGS which was then designated as the XM8. This 119.5 million dollar, 46 month, contract covered the delivery and test of six prototype vehicles by April 1996. News sources indicated that the competitive bids ranged from a high of 189 million dollars for the General Dynamics Land Systems-Teledyne Continental Motors vehicle to a low of 92 million dollars for the Hägglund entry. The Cadillac Gage Textron bid was reported to be 105 million dollars. Thus the selection was based upon both cost and the degree to which the vehicle met the expected operational requirements.

Below, the Army Chief of Staff, General Gordon Sullivan, checks out the driver's station in the XM8. At the right, the General is with William Highlander of United Defense at the rollout ceremony for the XM8 on 21 April 1994.

Above, the XM8 moves out during its test program. This vehicle has level 1 armor protection.

The first of the six XM8 prototypes was delivered to the Army at a rollout ceremony on 21 April 1994. This ceremony was held at the United Defense facility in San Jose, California. This was the former FMC plant prior to the formation of the United Defense Limited Partnership by FMC and BMY.

As delivered, the XM8 prototype was fitted with level 2 armor protection and a .50 caliber machine gun was installed at the commander's hatch. As usual, the development program had resulted in some weight increase. The air drop weight was now quoted as 36,900 pounds with a combat weight at level 1 protection of 39,800 pounds. The roll-on/roll-off weight with level 2 protection was now up to 44,000 pounds. The additional armor required to achieve level 3 protection brought the estimated weight up to 52,000 pounds.

All six prototypes were scheduled for delivery to the Army during April and May 1994. Engineering and service evaluations, including live fire tests, are expected to continue through 1997. The Army intends to make the decision regarding full scale production in March 1997. At the present time, the Army plans to field 237 armored gun systems to equip the 3/73rd Armor of the 82nd Airborne Division and the 2nd Armored Cavalry Regiment.

The level 2 and level 3 armor protection configurations on the XM8 are illustrated in the drawings at the right.

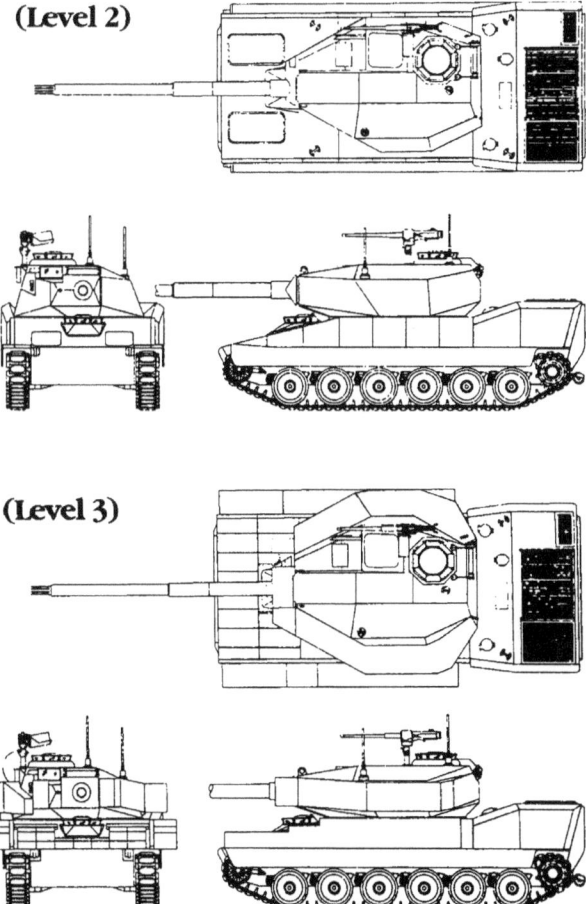

(Level 2)

(Level 3)

These two views show the XM8 with level 1 armor protection. Note that the muzzle brake is integral with the gun tube on the XM35 105mm cannon.

The XM8 in these photographs has been fitted with additional armor providing level 2 protection. Note the panels on the front and sides of the hull as well as the gun shield.

Level 3 armor protection is shown in these views of the XM8. The passive armor boxes can be seen attached to the sides of the hull and turret. Also, note the additional armor segments added to the upper front hull.

PART IV

SPECIALIZED LIGHT COMBAT VEHICLES

The pilot 90mm self-propelled gun T101 appears in the photographs on this page. The views at the bottom show details of the 90mm gun T125 and the T70 mount.

ANTITANK VEHICLES

Defense against attack by armored forces was always a serious problem for airborne troops, particularly during early stages of an airborne assault. A conference on antitank defense was held at Fort Monroe, Virginia on October 6-7 1948. This meeting established an urgent requirement for a self-propelled antitank gun that could be transported and landed during phase I of airborne operations. This weapon was to have ballistic characteristics equal to those of the medium tank gun using the same ammunition. On 13 April 1949, the Army Field Forces specified that the new antitank weapon be a tracked vehicle armed with the 90mm gun T119 then intended for the new medium tank T42. However, further investigation indicated that the weapon would have to be modified for installation on the lightweight tracked chassis. This was accomplished and the redesigned cannon was designated as the 90mm gun T125 for installation in the mount T70. This gun was ballistically identical to the tank cannon and used the same ammunition.

A contract was awarded to the Cadillac Motor Car Division of General Motors Corporation to build two pilot vehicles designated as the 90mm, full tracked, self-propelled gun T101. After evaluation and some modification of the

 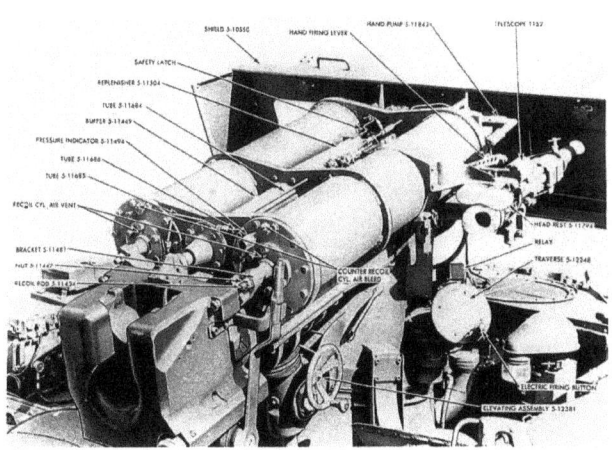

189

The production 90mm self-propelled gun M56 can be seen in these three photographs. The seat for the vehicle commander is visible on top of the radio in the bottom left view.

pilots, the vehicle was standardized as the 90mm, full tracked, self-propelled gun M56 with the 90mm gun M54. Although nicknamed the Scorpion, the M56 also was referred to as the SPAT (self-propelled antitank). The production vehicle differed from the pilots in several respects. The most obvious changes included the redesigned blast shield and the replacement of the muzzle brake on the cannon by a blast deflector or counterweight. Originally, the T101 was manned by a crew of three consisting of the driver, gunner, and loader. The driver and gunner were located behind the blast shield on the left and right respectively. The loader was seated at the left rear facing toward the rear. On the standardized M56, the crew was increased to four by the addition of a vehicle commander who also operated the

radio. A seat was provided for the commander on the left rear of the vehicle above the radio installation. The loader was relocated to a seat at the right front on top of the fender stowage box.

Combat loaded, the M56 weighed 15,500 pounds and did not have any armor. It was driven by a front mounted Continental AOI-402-5, air-cooled, gasoline engine with an Allison CD-150-4 transmission. This engine developed 200 gross horsepower at 3000 revolutions per minute. The maximum level road speed was 28 miles per hour and a 55 gallon fuel capacity gave it a cruising range of about 140 miles. An unusual feature of the M56 was the torsion bar suspension fitted with pneumatic tires on the four wheels per side. These tires were normally inflated to a pressure of 75 psi,

190

Above, the 90mm gun is locked in the travel position and the ammunition stowage rack is open at the top right. At the right, the travel lock for the cannon has been released. The band type tracks and pneumatic tires are clearly visible in all of these photographs.

but in an emergency, they could be run without pressure for up to 15 miles at 15 miles per hour. Tube over bar torsion springs were installed at the front and rear road wheel stations. The two intermediate stations were fitted with only solid torsion bars. The 20 inch wide, band type, tracks reduced the ground pressure to a little over four psi. Each of the two band tracks was assembled from eight sections. The M54 90mm gun was installed on the M88 gun mount located on top of the fuel tank in the center of the aluminum alloy hull. A 29 round 90mm ammunition rack was at the lower rear. A folding platform at the rear provided working space for the loader.

The M56 was designed for parachute delivery and its small size and low weight allowed it to be carried in a large cargo glider or a number of transport aircraft.

A top view of the M56 appears at the right. Below, the T101 pilot vehicle is being loaded into a C119 aircraft. Note the difference in the blast shield compared to the standard M56.

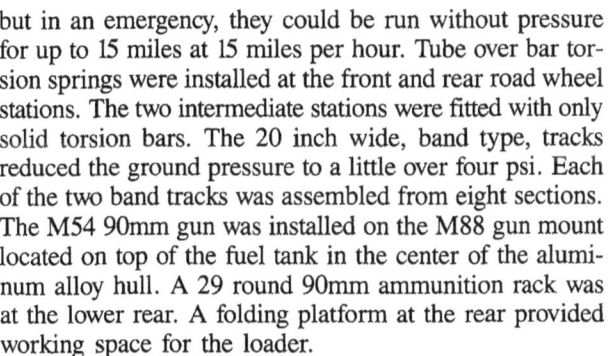

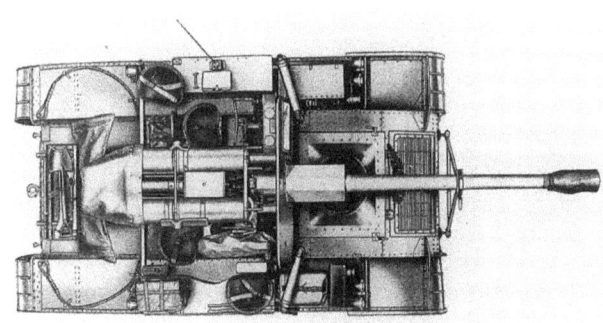

191

Above, the infantry tracked utility vehicles T55 and T56 are shown at the left and right respectively. They are essentially the same except for the length.

During August and September 1951, meetings were held at Detroit Arsenal and at the Tractor Division of the Allis-Chalmers Manufacturing Company regarding the development of a low cost, lightweight, highly mobile vehicle based upon a universal chassis. It was intended to adapt this vehicle to meet the requirements of several infantry missions. On 25 October 1951, the development and manufacture of five versions of the new vehicle, now dubbed the Ontos, was authorized and the following day, a contract was awarded to Allis-Chalmers for the project.

Two of the new vehicles were personnel carriers and were designated as the infantry tracked utility vehicles T55 and T56. These vehicles differed in length and they carried six or ten men respectively, including the driver, plus equipment. Another designation assigned was the multiple rifle and machine gun mount T164. This was a kit fitted with four recoilless rifles intended for installation on the basic vehicle. However, it was never completed. The multiple self-propelled rifle T165 was armed with six recoilless rifles and manned by a crew of three. The self-propelled rifle T166 was fitted with a single recoilless rifle using the ground

mount. The multiple self-propelled rifle T167 was to be armed with eight recoilless rifles, but it was never assembled. The original design specified 105mm recoilless rifles, however, they were replaced by the later 106mm rifle T170E1 before the vehicles were built. The latter were actually 105mm weapons also, but they were designated as 106mm to prevent confusion with the earlier rifle. All of the vehicles were assembled using ½ inch thick steel armor plate except for the floor which originally was $\frac{3}{16}$ inches, but later was increased to ¼ inch.

The first two vehicles completed were the six man personnel carriers. T55 number 1 was shipped to Aberdeen Proving Ground on 26 May 1952. Number 2 went to Fort Knox on 18 June. The first T166, armed with the single 106mm rifle, was shipped to Aberdeen on 1 July followed by the six rifle T165 number 1 on 18 July. The first T56 ten man personnel carrier went to Fort Knox on 30 July and the second was allocated to Aberdeen on 19 August. Only four personnel carriers were completed, two T55 and two T56. T55 number 2 was later converted to the T165E1 number 2 with six recoilless rifles.

Below are two views of the 106mm multiple self-propelled rifle T165. Note the single road wheels with pneumatic tires on this early pilot vehicle.

192

The 106mm self-propelled rifle T166, pilot number 1, appears above. Below are drawings of the proposed rifle arrangement on the T164 (left) and the T167 (right).

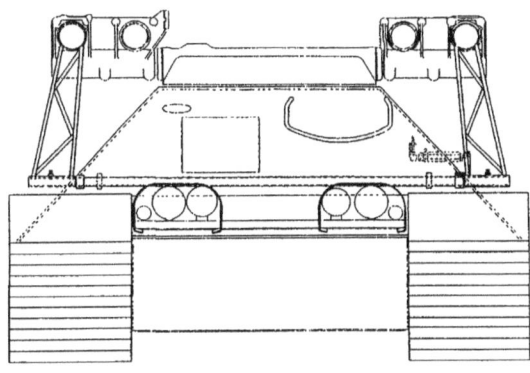
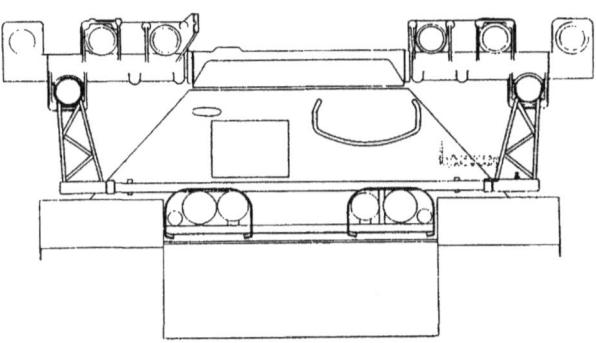

The lower hull configuration of the T55, T165, and T166 was essentially the same. The upper hull of the T55 also provided the basis for the other two vehicles. The T55 had a large double door hatch on top, a driver's hatch in the sloping front armor, and double doors at the rear. The ten man T56 was similar, but it was longer with two roof hatches. The top plate on the T165 had a circular opening

The photographs at the right and below show the 106mm multiple self-propelled rifle T165E2. Note that it is fitted with dual road wheels and solid rubber tires. Compare the length of the rear doors with those on the earlier T166 at the top of the page.

The eight round ammunition stowage compartment on the 106mm multiple self-propelled rifle T165E2 can be seen above closed (left) and open (right).

to mount the turret, but the driver's hatch and the rear double doors were the same as on the T55. The T166 was the same except for the roof plate which had a semicircular hatch on the left to permit access to the single recoilless rifle mounted on the right. This rifle had a .50 caliber spotting rifle installed coaxially on top. The turret on the T165 was fitted with six recoilless rifles, three on each mount at the left and right. Two .50 caliber spotting rifles were installed, one above each of the the two upper outboard recoilless rifles. The four rifle T164 design also included two .50 caliber spotting rifles. The same applied to the eight rifle T167 configuration. The T164, T165, and T167 designs also were armed with a .30 caliber machine gun between the recoilless rifles. The turret on these vehicles provided a hatch for the gunner. A stowage compartment for eight rounds of 106mm ammunition was located in the rear of the lower hull below the crew compartment on the T164, T165, T166, and T167. This compartment was accessible only from outside the vehicle and was protected by a steel cover below the double rear doors.

The power plant in all of the early Ontos vehicles consisted of the General Motors Model 302 gasoline engine. This was the same liquid-cooled engine used in the M135 truck. Located in the right front hull, it was coupled to the XT-90 transmission driving the band type tracks through the front mounted sprockets. The torsilastic suspension supported the vehicles on four wheels per side, five on the longer T56. The rubber torsion bushings were not susceptible to corrosion and this type of externally mounted suspension did not occupy any space within the hull. Originally, road wheels with pneumatic tires were installed, but these were replaced by solid tire dual road wheels on the final design. The latter rode on band type tracks with both center and outside guides. Each of the two 20 inch wide tracks was assembled from five sections.

Except for the driver, crew members in the T55 and T56 personnel carriers were seated facing inward along each side of the vehicle. The driver, in all cases, was located at the left front alongside the power plant. The gunner's seat in the T164, T165, and T167 was in the center and a floor cushion was provided for the loader. The first vehicles differed from those produced later. The threshold for the double rear doors was lowered and the doors lengthened to improve access to the vehicle. The original design required the removal of the upper hull in order to change the power pack. A new arrangement was introduced which allowed the power pack to be removed through the space for new air intake grills in the front armor. These grills replaced the original engine air intake louvers in the right side wall. The recoilless rifle mount on the original T166 was changed to incorporate an hydraulic powered arm to raise the cannon so that it could be easily dismounted for ground use. Only three T166 vehicles were completed, the first of which was later converted to T165E1 number 3.

Below, the chassis of the Ontos (Greek for Thing) vehicle appears at the left with the upper hull removed exposing the engine and transmission. At the right below is a view of the hydraulic powered arm on the modified gun mount of the T166.

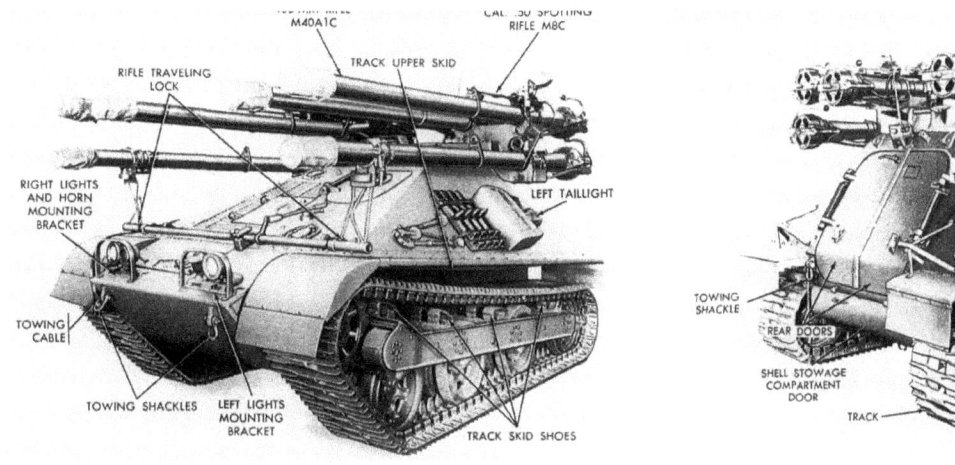

Above is the production 106mm multiple self-propelled rifle M50 with the rifles locked in the travel position. The 106mm recoilless rifle M40A1C appears below with the spotting rifle attached.

In late February 1953, procurement was authorized for 24 pilots of the six rifle T165 version of the Ontos. After some modification, these became the T165E1 and later testing by the Army and the Marine Corps resulted in additional changes. Pilots with serial numbers 10 through 24 were manufactured according to the revised specifications and designated as the T165E2. Additional tests produced further modifications before the vehicle was standardized as the multiple 106mm self-propelled rifle M50 and placed in production for the Marine Corps. On 12 August 1955, a contract was awarded to Allis Chalmers to manufacture 297 M50s for the Marine Corps. This production was completed on 30 November 1957.

On the standardized M50, the six M40A1C 106mm recoilless rifles were numbered 1 through 6 from left to right when viewed from the rear of the vehicle. All of the rifles could be dismounted for ground use, but numbers 2 and 5 were specifically designated for this purpose. These were the top outboard rifles on each side. Both were fitted with M8C .50 caliber spotting rifles and M92D elbow telescopes for ground use. M8C spotting rifles also were installed over M40A1C rifles 3 and 4, but the M92D telescopes were not provided. The latter two spotting rifles were part of the vehicle firing system for remote control operation from within the vehicle. All of the spotting rifles were magazine fed, semiautomatic, weapons firing a special .50 caliber tracer round which produced a puff of smoke on impact.

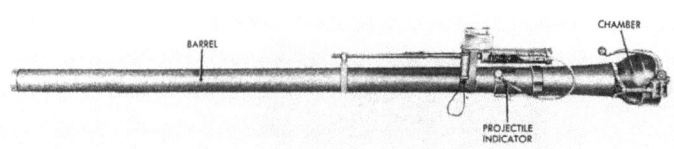

The .30 caliber machine gun was mounted between the 106mm rifles. It could be locked in position coaxial with the recoilless rifles and fired from inside the vehicle or it could be operated manually by the gunner standing in the open hatch.

The gunner's sighting device was the M20A3C periscope in the turret roof. The turret was manually operated in both traverse and elevation. It could be rotated 40 degrees to the left or right of center and elevated through a range of −10 to +20 degrees. The 106mm rifles could be loaded only from outside the vehicle, but an hydraulic system allowed the gunner to operate the breech mechanism of the rifles between the safe and fully closed positions from within the vehicle. In addition to the eight 106mm rounds in the lower hull stowage compartment, four rounds were carried in the right rear of the crew compartment bringing the total ammunition supply to 18, if the rifles were loaded. Originally, two types of ammunition were provided for the 106mm rifle. These were the high explosive plastic with tracer (HEP-T) M346 and the high explosive antitank (HEAT) M344. The first of these was a squash head projectile with its plastic explosive enclosed inside a thin metal case. It was fitted with a preengraved rotating band to engage the rifling in the cannon. Two indexing buttons on the shell were used to align the rotating band with the rifling grooves. The HEAT M344 round was a fin stabilized, shaped charge, projectile which did not have a rotating band. It was fitted with a rear bourrelet to support the projectile in the tube. Later, a flechette round was introduced for antipersonnel use.

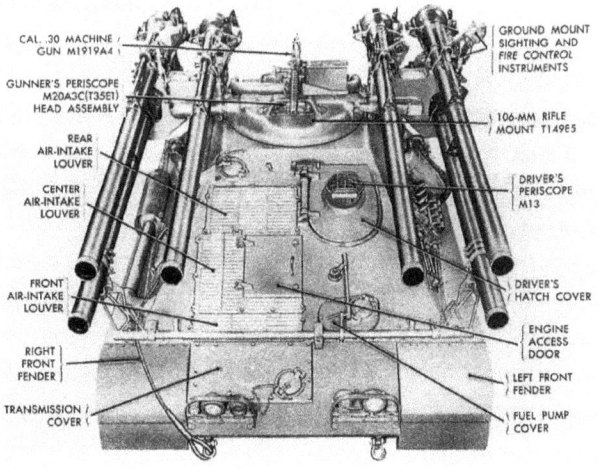

At the left is a top front view of the M50 with the various components identified.

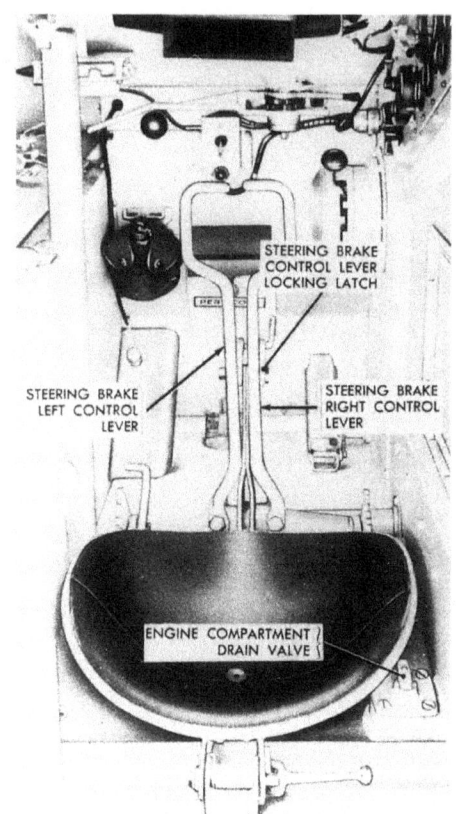

The interior of the M50 can be seen above through the rear doors. At the upper right is a view of the driver's controls and the inside ammunition rack is visible at the lower right.

With its three man crew, the M50 had a combat weight of about 9½ tons. The six cylinder General Motors SL12340 gasoline engine developed 145 gross horsepower at 3400 revolutions per minute. It drove the M50 through the Allison XT-90-2 transmission at a maximum level road speed of 30 miles per hour. The 47 gallon fuel tank provided a cruising range of about 150 miles. The normal fording depth was about two feet, but this could be increased to approximately five feet with the installation of a fording kit.

A later modification was the M50A1 fitted with the Chrysler HT-361-318 liquid-cooled V-8, gasoline, engine which developed 180 gross horsepower at 3450 revolutions per minute. It was coupled to the Allison XT-90-5 transmission. This version of the Ontos could be identified by the air intake louvers in the engine and transmission access doors. The M20A3G periscope replaced the earlier M20A3C periscope for the gunner and an M13A1C elevation quadrant was added. M92F elbow telescopes replaced the earlier M92D telescopes for use with the ground mount.

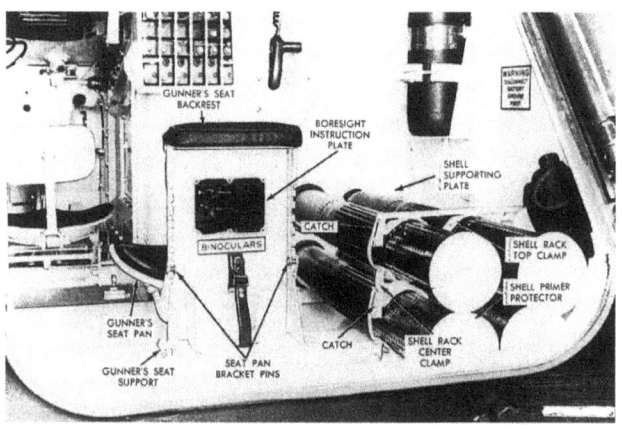

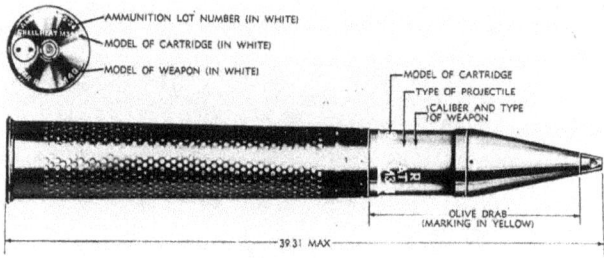

The 106mm HEAT round is above and the 106mm multiple self-propelled rifle M50A1 is at the left. Note the louvers in the engine and transmission access doors.

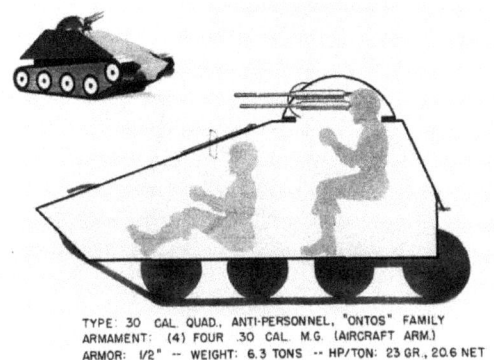

TYPE: 30 CAL. QUAD., ANTI-PERSONNEL, "ONTOS" FAMILY
ARMAMENT: (4) FOUR .30 CAL. M.G. (AIRCRAFT ARM.)
ARMOR: 1/2" -- WEIGHT: 6.3 TONS -- HP/TON: 23 GR., 20.6 NET

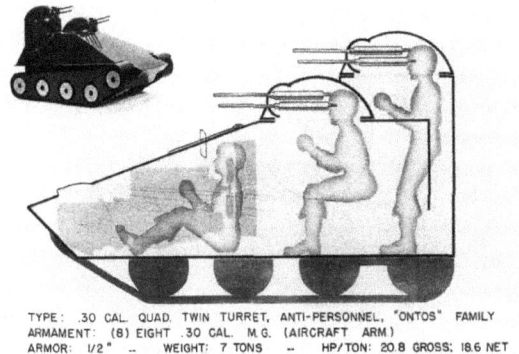

TYPE: .30 CAL. QUAD. TWIN TURRET, ANTI-PERSONNEL, "ONTOS" FAMILY
ARMAMENT: (8) EIGHT .30 CAL. M.G. (AIRCRAFT ARM.)
ARMOR: 1/2" -- WEIGHT: 7 TONS -- HP/TON: 20.8 GROSS; 18.6 NET

Above are the proposed Ontos assault vehicles armed with four (left) and eight (right) .30 caliber machine guns. The titanium alloy upper hull intended for the T55 Ontos is pictured at the right below.

The Ontos provided the basis for experiments with titanium armor. A welded titanium 6Al-4V alloy superstructure was fabricated for the T55 Ontos personnel carrier. It was ⅝ inches thick providing ballistic protection equal to the ½ inch thick steel armor with a weight reduction of 25 per cent.

A version of the Ontos also was proposed as a light assault vehicle at the second Questionmark Conference in September 1952. In two of these proposals, it was armed with four or eight .30 caliber machine guns.

At the same conference, three design concepts were presented for lightweight self-propelled antitank weapons. Two of these were armed with recoilless rifles with one using the chassis of the 90mm self-propelled gun T101. The third concept was a rather ambitious attempt to provide a very powerful antitank gun on a lightweight self-propelled mount. It proposed the installation of the 105mm gun T140 on a light chassis with an estimated combat weight of 16.7 tons.

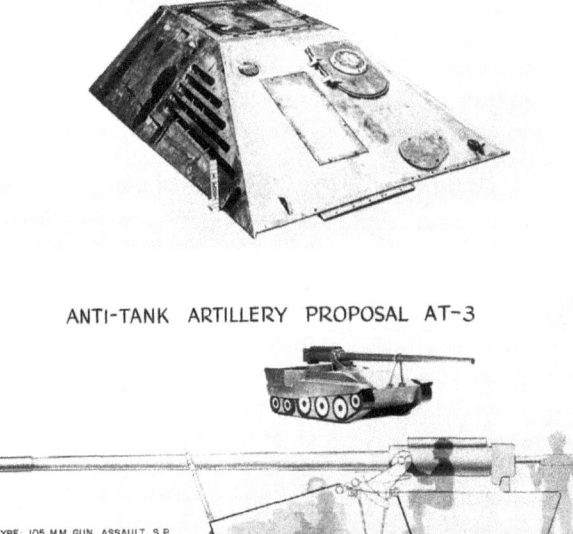

ANTI-TANK ARTILLERY PROPOSAL AT-3

TYPE: 105 MM GUN, ASSAULT, S.P.
ARMAMENT: 105 MM GUN, T-140
ARMOR: 1/2" PARTIAL
WEIGHT: 16.7 TONS
HP/TON: 25.7 GROSS; 18 NET

At the right is the AT-3 antitank proposal armed with the 105mm gun T140. Below, the AT-1 (left) and AT-2 (right) proposals are both armed with 105mm recoilless rifles.

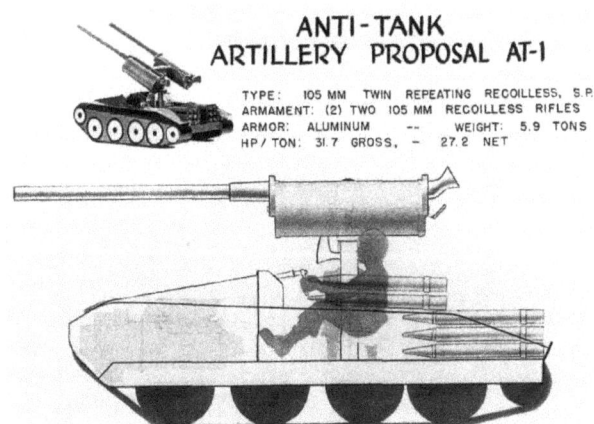

ANTI-TANK ARTILLERY PROPOSAL AT-1

TYPE: 105 MM TWIN REPEATING RECOILLESS, S.P.
ARMAMENT: (2) TWO 105 MM RECOILLESS RIFLES
ARMOR: ALUMINUM -- WEIGHT: 5.9 TONS
HP/TON: 31.7 GROSS, - 27.2 NET

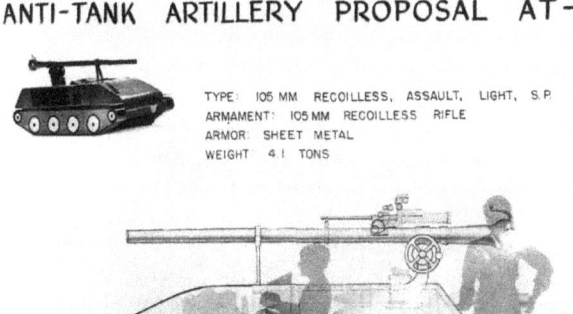

ANTI-TANK ARTILLERY PROPOSAL AT-2

TYPE: 105 MM RECOILLESS, ASSAULT, LIGHT, S.P.
ARMAMENT: 105 MM RECOILLESS RIFLE
ARMOR: SHEET METAL
WEIGHT: 4.1 TONS

197

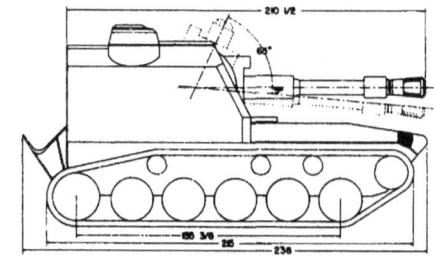
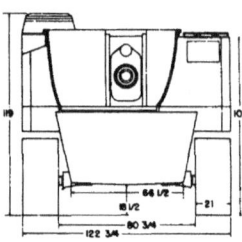

Above, a mock-up of the 155mm self-propelled howitzer T99 appears at the left and the dimensions of the vehicle are indicated in the sketch at the right.

SELF-PROPELLED ARTILLERY

Self-propelled artillery was employed with great success by the U. S. Army during World War II, particularly by armored units. Most of these weapons mounted the standard field artillery cannon on the open top chassis of the light and medium tanks. The latest such vehicles developed at the end of the war were based upon the then standard light tank M24. In January 1946, the Stilwell Board report specified the development of new self-propelled artillery for the postwar army. It also indicated that such vehicles should have overhead protection against artillery air bursts. During 1946-1947, several design concepts were developed to comply with the recommendations of the Stilwell Board. They included two self-propelled weapons based upon the chassis of the new T37/T41 light tank then under development. These two vehicles were originally designated as the 105mm howitzer motor carriage T98 and the 155mm howitzer motor carriage T99. Although they were developed in parallel, the T99 was the first to be officially authorized by OCM 31897 dated 19 September 1947. Later, it was redesignated as the 155mm self-propelled howitzer T99 by OCM 33807 on 30 April 1951. Two pilot vehicles were constructed at Detroit Arsenal. The T99 was armed with the T97 155mm howitzer in the T68 barbette mount with manual and hydraulic power traverse 30 degrees to the left and right of center. The elevation range was from −5 to +65 degrees, also with manual and hydraulic operation. A new "Ultimate" fire control system was proposed to minimize manual operations and to permit the central control of several howitzers. This system included a T28 out-of-level computer and a T30 drive controller. However, with the outbreak of war in Korea during June 1950, the program was greatly accelerated to provide production vehicles at the earliest possible date. To achieve this objective, a simplified "Alternate" fire control system was installed on pilot number 1 and specified for the

The photographs below show the 155mm self-propelled howitzer T99. It was originally fitted with a muzzle brake as in the view at the left.

198

The 155mm self-propelled howitzer T99E1 is above. Note the additional cooling louvers in the front hull compared to the T99.

production vehicles. The "Ultimate" fire control system was installed on pilot number 2 for evaluation and further development.

As frequently happened during this period, the crash development program encountered numerous problems. With the advantage of hindsight, it is easy to conclude that a slower, systematic, development and test program would have been preferable. However, at that time it was widely believed that the war in Korea was the opening phase of a worldwide conflict and that large quantities of new weapons, even if imperfect, would be required as soon as possible.

The T41 light tank, whose chassis provided the basis for the self-propelled howitzers, was being modified as mentioned earlier during the description of its development program. The version that was rushed into production was the T41E1. To maintain commonality between the light tank and the self-propelled howitzer, the latter was modified and redesignated as the 155mm self-propelled howitzer T99E1.

With a five man crew, the T99E1 had a combat weight of 30 tons and, like the light tank, it was powered by the Continental AOS-895-3 engine with the Allison CD-500-3 transmission. Although it shared power train and suspension components with the light tank, the arrangement was different on the self-propelled howitzer. The power train and drive sprockets were in front leaving the rear hull free for the howitzer and its crew. The suspension also was fitted with a 28 inch diameter trailing idler in addition to the five standard road wheels per side. The vehicle had a maximum level road speed of 35 miles per hour and a cruising range of about 75 miles. Minor modifications changed the howitzer and mount designations on the production vehicle to T97E1 and T68E1 respectively. The T97E1 howitzer used separated ammunition with a metal cartridge case, unlike the bag charges provided for the standard 155mm field piece. A total of 30 rounds of ammunition for the howitzer were stowed on the vehicle. A .50 caliber machine gun mounted on the cab roof was the only secondary armament.

As mentioned before, tests of the production pilots revealed the need for numerous changes. The T97E1 howitzer was considered to be unsatisfactory and the artillery requested that it be replaced by a new howitzer using bag

An experimental lightweight self-propelled 155mm howitzer is shown below using the chassis of the light tank. This austere approach to the design of the self-propelled howitzer eliminated all armor protection for the crew.

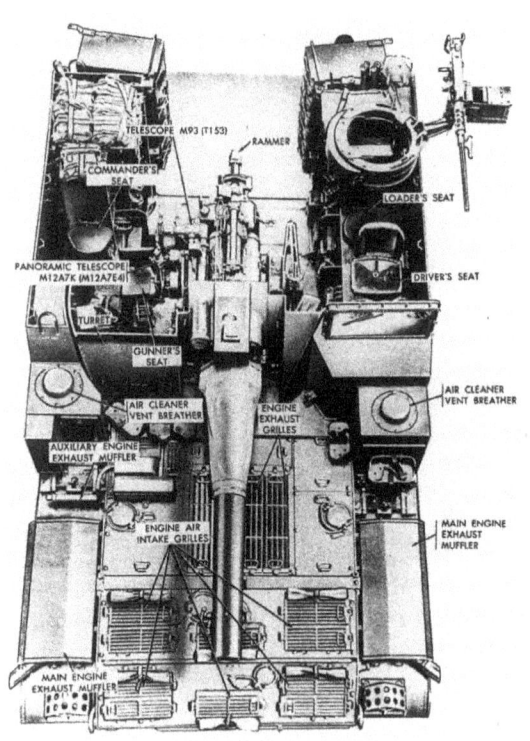

The external components on the 155mm self-propelled howitzer M44 are identified in the views above and at the right.

type propellant charges. The user also complained about the fumes in the enclosed fighting compartment. The "Alternate" fire control system required considerable modification and one T99E1 was fitted with the standard World War II fire control equipment as a backup in case the changes were unsuccessful. As a result of these problems, the production of the T99E1 was halted after 250 were completed and the vehicle was redesigned. The new version was designated as the 155mm self-propelled howitzer T194. The T194 was an open top vehicle eliminating the overhead armor protection. It was armed with the 155mm howitzer T186E1 in the mount T167. The new howitzer used bag charge propellant and space was available to stow 24 rounds on the vehicle. The .50 caliber machine gun was installed on a ring mount at the left side just to the rear of the driver's seat. Like the T99E1, the T194 used the track and suspension components of the light tank and had a 28 inch diameter dual trailing idler in addition to the five standard dual road wheels per side. Also, the rear track support roller was larger in diameter than the front three.

Below, the M44 is shown with the full canvas covers (left) and with only the supporting bows installed (right).

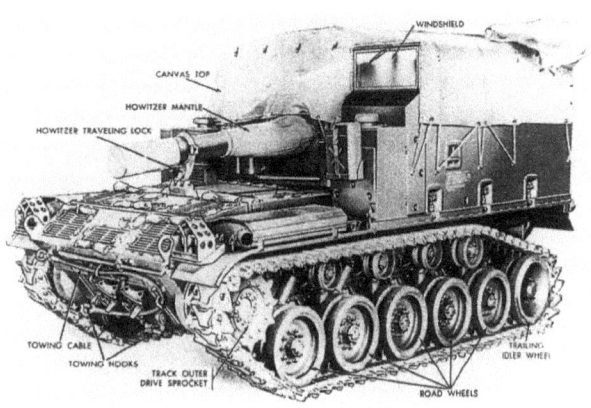

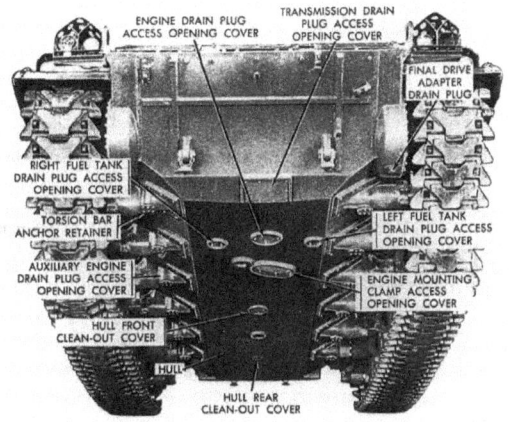
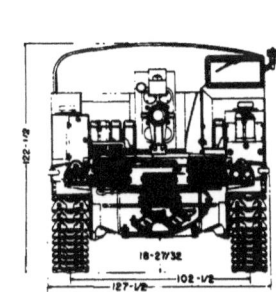
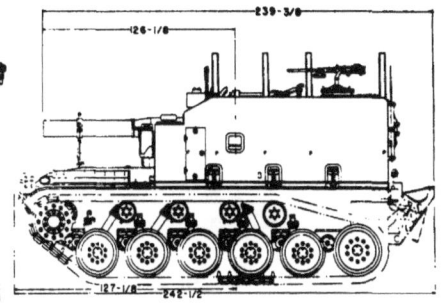

Above, the bottom front of the M44 hull can be seen at the left and a sketch at the right shows the dimensions of the vehicle.

Production was resumed using the new configuration and the first 250 T99E1 vehicles were rebuilt to the T194 standard. Subsequently, the T194 was standardized as the 155mm self-propelled howitzer M44 with the 155mm howitzer M45. With the installation of the Continental AOSI-895-5 fuel injection gasoline engine, the vehicle was redesignated as the M44A1.

OCM 32032, dated 19 February 1948, authorized the development of the 105mm howitzer motor carriage T98 and approved the procurement of two pilot vehicles. Later, on 19 July 1951, OCM 33806 changed the designation to the 105mm self-propelled howitzer T98. The two pilots were built at Detroit Arsenal and tests of T98 number 1 began at Aberdeen Proving Ground during October 1950. The T98 was armed with the 105mm howitzer T96 in mount T67 installed in a turret at the rear of the chassis. This turret could be traversed 60 degrees to the left or right of center either manually or by hydraulic power. The howitzer was fitted with a muzzle brake and it could be elevated through a range of −10 to +65 degrees also manually or by hydraulic power.

Originally, the "Ultimate" fire control system was intended for installation in the T98. However, due to the need for immediate production, the simplified "Alternate" fire control system was selected for the production vehicles and installed in pilot number 1. Pilot number 2 received the "Ultimate" system and was shipped to Aberdeen for evaluation.

The hull and turret of the T98 was a welded assembly of ½ inch thick steel armor plate. The front mounted Continental AOS-895-3 gasoline engine drove the vehicle through the Allison CD-500-3 transmission. The five men in the crew were all located in the turret with the driver at the left front. The gunner's seat was at the right front alongside the howitzer. A cupola on the right rear of the turret roof was provided for the section chief or vehicle commander. A .50 caliber machine gun was mounted on the roof adjacent to the cupola.

Below is the 105mm self-propelled howitzer T98 as originally assembled. Test instrumentation has been installed on the front hull.

201

EXHAUST   HEATER

18"

CARBON MONOXIDE CONCENTRATION
130% OF FATAL RATIO

The close proximity of the engine exhaust to the air intake for the personnel heater can be seen in the photograph of the original T98 at the left.

The tests at Aberdeen revealed a number of problems. For example, the air intake for the personnel heater was adjacent to the end of the engine exhaust pipe resulting in a high concentration of carbon monoxide gas inside the turret when the heater fan was in operation. This and other test results required several modifications. The inside diameter of the turret ring was increased from 69 to 73 inches to provide additional space and the outlet duct for the turret ventilation blower was redesigned. The intake for the personnel heater was relocated to prevent the exhaust fumes from entering the turret. The front hull over the engine compartment was redesigned and the modified vehicle was redesignated as the 105mm self-propelled howitzer T98E1 armed with the 105mm howitzer T96E1 on mount T67E1. The muzzle brake on the original howitzer was replaced by a counterweight. Hydraulic power was retained for both elevation and traverse. Initially, the suspension and tracks remained the same as on the T98 with five dual road wheels, a 28 inch diameter dual trailing idler, and three dual track support rollers per side. Later, a fourth track support roller was added on each side. Hydraulic double acting shock absorbers were mounted at road wheel stations 1, 2, and 5 as well as on the trailing idler. The latter was a reservoir type shock absorber and was not interchangeable with the others.

Production of the T98E1 began in January 1951 for a run of 684 self-propelled howitzers. Tests on the early production vehicles revealed additional deficiencies. These were corrected by several Industrial Modification Programs and tests of the modified vehicles during the Summer of 1955 indicated that they were suitable for issue to the troops. OCM 36028, dated 15 November 1955, standardized the modified T98E1 as the 105mm self-propelled howitzer M52.

Modifications incorporated into the M52 included a new commander's cupola and mount for the .50 caliber machine gun, elimination of the power elevation and traverse systems, elimination of the auxiliary generator and engine although the controls were retained, elimination of the shock absorber on the trailing idler, and the installation of a larger rear track support roller.

When the Continental AOSI-895-5 engine with fuel injection was approved for use on the M41A2 and M41A3 tanks, it also was released for installation on the other vehicles of the light tank family. When this power plant was installed in the M52, it was redesignated as the 105mm self-propelled howitzer M52A1.

At the left are front and rear views of the T98 as originally constructed. The test instrumentation is clearly visible on the upper front hull.

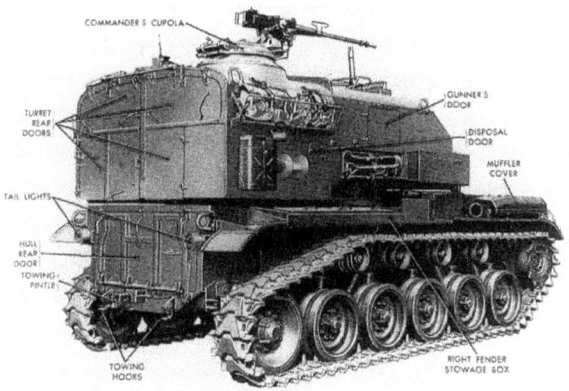

Components of the 105mm self-propelled howitzer M52 are identified in the photographs above and at the right. Note that the muzzle brake on the T98 has been eliminated.

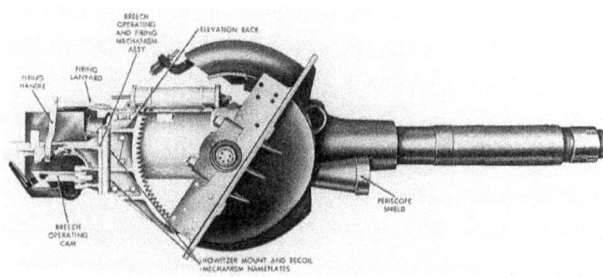

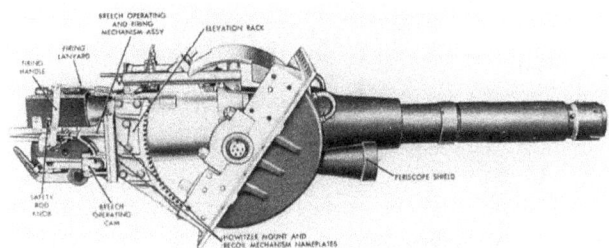

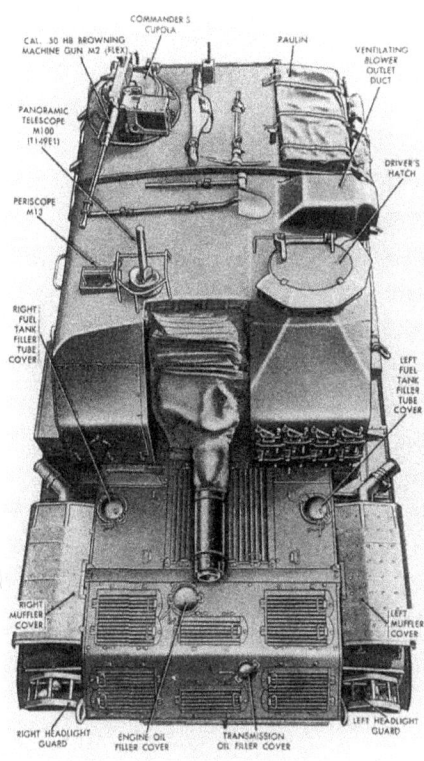

Above, the 105mm howitzer M49 is shown in the early (upper) and late (lower) mounts. The internal arrangement of the M52 can be seen in the sectional drawing below.

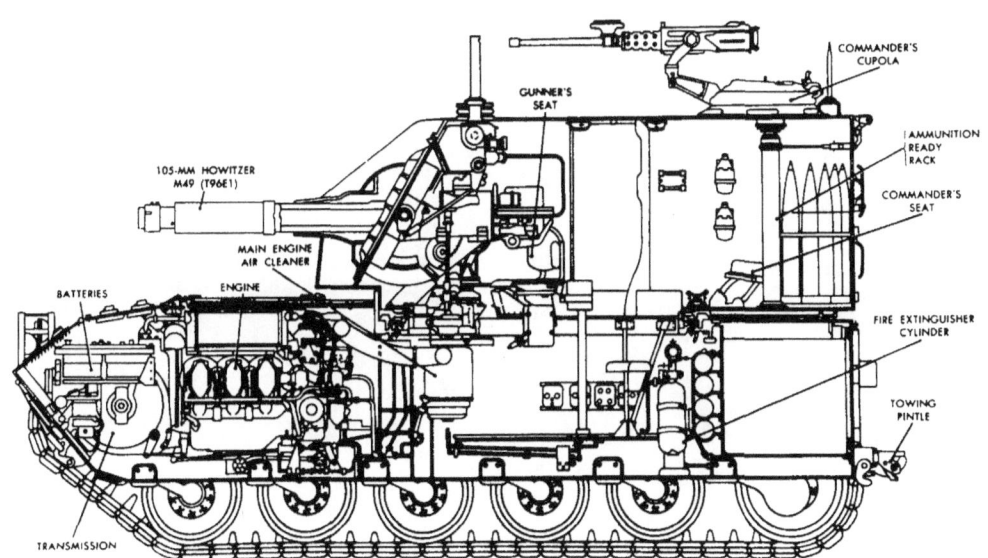

Note that the engine exhaust pipe has been extended and the personnel heater air intake relocated on the T98E1 above to eliminate the carbon monoxide problem. Below, the T98E1 has been fitted for deep water fording.

LIGHT ARTILLERY
PROPOSAL LA-3

TYPE: 105 MM HOW., S.P., AIRBORNE - PHASE I, T-101
ARMAMENT: 105 MM HOW., T96E1
ARMOR: ALUMINUM   WEIGHT: 8 TONS
HP/TON: 23.6   GROSS: 20 NET

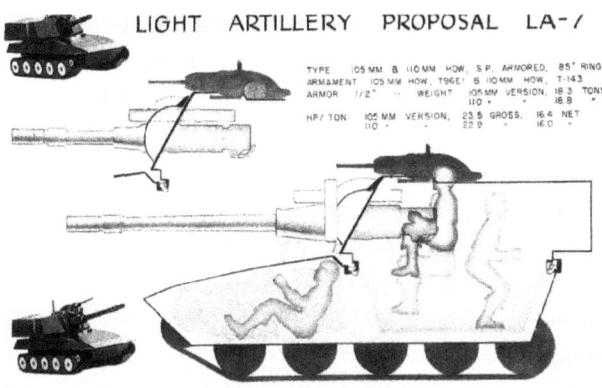

LIGHT ARTILLERY PROPOSAL LA-7

TYPE: 105 MM & 110 MM HOW., S.P. ARMORED, 85" RING
ARMAMENT: 105 MM HOW., T96E1 & 110 MM HOW, T-143
ARMOR: 1/2"   WEIGHT: 105 MM VERSION, 18.3 TONS
                              110 "   18.8 "
HP/TON: 105 MM VERSION, 23.5 GROSS, 16.4 NET
         110 "   22.9 "   16.0 "

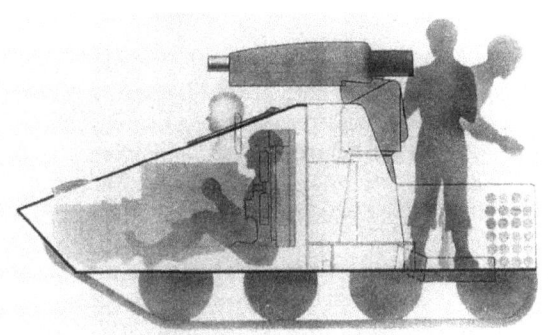

Above are two of the lightweight self-propelled artillery proposals from Questionmark II. At the left is a sectional view of the 75mm pack howitzer mounted on the Ontos chassis.

At the Questionmark II Conference in September 1952, fourteen design concepts were presented for light and medium self-propelled artillery, all of which had combat weights less than 28 tons. One of these mounted the 75mm pack howitzer M1A1 on the Ontos chassis and another installed the 105mm howitzer T96E1 on the unarmored chassis of the 90mm self-propelled gun T101. Two of the concepts were wheeled vehicles used as mounts for the new 110mm and 156mm howitzers. One design considered the 105mm howitzer and the 110mm howitzer as alternate armament in a turret mount on a tracked chassis. The maximum protection on any of the vehicles was ½ inch thick steel armor.

The study of new design concepts for self-propelled artillery continued with five proposed configurations presented during the Questionmark III Conference in June 1954. Two of these mounted cannon on wheeled chassis. Among the tracked vehicles, the new 110mm howitzer T203 provided the armament on proposals TS-12 and TS-27. Both designs required the cannon to be served from the ground at the rear of the vehicle. The TS-13 concept installed the 156mm howitzer T202 on a tracked chassis, but like the two 110mm howitzer proposals, the cannon was served from the ground at the rear of the vehicle. Also like the other proposals, protection was limited to ½ inch thick steel armor in front and a nylon blanket cover at the rear for the howitzer crew.

Detroit Arsenal presented eight self-propelled artillery concepts at the Questionmark IV Conference in August 1955. At that time, the greatest emphasis was on providing vehicles of all types that could be transported by air. Three of the eight proposals were wheeled vehicles and of the remaining five, two were armed with recoilless rifles and two with rocket launchers. The last concept was armed with the 156mm howitzer converted to use liquid propellant. The estimated weight of these vehicles ranged from 10 to 12 tons. Although some of these studies were impractical considering the state of the art at that time, they provided valuable information to the development programs then in progress.

Sectional views of three proposals from Questionmark III are shown here. At the left is the TS-12 concept. Below, the TS-27 appears at the left and the TS-13 is at the right.

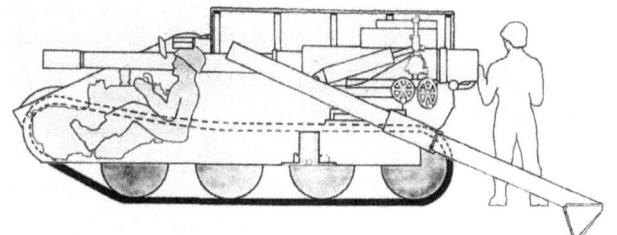

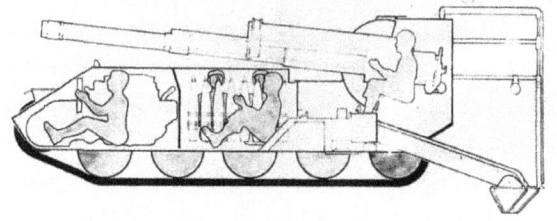

Concept drawings are shown here for the moritzer armed with the 105mm howitzer (below) and a 105mm liquid propellant cannon (right).

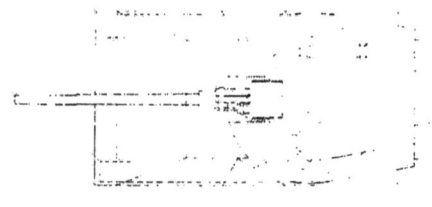

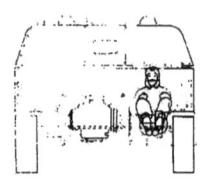 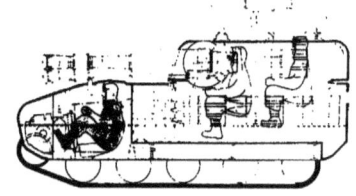

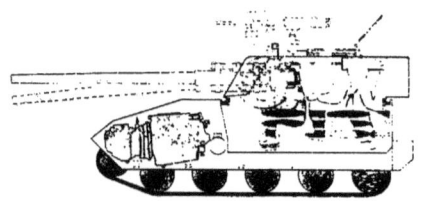

Another concept of a self-propelled artillery weapon suitable for use during phase I of airborne operations was dubbed the moritzer. Under study during 1956, it was intended to combine the high angle, short range, fire of the mortar with the longer range and direct fire capability of the howitzer. The characteristics outlined for the moritzer specified a minimum range of 400 yards and a maximum range of 12,500 to 16,000 yards. It also was to be capable of direct fire support. The unstowed weight was not to exceed 16,000 pounds and the armor was expected to provide protection against .30 caliber armor piercing ammunition at ranges above 200 yards. One moritzer concept drawing showed the vehicle armed with a 105mm howitzer, but that weapon was not expected to meet the full range requirements. Some earlier drawings showed a 105mm liquid propellant cannon. Since suitable armament was not available at that time, the moritzer project was suspended pending the results of future weapon development.

A meeting was convened in January 1952 to consider future requirements in self-propelled artillery. A recommendation from this meeting was to develop lightweight, self-propelled, mounts for the new 110mm and 156mm howitzers then under development. Early proposals for such self-propelled howitzers were rejected by the Continental Army Command (CONARC), but in September 1953, a design concept was selected for development as the 110mm self-propelled howitzer T195. The following May, a design was approved for development as the 156mm self-propelled howitzer T196. The two vehicles were to be armed with the 110mm howitzer T203 and the 156mm howitzer T202 respectively. After further modification, mock-ups were constructed of both vehicles which were reviewed during a meeting at Detroit Arsenal in March 1954. At this stage, the two vehicles differed in several respects. Both mounted the cannon in a turret or cab with 360 degrees rotation, but the inside diameters of the turret rings were 80 inches

 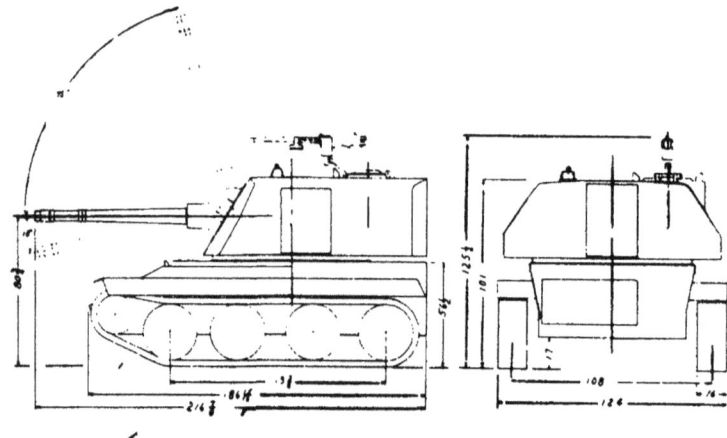

Mock-up photographs and dimensional sketches can be seen here for the 110mm self-propelled howitzer T195 (upper) and the 156mm self-propelled howitzer T196 (lower).

 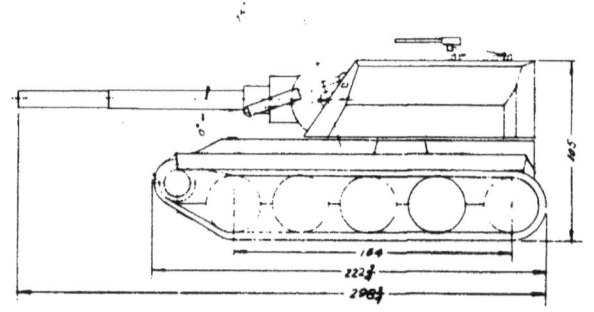 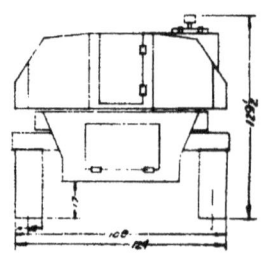

Above and at the right are views of the 105mm self-propelled howitzer T195 armed with the 105mm howitzer T252.

for the T195 and 85 inches for the T196. Both had an elevation range of −10 to +75 degrees and were protected by ½ inch thick steel armor on the top and all sides. The crew on both vehicles consisted of five men and the combat weight was estimated to be 38,500 pounds for the T195 and 50,000 pounds for the T196. The driver was in the left front hull alongside the power plant and the remainder of the crew rode in the turret. Each vehicle had a flat track suspension without support rollers and the T195 had four road wheel stations per side compared to five for the T196. On both vehicles, the road wheels were 30 inches in diameter and the rear wheels served as trailing idlers. The power pack in both cases was the Continental AOI-628-1 gasoline engine with the Allison XT-300 transmission. The front mounted sprockets drove 16 inch wide tracks.

In January 1956, the Chief of Research and Development, Office of the Chief of Staff, directed that the development of the 110mm and 156mm howitzers and their ammunition be terminated. Thus the T195 and T196 were modified to use the 105mm howitzer and the 155mm howitzer. In addition, changing specifications required extensive redesign of both vehicles. Directives of the Ordnance Tank Automotive Command (OTAC) specified that all vehicles in this weight class conform to the following requirements. 1. Use the AOI-628 engine, 2. Use the XT-300 transmission, 3. Adapt the suspension from the T113 personnel carrier to the requirements of the new vehicle, 4. Use a 15 inch wide track. The first two items were no problem as they were already incorporated into the design. The suspension and tracks, however, required major changes. Design studies were prepared for a variety of configurations using steel and aluminum armor with turret ring diameters ranging from 80 to 100 inches. As a result of these studies, it was concluded that a single basic chassis design could serve for both weapons. Mock-ups of the new T195 and T196 were reviewed during a meeting at OTAC in October 1956. The

proposed vehicles were protected by 1¼ inch thick 5083 aluminum alloy armor and had estimated weights of 32,500 pounds for the T195 and 35,827 pounds for the T196. They were to be armed with the 105mm howitzer T252 and a modified version of the 155mm howitzer T186E1 (later the T255) respectively. Stowage space was provided for 103 rounds of 105mm ammunition in the T195 and 32 rounds of 155mm ammunition in the T196. Both vehicles had a 100 inch diameter turret ring, but only manual traverse and elevation were provided. The turret could be traversed 360 degrees and the elevation range was −10 to +75 degrees. The T195 and T196 were manned by crews of five and six men respectively. In each case, the driver remained in his position in the left front of the hull and the other crew members rode in the turret. A .50 caliber machine gun was installed at the commander's hatch on the turret roof. As specified, the power pack consisted of the Continental AOI-628-1 engine with the Allison XT-300 transmission. The flat track, torsion bar, suspension had seven 22 inch diameter dual road wheels per side with the rear wheels acting as trailing idlers. The tracks were 15 inches wide as called for in the OTAC directive.

At the right is the 155mm self-propelled howitzer T196 at Detroit Arsenal on 1 April 1960.

Above, the 105mm self-propelled howitzer T195E1 appears at the left and the 155mm self-propelled howitzer T196E1 is at the right. Note the front hull changes on the T195E1 compared to the T195.

After the detailed design work was completed, four T195 and four T196 vehicles were authorized for construction. The first T195 was completed in August 1958 and tests began in September at the Erie Ordnance Depot and in December at Aberdeen Proving Ground and Fort Knox.

The test program revealed serious problems with the suspension system. There were numerous failures of torsion bars and other components. The pilots were returned to Detroit Arsenal and the suspension was rebuilt to use 24 inch diameter dual road wheels and an adjustable idler was added at the rear of each track. The pilot 155mm howitzer T196 was fitted with this modified suspension when it arrived for test at Aberdeen Proving Ground on 3 October 1959. In addition to being armed with the T255 155mm howitzer, the T196 also differed from the T195 by the installation of spades on the rear hull to provide stability when firing. During this same period, OTAC was directed to install diesel engines in the pilot vehicles which required more redesign work. With the new power plant, the two vehicles were redesignated as the 105mm self-propelled howitzer T195E1 and the 155mm self-propelled howitzer T196E1. Two pilots of each type were reworked to the new configuration. These vehicles were powered by the 8V71T, two stroke cycle, diesel engines produced by the Detroit Diesel Division of General Motors Corporation. This engine was coupled to the XTG-411-2A transmission from the Allison Division of General Motors.

The T195E1 and T196E1 pilots were submitted to extensive tests which, unfortunately, revealed serious failures of critical suspension and final drive components. As a result, CONARC terminated the tests for the T196E1 in November 1960 and those for the T195E1 in December. Plans to start production were suspended until these problems could be corrected. In February 1961, a new pilot program was approved which called for modification of the original pilots and the fabrication of two new T195E1s and two new T196E1s. After numerous changes, followed by engineering and service tests, the T195E1 and the T196E1 were released for limited production in December 1961. Subsequently, the two vehicles were standardized in July 1963 as the 105mm self-propelled howitzer M108 and the 155mm self-propelled howitzer M109. The initial production was carried out by the Cadillac Motor Car Division of General Motors at the Cleveland Tank Plant.

The production vehicles differed from the early T195E1 and T196E1 pilots in several respects. The early T118E1 tracks were replaced by the T137 track with replaceable rubber pads. The armament was changed on the production vehicles, particularly on the M108. The 105mm howitzer T252 on the early vehicles was replaced by the 105mm howitzer XM103 which was later standardized as the M103 in the mount M139. This cannon had a longer tube than the T252. The production M109s were armed with the 155mm howitzer T255E4 in mount XM127. Later, these were

Below, the early versions of the 105mm self-propelled howitzer M108 and the 155mm self-propelled howitzer M109 are shown at the left and right respectively.

The 105mm self-propelled howitzer M108 appears in the two upper photographs at the right.

standardized as the M126 and M127 respectively. The M109 also was provided with hydraulic power for the turret traverse and the howitzer elevation. A flotation device was available for both the M108 and the M109. It consisted of eight metal enclosed, neoprene impregnated, nylon bags for installation on the hull sides and one additional bag for attachment to the front as well as water barriers, a selector valve, a turboblower, and the necessary hoses and fittings.

Production of the M108 started in October 1962 and continued into 1963. The first M109 was delivered in November 1962 beginning a production program that was to last through many modifications.

The M108 and M109 vehicles were the same size except for the overhang of the spades and the 155mm howitzer on the latter. Their combat weights were 46,221 pounds and 52,461 pounds respectively. However, the firepower of the 155mm howitzer on the M109 greatly exceeded that of the 105mm weapon on the M108. As a result, additional production of the M108 was cancelled and future development centered on the M109. A total of 2111 M109s (1961 for the Army and 150 for the Marine Corps) were produced at the Cleveland Tank Plant through fiscal year 1969. As mentioned before, the first production was by the Cadillac Motor Car Division of General Motors. In the third year of production, the Cleveland Tank Plant was transferred to the Chrysler Corporation and after that, it was operated by the Allison Division of General Motors.

The 155mm self-propelled howitzer M109 can be seen in the two lower photographs at the right. The hull components on the M108 and the M109 are identified in the drawing below.

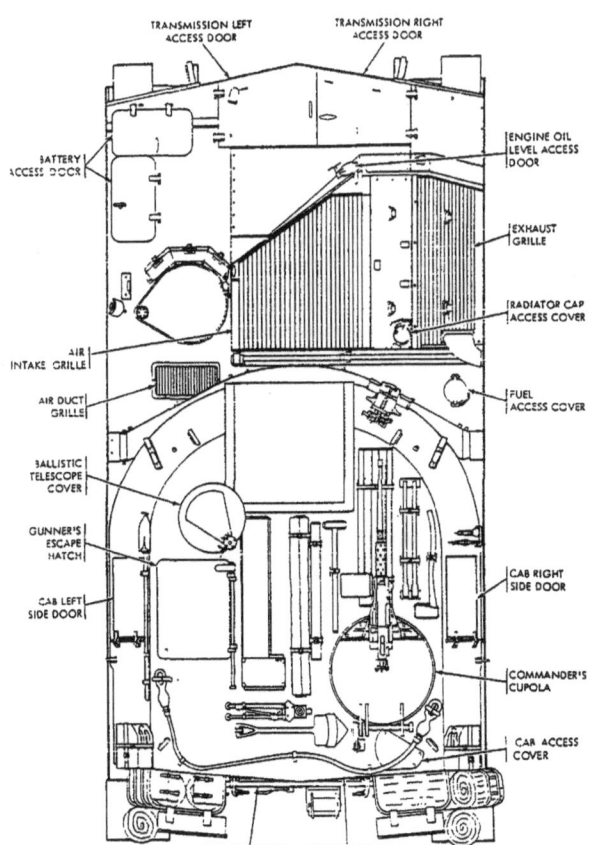

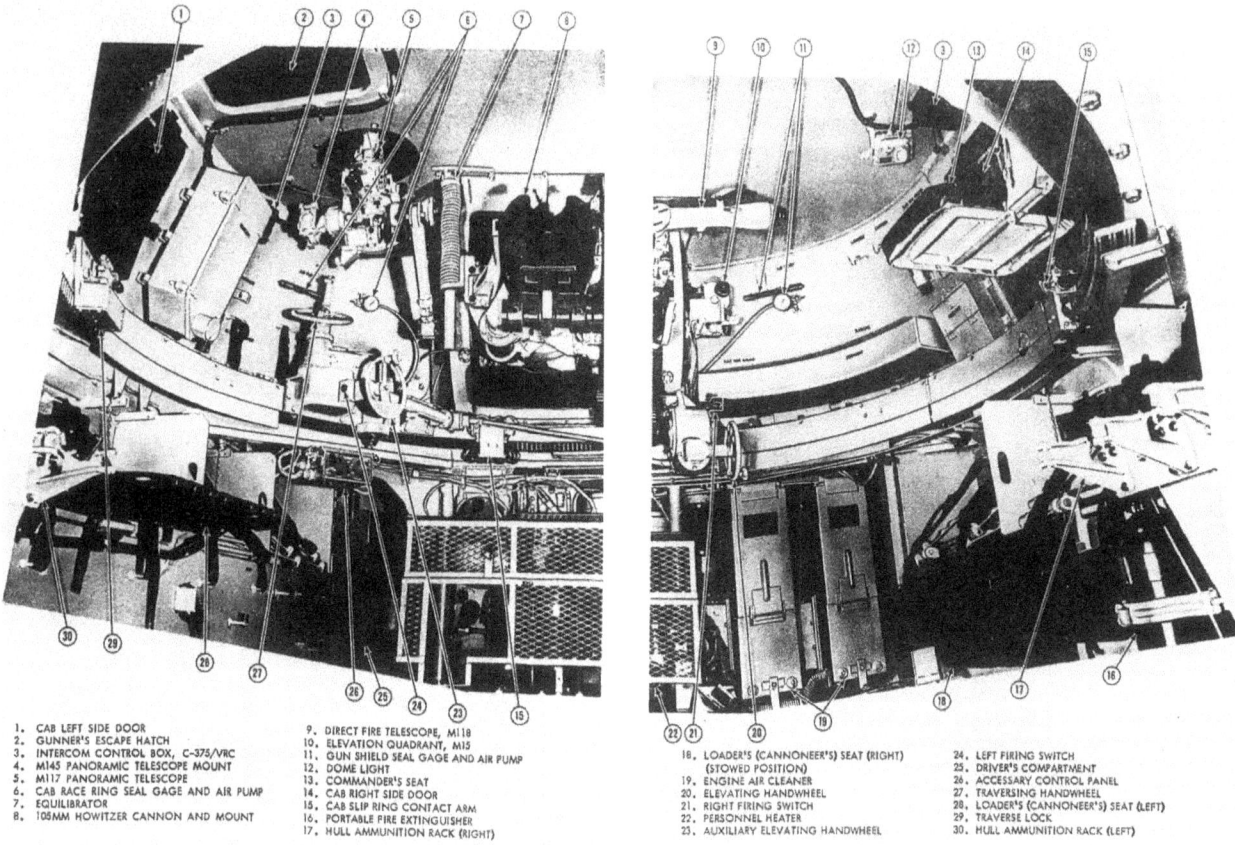

1. CAB LEFT SIDE DOOR
2. GUNNER'S ESCAPE HATCH
3. INTERCOM CONTROL BOX, C-375/VRC
4. M145 PANORAMIC TELESCOPE MOUNT
5. M117 PANORAMIC TELESCOPE
6. CAB RACE RING SEAL GAGE AND AIR PUMP
7. EQUILIBRATOR
8. 105MM HOWITZER CANNON AND MOUNT
9. DIRECT FIRE TELESCOPE, M118
10. ELEVATION QUADRANT, M15
11. GUN SHIELD SEAL GAGE AND AIR PUMP
12. DOME LIGHT
13. COMMANDER'S SEAT
14. CAB RIGHT SIDE DOOR
15. CAB SLIP RING CONTACT ARM
16. PORTABLE FIRE EXTINGUISHER
17. HULL AMMUNITION RACK (RIGHT)

18. LOADER'S (CANNONEER'S) SEAT (RIGHT) (STOWED POSITION)
19. ENGINE AIR CLEANER
20. ELEVATING HANDWHEEL
21. RIGHT FIRING SWITCH
22. PERSONNEL HEATER
23. AUXILIARY ELEVATING HANDWHEEL
24. LEFT FIRING SWITCH
25. DRIVER'S COMPARTMENT
26. ACCESSARY CONTROL PANEL
27. TRAVERSING HANDWHEEL
28. LOADER'S (CANNONEER'S) SEAT (LEFT)
29. TRAVERSE LOCK
30. HULL AMMUNITION RACK (LEFT)

The left front and right front of the M108 turret interior are shown above. The inside rear of the same turret appears below.

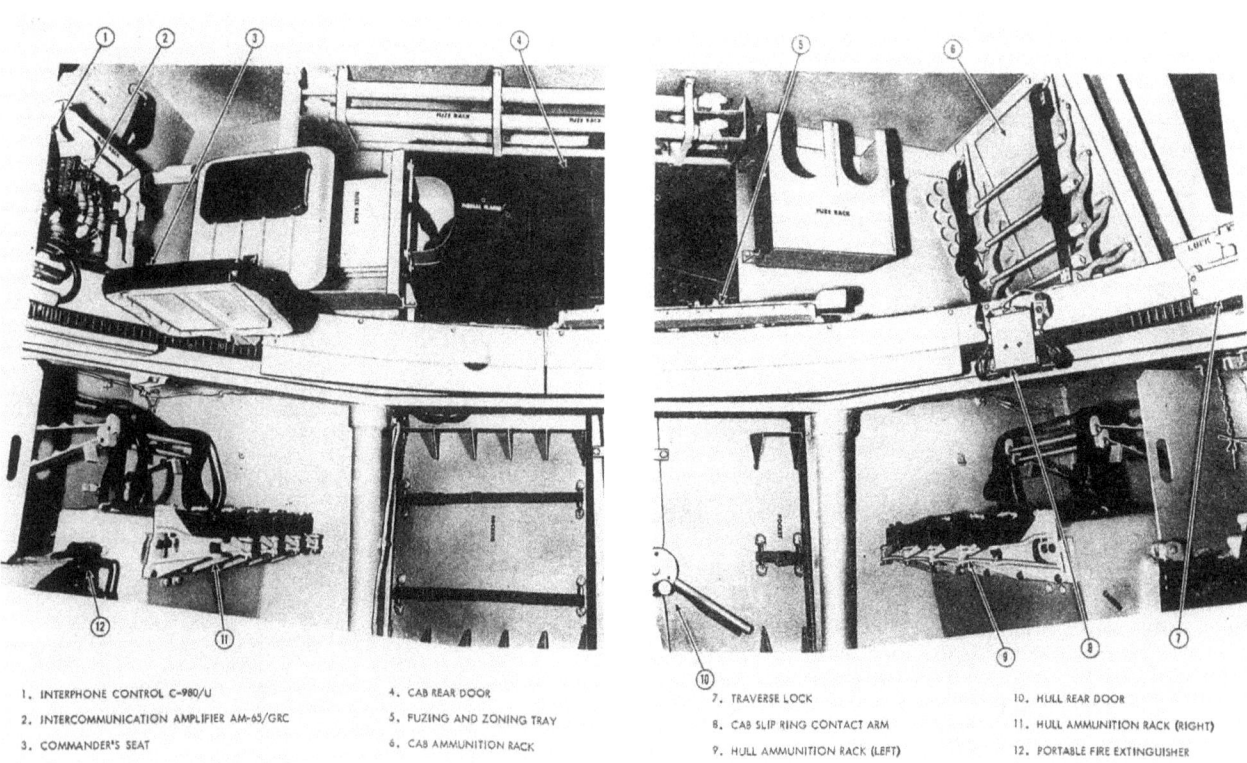

1. INTERPHONE CONTROL C-980/U
2. INTERCOMMUNICATION AMPLIFIER AM-65/GRC
3. COMMANDER'S SEAT
4. CAB REAR DOOR
5. FUZING AND ZONING TRAY
6. CAB AMMUNITION RACK

7. TRAVERSE LOCK
8. CAB SLIP RING CONTACT ARM
9. HULL AMMUNITION RACK (LEFT)
10. HULL REAR DOOR
11. HULL AMMUNITION RACK (RIGHT)
12. PORTABLE FIRE EXTINGUISHER

210

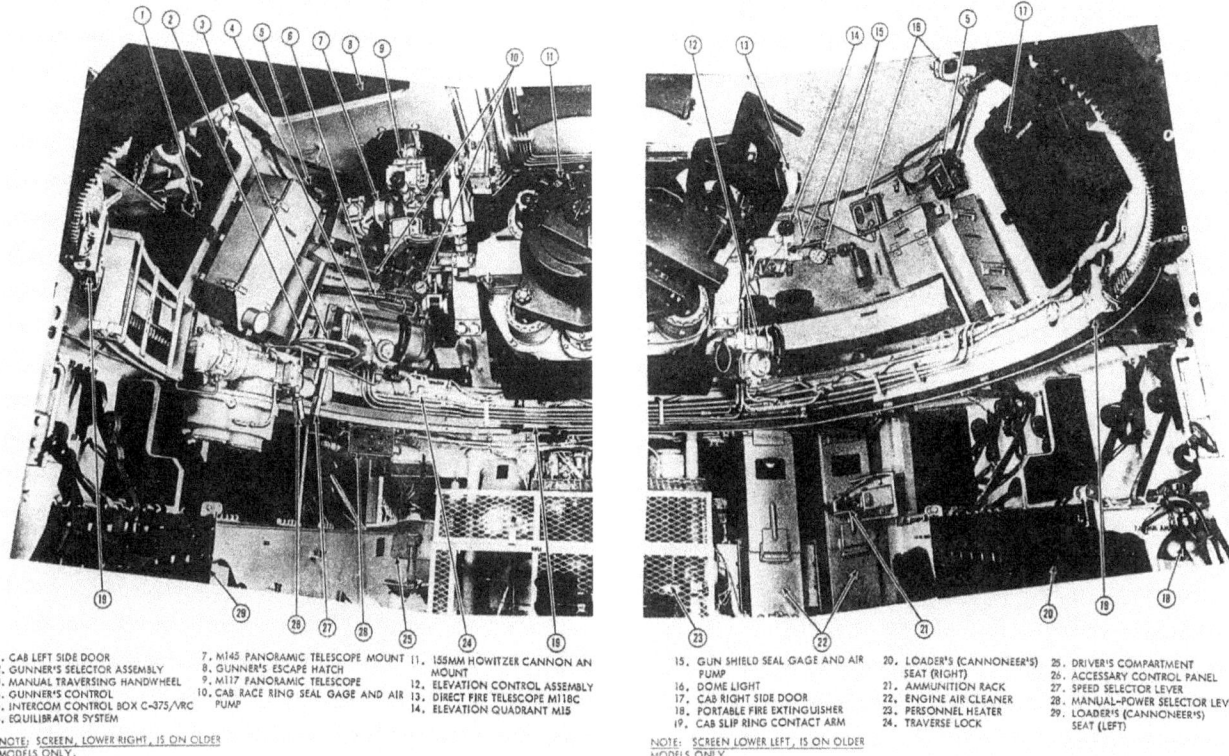

1. CAB LEFT SIDE DOOR
2. GUNNER'S SELECTOR ASSEMBLY
3. MANUAL TRAVERSING HANDWHEEL
4. GUNNER'S CONTROL
5. INTERCOM CONTROL BOX C-375/VRC
6. EQUILIBRATOR SYSTEM

7. M145 PANORAMIC TELESCOPE MOUNT
8. GUNNER'S ESCAPE HATCH
9. M117 PANORAMIC TELESCOPE
10. CAB RACE RING SEAL GAGE AND AIR PUMP

11. 155MM HOWITZER CANNON AN MOUNT
12. ELEVATION CONTROL ASSEMBLY
13. DIRECT FIRE TELESCOPE M118C
14. ELEVATION QUADRANT M15

NOTE: SCREEN, LOWER RIGHT, IS ON OLDER MODELS ONLY.

15. GUN SHIELD SEAL GAGE AND AIR PUMP
16. DOME LIGHT
17. CAB RIGHT SIDE DOOR
18. PORTABLE FIRE EXTINGUISHER
19. CAB SLIP RING CONTACT ARM

20. LOADER'S (CANNONEER'S) SEAT (RIGHT)
21. AMMUNITION RACK
22. ENGINE AIR CLEANER
23. PERSONNEL HEATER
24. TRAVERSE LOCK

25. DRIVER'S COMPARTMENT
26. ACCESSARY CONTROL PANEL
27. SPEED SELECTOR LEVER
28. MANUAL-POWER SELECTOR LEVER
29. LOADER'S (CANNONEER'S) SEAT (LEFT)

NOTE: SCREEN LOWER LEFT, IS ON OLDER MODELS ONLY.

Above, the left front and right front of the M109 turret interior can be seen at the left and right respectively. Rear views of the turret interior are below.

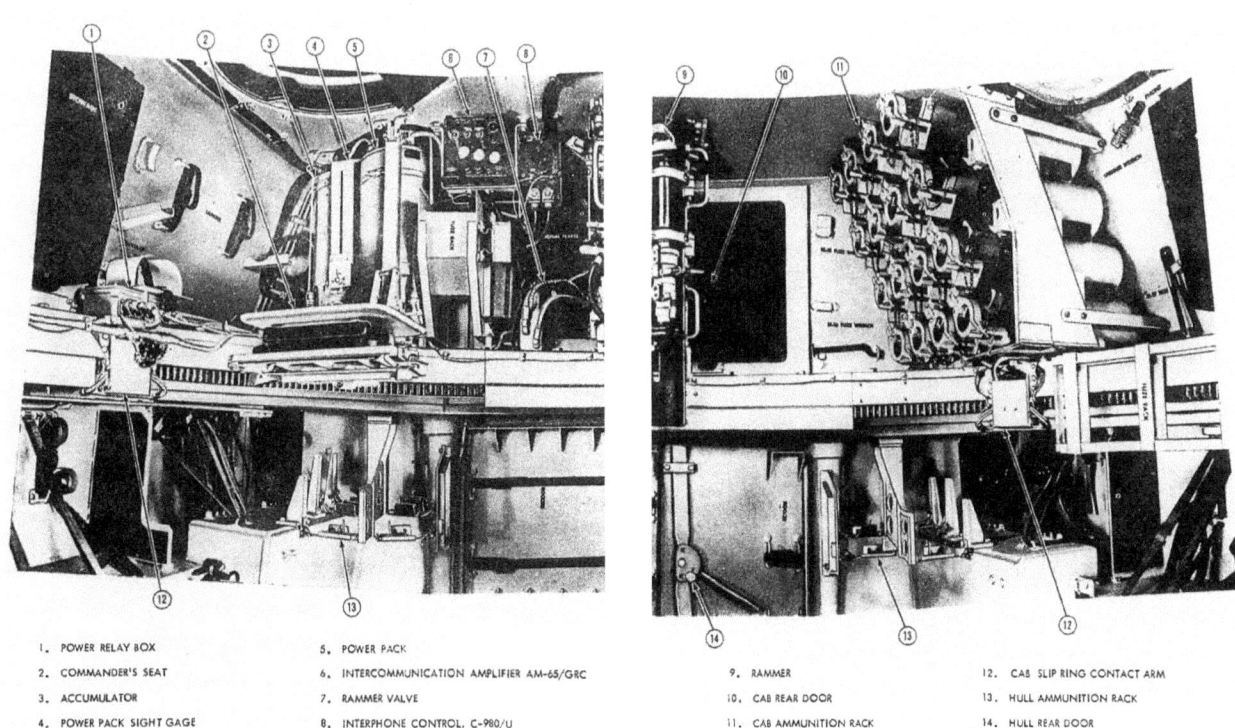

1. POWER RELAY BOX
2. COMMANDER'S SEAT
3. ACCUMULATOR
4. POWER PACK SIGHT GAGE

5. POWER PACK
6. INTERCOMMUNICATION AMPLIFIER AM-65/GRC
7. RAMMER VALVE
8. INTERPHONE CONTROL, C-980/U

9. RAMMER
10. CAB REAR DOOR
11. CAB AMMUNITION RACK

12. CAB SLIP RING CONTACT ARM
13. HULL AMMUNITION RACK
14. HULL REAR DOOR

211

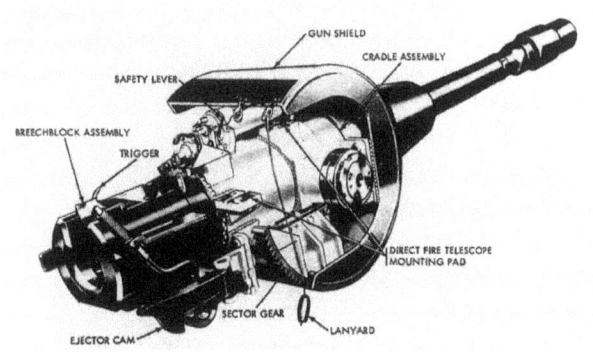

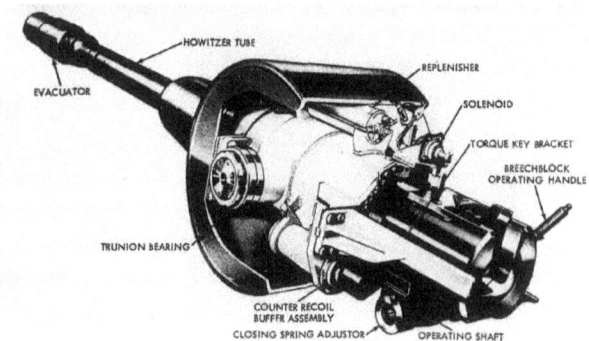

The 105mm howitzer M103 is shown above in mount M139. Below, the 155mm howitzer M126 is installed in mount M127.

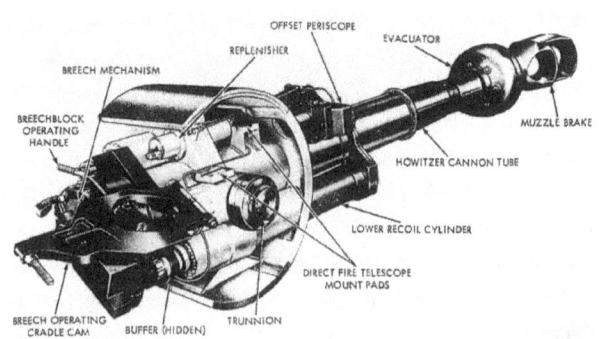

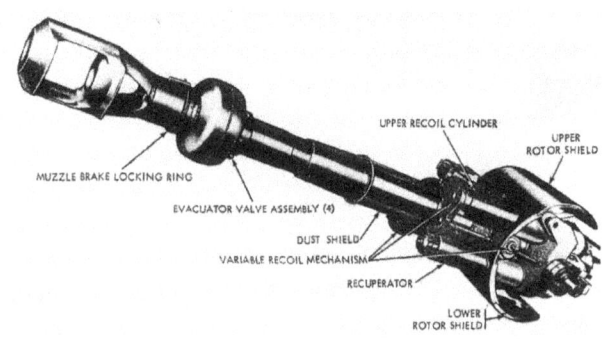

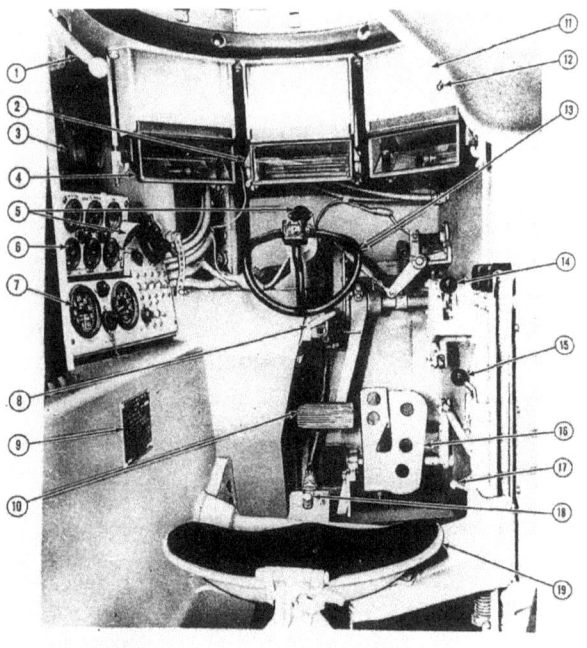

1. HATCH CONTROL
2. DOME LIGHT (HIDDEN)
3. AUXILIARY POWER RECEPTACLE
4. DRIVER'S PERISCOPE (3)
5. MASTER WARNING LIGHT
6. PORTABLE INSTRUMENT PANEL
7. MAIN INSTRUMENT PANEL
8. BRAKE LOCK
9. VEHICLE IDENTIFICATION PLATE
10. BRAKE PEDAL
11. HATCH LOCKING HANDLE
12. FUEL SHUT-OFF
13. STEERING WHEEL
14. TRANSMISSION SHIFT LEVER
15. THROTTLE CONTROL LEVER
16. ACCELERATOR PEDAL
17. SEAT VERTICAL CONTROL LEVER
18. HEADLIGHT DIMMER SWITCH
19. DRIVER'S SEAT

The driver's controls on the M108 and M109 appear at the left. Above, the M109 can be seen with the flotation barrier installed.

212

Above, the 155mm self-propelled howitzer M109 is at the left and the M109A1 is at the right. Below, the drawings show the 155mm self-propelled howitzer M109A2.

The M109 was standardized under the condition that the maximum range of the 155mm howitzer would exceed the 14,500 meter (15,857 yards) range of the 155mm self-propelled howitzer M44A1. This extended range was achieved through the development of the XM119 super propelling charge. Unfortunately, the high muzzle blast pressure of the XM119 charge fired in the short tube howitzer damaged vehicle components and produced undesirable physiological effects on the crew. To correct this situation, the long tube 155mm howitzer XM185, then under development for the proposed 155mm self-propelled howitzer XM179, was installed in the M109. When used with the longer tube, the powerful XM119 propelling charge did not produce undesirable effects on the vehicle or crew. Only minor turret modifications and a new travel lock were required for the installation of the XM185 howitzer. When armed with the new weapon, the vehicle was redesignated as the 155mm self-propelled howitzer M109E1. Three M109E1s completed service tests at Fort Sill in April 1969 and the new version was standardized in October 1970 as the 155mm self-propelled howitzer M109A1 with the 155mm howitzer M185. Conversions of the M109 into the M109A1 began in 1972 and the converted vehicles were issued to the troops in 1973. When new production of the self-propelled howitzer began in 1974 at Bowen-McLaughlin-York (later BMY Combat Systems), they were designated as the M109A1B.

The experience of the field artillery units operating with the M109s resulted in the usual requests for numerous modifications. These were incorporated in a new version designated as the 155mm self-propelled howitzer M109A2.

The improvements included a new M178 howitzer mount with a better recoil system and a redesigned rammer. The turret bustle was lengthened to carry 22 rounds of ammunition. This increased the total of 155mm rounds from 28 to 36. A ballistic cover was provided for the panoramic sight and the hydraulic system was simplified. Engine warning devices were installed and the hatch and door latches were redesigned. The flotation equipment was removed eliminating the swimming capability. The Army purchased 823 new production M109A2s during the fiscal years 1976 through 1985. Many of the M109A1s and the M109A1Bs were converted to the new standard and redesignated as the 155mm self-propelled howitzer M109A3.

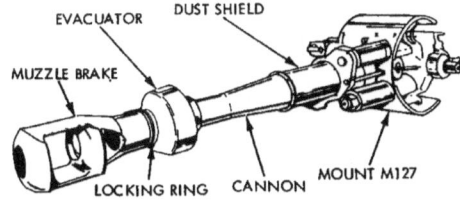

Above is the 155mm howitzer M126 in mount M127. Below, the 155mm howitzer M185 is installed in mount M178.

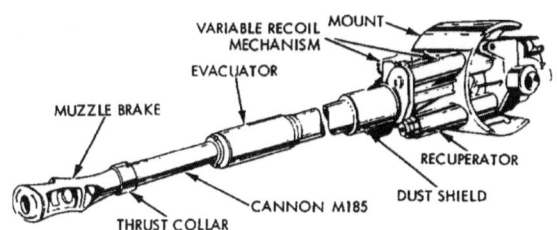

213

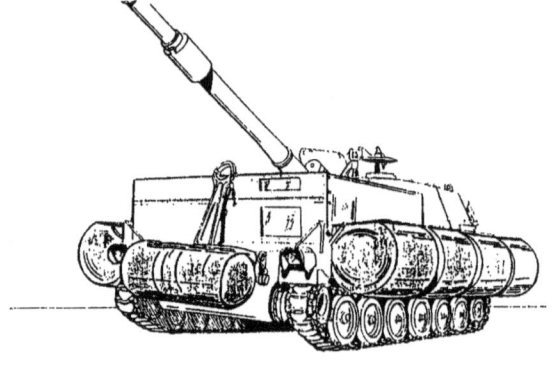

Below are two views of the 155mm self-propelled howitzer M109A2. At the left is a sketch of the M109A1 with the flotation kit which was eliminated on the M109A2.

90.79 C. of G.
FROM REAR FACE
OF BREECH RING
11.2 DIA
10.25 DIA
8. DIA
272.12 OVERALL LENGTH
(LENGTH OF TUBE 238.05)

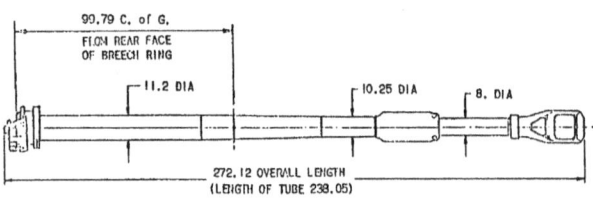

Additional photographs of the 155mm self-propelled howitzer M109A2 appear at the right and below. Above is a drawing of the 155mm howitzer M185.

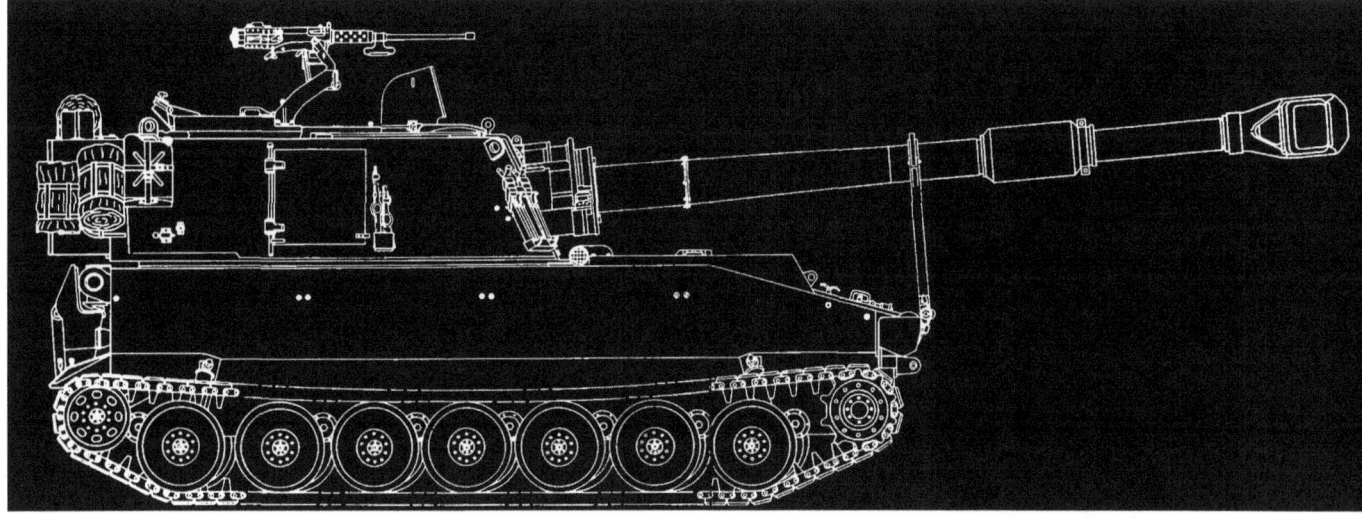

A side view of the 155mm self-propelled howitzer M109A2 is shown in the drawing above. Note the turret bustle extension for ammunition stowage and the ballistic cover for the panoramic sight.

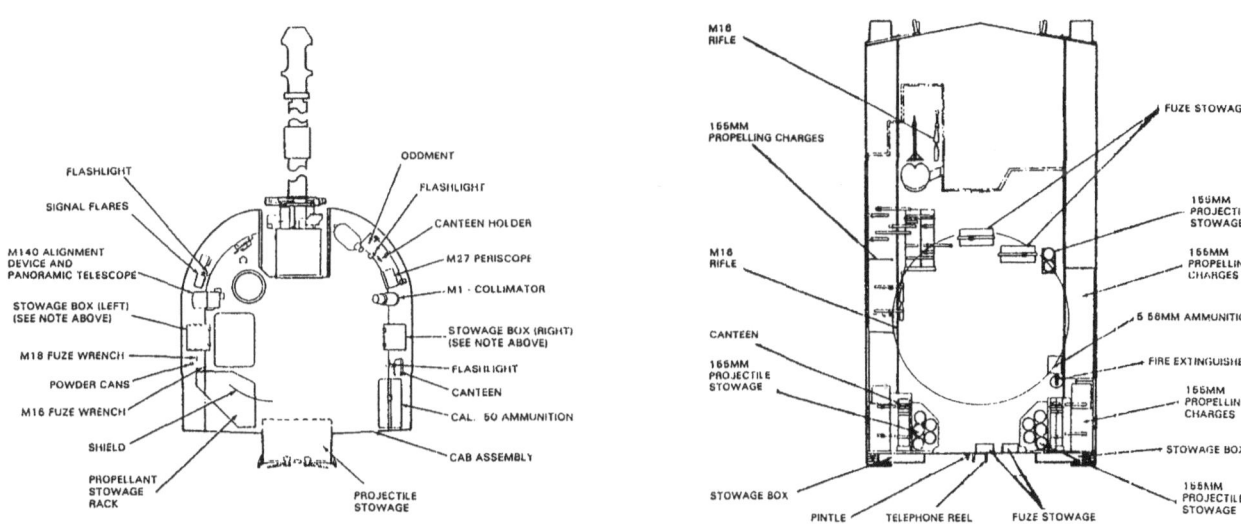

Above, the turret stowage is shown at the left and the hull stowage can be seen at the right for the M109A2. Below, an artist's concept of the M109A2 is at the left and the self-propelled howitzer is firing at the right.

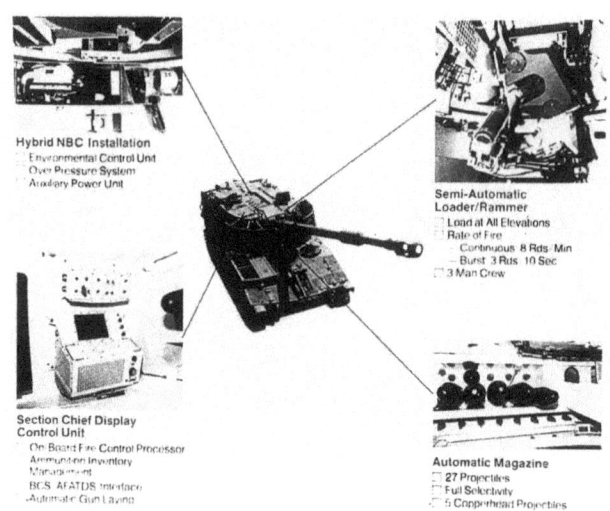

Above, the second prototype 155mm self-propelled howitzer M109E4 is at the left and some of the features evaluated on this vehicle are shown at the right.

During 1979-80, concept studies for a new vehicle were carried out under the project for the enhanced self-propelled artillery weapon system (ESPAWS). However, this program was replaced in the early 1980s by the division support weapon system (DSWS). The latter project also was cancelled and interest shifted to further upgrades of the basic M109. These included improvements in reliability and the installation of nuclear, biological, chemical (NBC) protection for the crew. The modifications were developed under the howitzer extended life program (HELP) in the early 1980s. Prototypes, designated as the M109E4, were delivered in late 1983 incorporating these features. M109A2 and M109A3 vehicles converted to the new standard were designated as the 155mm self-propelled howitzer M109A4.

The howitzer improvement program (HIP) which began in 1985 introduced the most extensive modernization project for the M109 series. Several armament options were considered under this program. The least expensive of these featured the 155mm howitzer XM284 and was referred to as the modified armament system (MAS). This weapon, based upon the 39 caliber M185 howitzer, had a stronger, redesigned, torque key to prevent rotation in the mount during firing. This allowed the use of the Zone 8s M203 propelling charge which provided a maximum range of 30 kilometers with the M549A1 rocket assisted projectile.

At the right is a prototype M109A3E2 armed with the 155mm howitzer XM284.

The advanced armament system (AAS) study investigated two additional cannon. These were the 155mm howitzers XM283 and XM282. The XM283 was a 39 caliber weapon like the XM284, but it was based upon the M199 cannon installed in the M198 towed howitzer. It used a modified muzzle brake which reduced the overpressure, but the internal ballistics were identical to the XM284. The 155mm howitzer XM282 was a much heavier weapon, 58 calibers in length, with the chamber capacity increased to 1700 cubic inches. The combustible case propelling charge XM244 developed for this weapon was expected to provide a maximum range of about 45 kilometers using the XM864 base-bleed projectile. All of the standard 155mm ammunition also could be used with the XM282.

The chassis of an M108 105mm self-propelled howitzer was fitted with a new turret to serve as a mock-up for the vehicle fitted with the MAS XM284. Another M108 was refitted for use as a vulnerability test bed. The HIP program called for the conversion of 11 M109s into prototypes using the various armament systems. Four of these, fitted with the XM284 howitzer, were designated as the

Above, another prototype from the howitzer improvement program (HIP) is at the left and the human factors howitzer test bed (HFHTB) is at the right. The figure below illustrates the large number of components evaluated under the HFHTB program.

M109A3E2. Five vehicles were allocated for use with the AAS armed with the XM282/283 howitzers under the designation M109A3E3. Two vehicles, designated as the M109A1C, retained the original M185 cannon in an improved M178 mount. This version, intended for the Israeli Defense Force, featured a new turret with increased ammunition stowage and an auxiliary power supply. Like the M109A3E3, it also included an improved suspension system with hydraulic bump stops, high strength torsion bars, and upgraded shock absorbers. Many new components also were evaluated by the Human Engineering Laboratory (HEL) using the human factors howitzer test bed (HFHTB).

TURRET GROUP

25 MM CHAINGUN
STAGET SIGHT W/LASER RANGEFINDER
IMPROVED COMMANDER'S HATCH
LOW PROFILE ANTENNA
INERTIAL REFERENCE UNIT
AUTOMATIC PRIMER SYSTEM
FREQUENCY HOPPING RADIOS
AUTOMATIC FUZE SETTER
LOADER/RAMMER ASSIST
EQUILIBRATOR (MECHANICAL)
IMPROVED INTERCOM W/WARNING
GLOBAL POSITIONING SYSTEM
BALLISTIC COMPUTER
VIDEO MONITOR
CHIEF OF SECTION STATION
GUNNER'S STATION
VIDEO MONITOR
DATA LINKS
BALLISTIC COMPUTER DISPLAY

HFHTB

CHASSIS GROUP

IMPROVED DRIVER'S HATCH
IMPROVED DRIVER'S SEAT
AUTOMATIC TRAVEL LOCK
AZIMUTH DRIVE (ELECTRICAL)
GAS TURBINE ENGINE
AUTOMATIC SPADES
AIR CYCLE UNIT
NBC FILTERS
NATO SLAVE
NBC SYSTEM

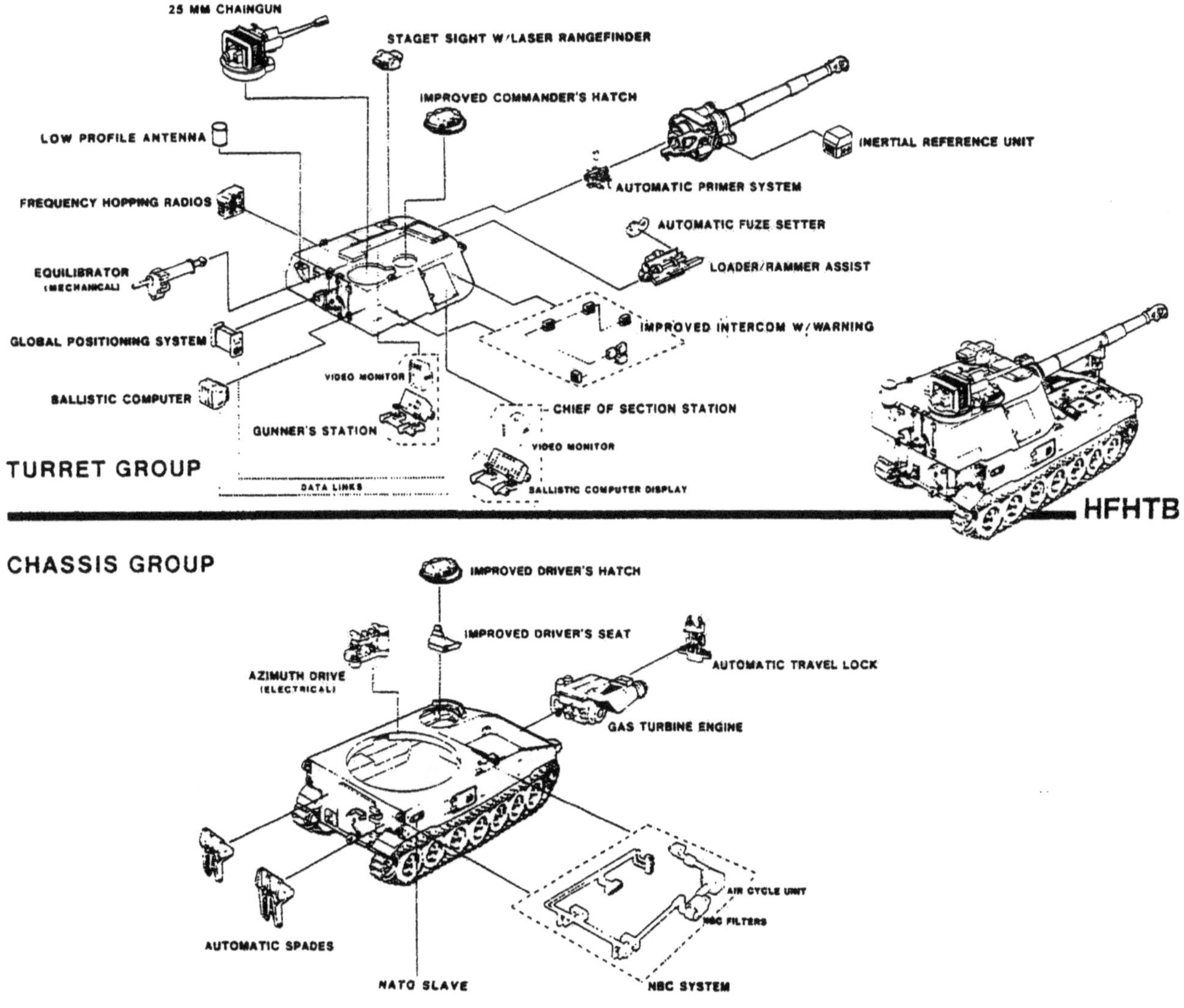

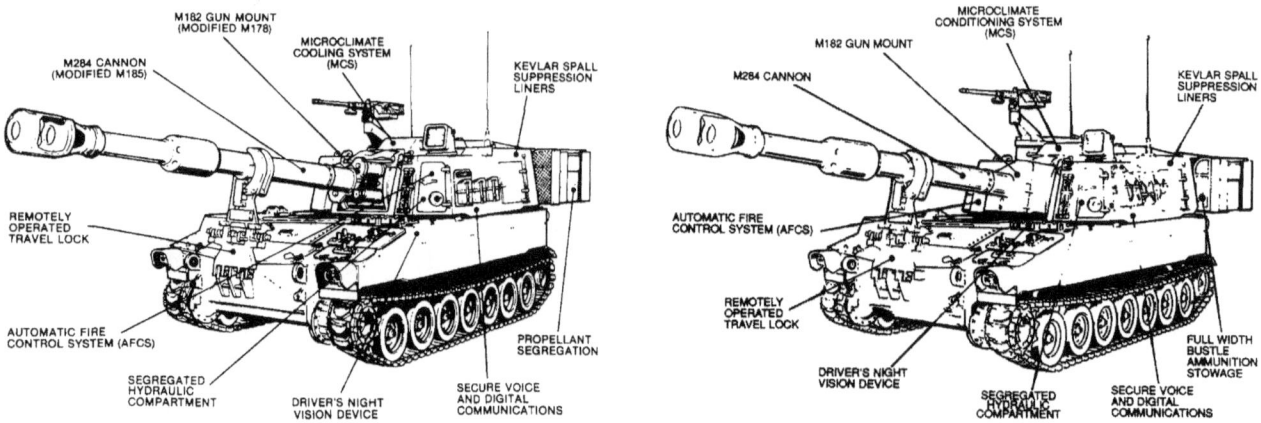

The sketches above show the early (left) and late (right) configurations of the 155mm self-propelled howitzer M109A6. Note the changes to the M182 mount on the later vehicle. Below are drawings of the M284 cannon (left) and the M182 mount (right).

M284 Cannon (Modified M185) (30 KM Range Capability)

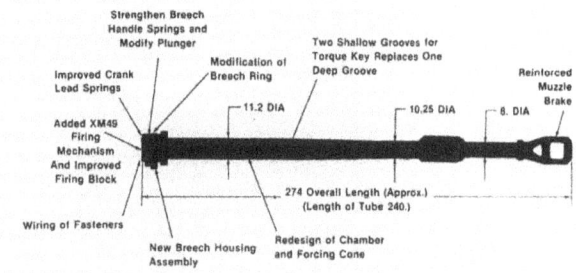

M182 GUN MOUNT (MODIFIED M178)

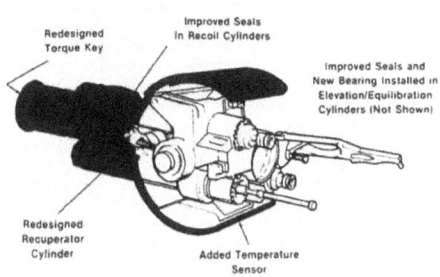

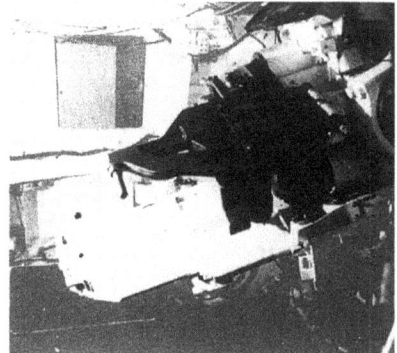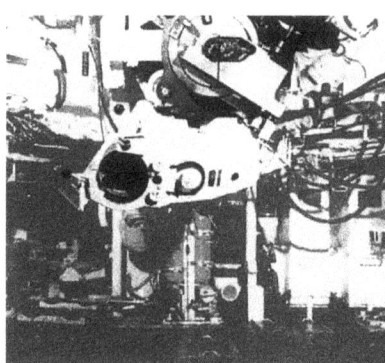

Above are three views of the semiautomatic loader in the 155mm self-propelled howitzer M109A6. Below, from left to right, are the segregated hydraulic system, radio, and environmental control system in the M109A6.

Above is the production 155mm self-propelled howitzer M109A6. The stowage bins on the sides of the turret can be swung to the rear to reduce the vehicle width for transportation.

After test and evaluation, the modified armament system was selected for installation in the production vehicle and it was standardized as the 155mm howitzer M284 in the mount M182. This armament, as well as an NBC system with micro-climate cooling and the improved suspension system, was applied to the new production vehicle. After standardization, it was designated as the 155mm self-propelled howitzer M109A6 and nicknamed Paladin. The Paladin also featured a new aluminum armor turret with aramid spall liners. The bustle extended to the full width of the new turret providing stowage space for the propellant charges and it incorporated steel armor for additional protection. The 155mm ammunition stowage was increased to 39 rounds. The automatic fire control system included inertial navigation and positioning, automatic gun laying, and full ballistic computation. Power assisted semiautomatic loading allowed the firing of short rapid bursts. The howitzer travel lock was remotely operated, reducing the time required to bring the weapon into action. The cannon could be moved to a preset elevation and deflection by the touch of a button. These features allowed the Paladin to take up a firing position and fire the first shot in less than 60 seconds.

The first production M109A6 was delivered in April 1992 and the program at that time called for the production of 824 vehicles. The first 104 were to be built under a low rate initial production contract at BMY Combat Systems.

Additional views of the M109A6 appear here. The full width turret bustle is obvious above. Below, the howitzer is firing at the left and the cannon is fully elevated at the right. Note the field artillery ammunition support vehicle (FAASV) in the latter photograph.

At the right is the 155mm self-propelled howitzer M109A3G modified in Germany and armed with the Rheinmetall howitzer.

In the Spring of 1992, more than 2400 M109s of various models were in the Army inventory. More than 1900 of these were in active and reserve units. To improve the performance of this fleet, a less expensive upgrade was planned to introduce many of the Paladin features. The upgraded vehicle, intended mainly for reserve units, was designated as the 155mm self-propelled howitzer M109A5. They were fitted with the new M284 howitzer in the M182 mount and were equipped with the NBC protection system first installed in the M109A4.

The M109 in various versions is widely employed in armies around the world. In Germany, the original M109s were designated as the M109G and the howitzer was fitted with a sliding breech mechanism developed by Rheinmetall. In 1983, kits were purchased to convert 586 vehicles to the M109A3G configuration armed with a new long tube Rheinmetall howitzer using the original sliding breech mechanism.

Because of the widespread use of the M109 series, there was considerable interest in providing improved versions of the vehicle to the large number of users. Many of these improvements were developed under the HIP project and work continued to further extend the capability of the M109 series. During 1983, BMY initiated a program to develop a new turret that could be installed to upgrade M109s then in service. Referred to as the International Turret, it mounted a 155mm howitzer with a 45 caliber tube developed by Royal

Ordnance in Britain. This cannon used a sliding breech-block with the Crossley pad obturating mechanism. It was fitted with a semiautomatic loader produced by the Emerson Electric Company which permitted a firing rate of eight rounds per minute and was capable of firing three round bursts in 15 seconds. The maximum range of this howitzer was 38 kilometers (41,557 yards). A Kevlar liner improved the crew protection in the turret and the 36 propellant charges in the turret bustle were separated from the crew and vented to the outside if ignition should occur. The estimated weight of the M109 with the International Turret was 57,300 pounds. Depending upon the customer's requirements, a full solution fire control system with automatic gun laying could be provided. Other upgrades for the M109 series were proposed by the FMC Corporation, Rheinmetall Defense Engineering, and United Technologies.

Below is an artist's concept of a 155mm self-propelled howitzer fitted with the International turret proposed by BMY. Note the extreme length of the 45 caliber cannon.

220

The prototype 8 inch self-propelled howitzer T236 appears at the right. The only armor on the vehicle was around the driver's compartment although some protection could be obtained through the use of nylon covers.

As described earlier, the Questionmark IV Conference in August 1955, placed major emphasis on the development of vehicles suitable for air transportation. The objective was to have all army equipment meet this requirement. In line with this policy, a feasibility study was approved during a meeting at the Office, Chief of Ordnance in January 1956. This study was to determine the design requirements to allow the air transportation of heavy artillery. A subsequent meeting at the Pacific Car and Foundry Company in September 1956 approved their proposal for a lightweight, self-propelled, artillery weapon. A contract was then awarded to Pacific Car and Foundry for the design and construction of six pilot vehicles. They consisted of two 175mm self-propelled guns T235, three 8 inch self-propelled howitzers T236, and one 155mm self-propelled gun T245. The mount and chassis were to be the same in all three cases with the weapons interchangeable. Later, in 1957, the program was extended to include the development of an unarmored wrecker/recovery vehicle T119 and an armored wrecker/recovery vehicle T120 using the same chassis.

The basic chassis was a welded assembly of high strength alloy steel and steel armor. The Continental AOI-628-3 gasoline engine and the Allison XTG-410-2 transmission were located in the right front hull. Engine intake air passed through filters in the left side of the hull and the engine exhaust went through the muffler and out through ports in the right side of the hull. The driver was located in the left front hull alongside the power plant. The hull front and the driver's compartment consisted of ½ inch thick steel armor, the only armor on the vehicle. The front mounted sprockets drove the 18 inch wide single pin tracks.

The flat track, torsion bar, suspension supported the vehicle on five dual road wheels per side. The rear road wheels served as trailing idlers.

The cannon was installed on the turret mount near the rear of the vehicle. The weapon was pointed forward and could be traversed 30 degrees to the left or right of center. The elevation range was 0 to +65 degrees. Four seats were provided, one for a gunner on the left side of the turret and three for another gunner and two loaders on the right side of the turret. The remaining eight members of the 13 man crew rode in an M548 cargo carrier with the main ammunition supply. Two rounds of 155mm, 175mm, or 8 inch ammunition were carried on the self-propelled mount itself. Hydraulic power was provided to traverse and elevate the cannon, to operate the rammer and winches, to operate the spades, and to lock out the suspension stabilizing the vehicle when firing. The combat weights were 59,200 pounds for the T235, 55,500 pounds for the T236, and 54,800 pounds for the T245. All of these weights could be reduced by 5500 pounds for air transportation.

Above is a side view of the suspension system used on the lightweight self-propelled weapons. Below, the 175mm self-propelled gun T235 is at the left and the 155mm self-propelled gun T245 appears at the right.

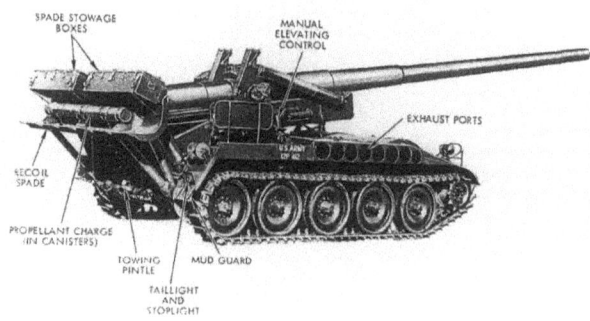

Various features of the 175mm self-propelled gun M107 are identified in the photographs above and below. At the bottom of the page are two views of the 8 inch self-propelled howitzer M110.

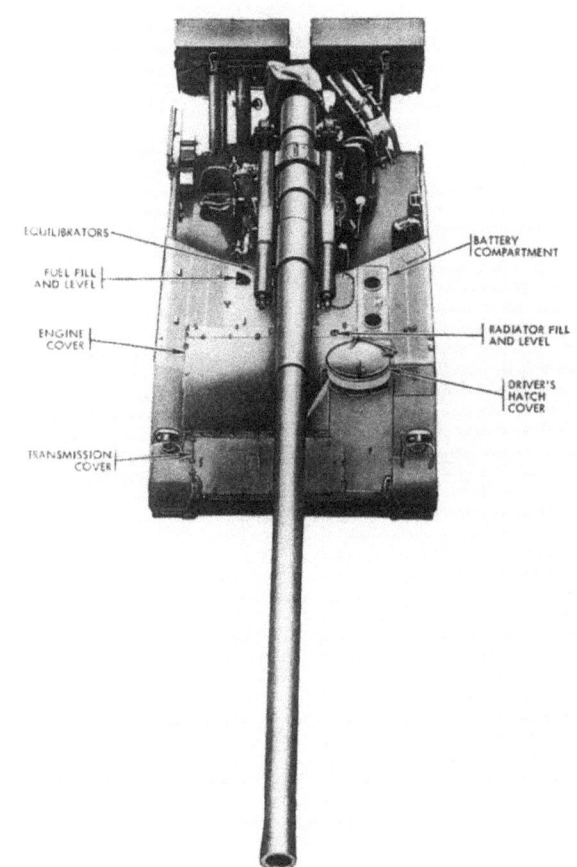

With the establishment of the policy in 1959 requiring the use of diesel engines, one pilot each of the T235 and T236 was modified and retrofitted with the General Motors 8V71T diesel engine and the Allison XTG-411-2 transmission. The 450 gallon fuel tank was replaced by a 300 gallon tank. Despite the smaller tank, the vehicle cruising range remained the same at about 450 miles because of the greater fuel economy of the diesel engine. The Continental AVDS-750 and the Caterpillar LDS-750 diesel engines were installed experimentally in the T236. With the 8V71T engine, the vehicles were redesignated as the 175mm self-propelled gun T235E1 and the 8 inch self-propelled howitzer T236E1. The T245 armed with the 155mm gun was dropped from further consideration. Service tests of the T235E1 were successfully completed at Fort Knox in March 1961. Both the T235E1 and the T236E1 were standardized by OTCM 37684 on 9 March 1961 and designated as the 175mm self-propelled gun M107 and the 8 inch self-propelled howitzer M110. They were armed with the 175mm gun M113 (T256E3) and the 8 inch howitzer M2A2 (M2A1E1) respectively. Both weapons were installed in the M158 mount using the same loader and rammer. The cannon was retracted in the travel position to reduce the barrel overhang at the front of the vehicle. A contract was awarded to the Pacific Car and Foundry Company in June 1961 for a limited production run.

Deliveries of the production vehicles began in 1962 with the first M107 battalion being formed at Fort Sill in January 1963. Later, FMC Corporation and Bowen-McLaughlin-York (later BMY Combat Systems) entered the production program. By 1977, interest had shifted to the use of the new

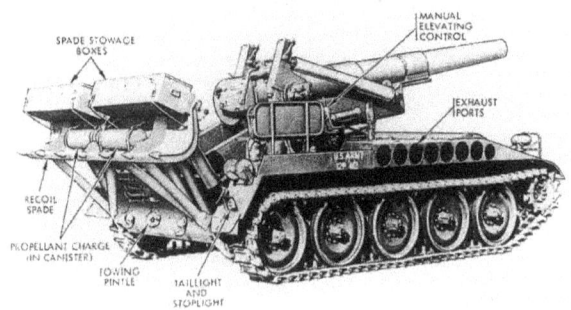

The 8 inch self-propelled howitzer M110 appears here with the cannon fully elevated and the spade lowered into the firing position.

long barrel 8 inch howitzer and by 1981, all of the M107s in the U.S. Army and Marine Corps were converted to the M110A2 configuration.

Production of the original 8 inch self-propelled howitzer M110 began at Pacific Car and Foundry Company in 1962 parallel with the M107 production. The first M110 battalion also was formed at Fort Sill in early 1963. Later, both FMC and BMY entered the production program.

In 1969, a development program was initiated to increase the range of the M110. This resulted in the installation of the new 8 inch howitzer M201 with a much longer tube. The new vehicle was equipped with the M139, direct fire, elbow telescope and it was standardized as the 8 inch self-propelled howitzer M110A1 in March 1976. The M110A1 entered service in 1977 to replace both the M110 and the M107. However, the M110A1 was limited to zone 8 of the M188A1 propelling charge because of excessive recoil forces with zone 9. This resulted in the installation of a double baffle muzzle brake and the redesignation of the vehicle in 1978 as the 8 inch self-propelled howitzer M110A2. With the muzzle brake, the howitzer became the M201A1. It could now fire zone 9 of the propelling charge M188A1.

At the left is the 8 inch self-propelled howitzer M110A1 armed with the 8 inch howitzer M201. Below is an artist's concept of the 8 inch self-propelled howitzer M110A2 with the M201A1 howitzer fitted with a muzzle brake.

The 8 inch self-propelled howitzer M110A2 appears in these photographs. The cannon and the recoil spade are in the travel position.

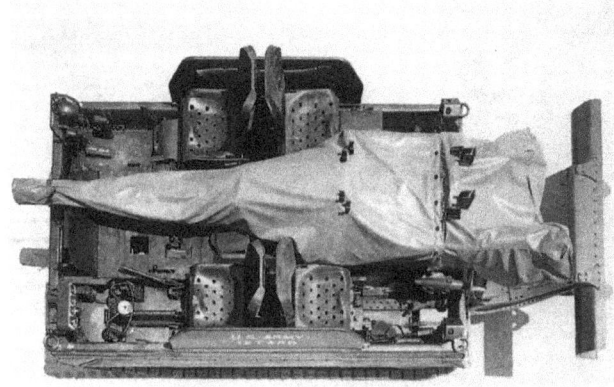

The photographs above and below show the 105mm self-propelled howitzer XM104. Note the small size of this lightweight weapon. The cannon and spade are secured in the travel position in the views above. Below, the howitzer is ready to fire.

Another effort to provide lightweight artillery for the airborne forces was the 105mm self-propelled howitzer XM104. Armed with the 105mm howitzer XM103, it had a combat weight of 8600 pounds, but this could be reduced to 7200 pounds for air transportation. Powered by a four cylinder, 66 horsepower, liquid-cooled engine with a five speed synchromesh transmission, it had a maximum road speed of 35 miles per hour. Two test rigs and six pilots were authorized for the XM104 and the complete vehicle was tested at Aberdeen in early 1963. However, the Department of the Army concluded that there was no requirement for the XM104 and the program was terminated.

Other projects to develop a lightweight 155mm self-propelled howitzer included the XM62 and the XM138. The XM62 reached the mock-up stage in May 1962 and a test rig was assembled for the XM138 in early 1966. This test rig was unarmored and had an empty weight of 22,900 pounds. With the concentration on the development of the M109 series, these projects were terminated.

Below, the mock-up of the 155mm self-propelled howitzer XM62 appears at the left and the 155mm self-propelled howitzer XM138 is at the right. Note the low silhouette of the latter test rig.

225

Above is the multiple caliber .60 machine gun motor carriage T100 after assembly by the Sperry Gyroscope Company. Note the track tension idler is retained on the T37 chassis at the left, but it has been eliminated at the right.

ANTIAIRCRAFT VEHICLES

Near the end of World War II, the .60 caliber machine gun was under development as an aircraft weapon. On 24 June 1948, OCM 32262 initiated a project to use this machine gun in a multiple mount as a short range antiaircraft weapon. Radar controlled, it mounted four T17E5 .60 caliber machine guns and was nicknamed Stinger. The T41 light tank was selected as the most suitable chassis to provide a self-propelled mount for the Stinger. A meeting was held at Detroit Arsenal to determine the necessary modifications to the T41 light tank chassis to meet the requirements of the Stinger installation. On 16 March 1950, OCM 33209 designated the new vehicle as the multiple caliber .60 machine gun motor carriage T100. Since it was the most readily available, the chassis from one of the T37 light tank pilots was reworked at Detroit Arsenal for use as the pilot T100. This chassis was shipped to the Sperry Gyroscope Company, Great Neck, Long Island, New York on 17 September 1951 for the installation of the turret and fire control system.

Below is the .60 caliber aircraft machine gun T17E3 similar to the T17E5 installed on the T100. At the right is the twin 40mm self-propelled gun T141.

Combat loaded with its crew of three, the T100 weighed about 22 tons. The unarmored turret with the four .60 caliber machine guns rotated 360 degrees and the elevation ranged from −10 to +90 degrees. A total of 4000 rounds of ammunition were provided for the .60 caliber machine guns.

Unfortunately, the .60 caliber machine gun itself was a major problem. It had numerous failures during the development program and did not provide reliable performance. Interest also was shifting toward the use of heavier caliber antiaircraft weapons and as a result, the T100 program was terminated.

On 30 August 1951, OCM 33869 authorized the design and development of a new self-propelled antiaircraft weapon to be used with a separate radar fire control unit.

226

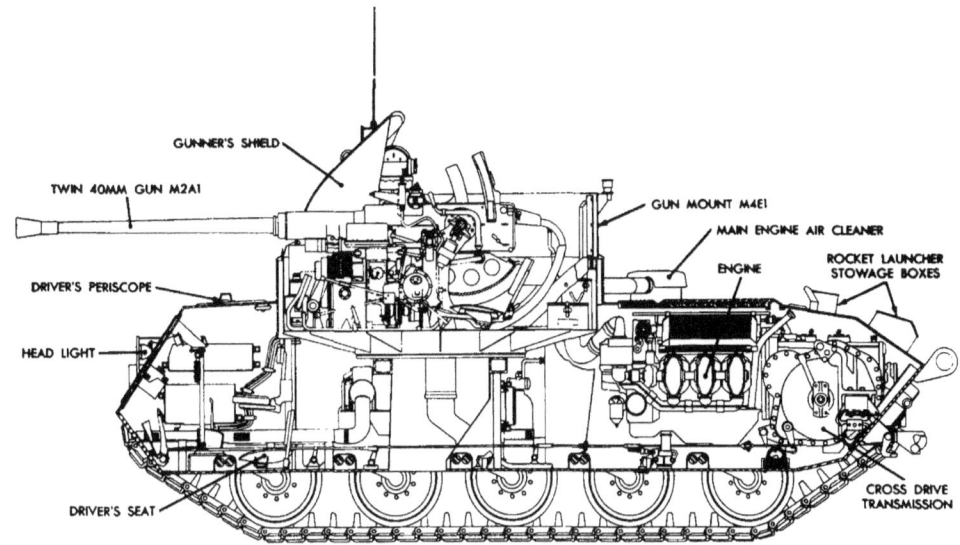

The twin 40mm self-propelled gun T141 is shown in the sectional drawing above and in the photograph below at the right. Note the conical flash hiders on this early vehicle.

Designations were assigned for the twin 40mm self-propelled gun T141 (interim vehicle), the twin 40mm self-propelled gun T141E1 (ultimate vehicle), and the 40mm gun fire control carrier T53. However, OCM 34340 cancelled the T141E1 gun and the T53 fire control carrier on 9 May 1952 because of complexity as well as the time and cost required for development.

In the meantime, work progressed rapidly on the T141 and it was rushed into production with the first vehicle being delivered by the Cadillac Motor Car Division at the Cleveland Tank Plant in August 1951. Obviously, the project was well underway before the official Ordnance Committee action. ACF Industries, Incorporated also entered the production program and delivered their first T141 in April 1952. As with the early light tank production, tests revealed the need for numerous modifications.

The general configuration of the T141 followed that of the light tank. The driving compartment was in front with a door in the hull front and two roof hatches. The gun mount was in the center with 360 degree traverse using both power and manual operation. Power elevation ranged from −3 to

+85 degrees. The manual elevation range was −5 to +87 degrees. The engine compartment was in the rear with the Continental AOS-895-3 engine driving the rear mounted sprockets through the Allison CD-500-3 transmission.

Originally, the auxiliary generator engine was not provided with a muffler and its exhaust passed out through the exhaust deflector on the hull roof. Later, a small muffler

Below is the late production twin 40mm self-propelled gun M42. The three prong flash suppressors have replaced the early conical flash hiders on the 40mm guns.

227

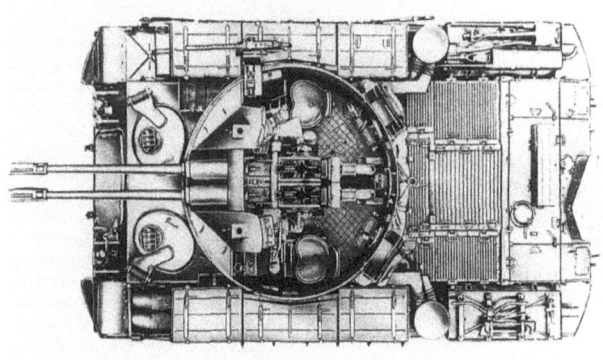

Above are additional views of the late production twin 40mm self-propelled gun M42. Note that the doors in the hull roof between the hatches for the driver and the vehicle commander have been eliminated.

was installed on the front of the tool stowage rack. Even later, a large muffler was fitted that extended to the rear of the tool stowage rack. This version also was retrofitted to the earlier vehicles that did not have an auxiliary generator engine muffler.

On the early T141s, the hull roof between the hatches for the driver and the vehicle commander was split to form front and rear overhead doors. These doors could be opened along with the hatches to provide an unobstructed path for passing ammunition from the hull stowage racks to the gun mount. On the later vehicles, the overhead doors were eliminated and replaced by a steel plate welded to the hull between the two hatches. Also, the automotive improvements introduced during the light tank production program were applied to the T141.

The T141 was manned by a crew of six. The driver and the vehicle commander rode in the front hull on the left and right respectively. The commander also operated the radio. The four man gun crew consisted of the gunner, sight setter, and two loaders. Seats were provided for the gunner and sight setter on the left and right sides of the gun respectively. Folding seats were installed in the rear of the mount for the loaders. On the early vehicles, the 40mm automatic dual gun M2A1 was fitted with conical flash hiders. Later,

these were replaced by three prong flash suppressors. A reinforcing ring was added to each flash suppressor at a later date. Although the radar controlled T141E1 and the fire control carrier T53 were cancelled prior to development, a radar fire control system was installed experimentally on the T141.

On 22 October 1953, OCM 35012 standardized the T141 as the twin 40mm self-propelled gun M42. The same action reclassified the twin 40mm self-propelled gun M19A1 as Limited Standard. The new vehicle was widely known as the Duster. With the later introduction of the fuel injection AOSI-895-5 engine, it was reclassified as the M42A1 and the M42 became Limited Standard.

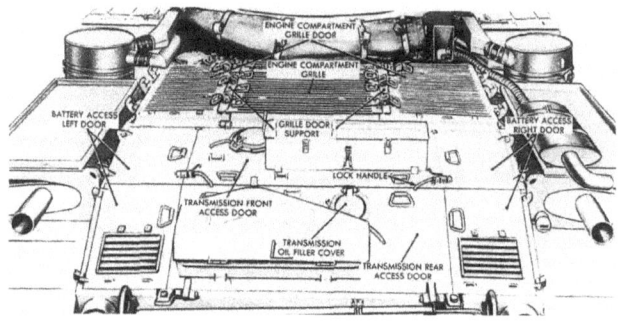

The rear deck and engine compartment covers on the M42 can be seen above at the right. Below, the components of the twin 40mm gun mount are identified.

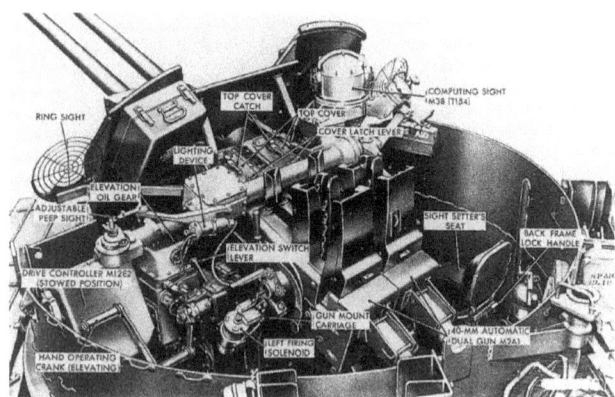
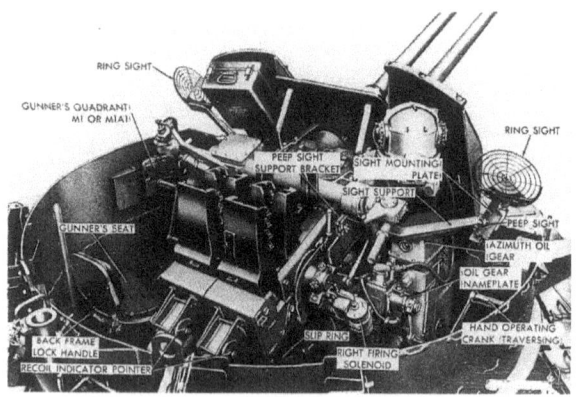

228

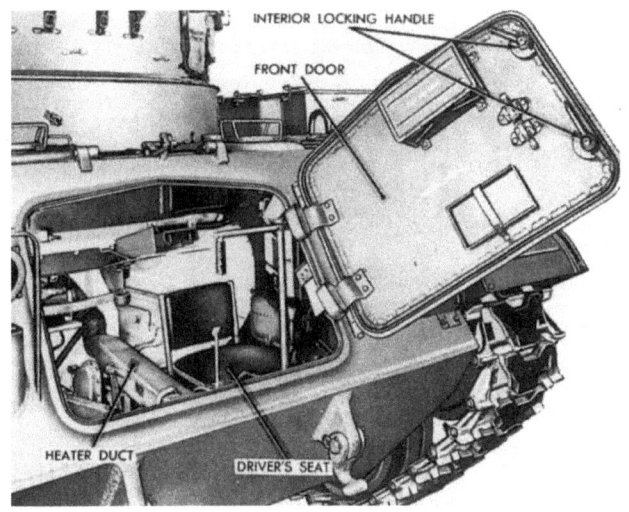

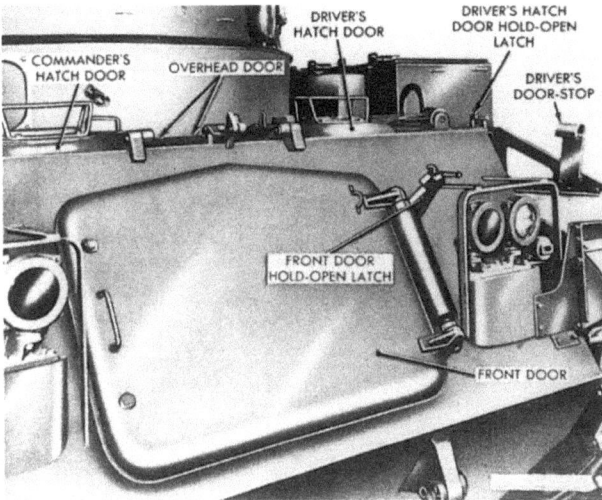

The top photographs show the front hull door open and closed. The overhead doors between the hatches for the commander and the driver can be seen in the open position below.

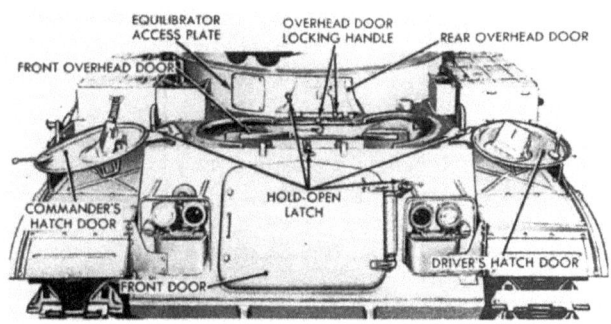

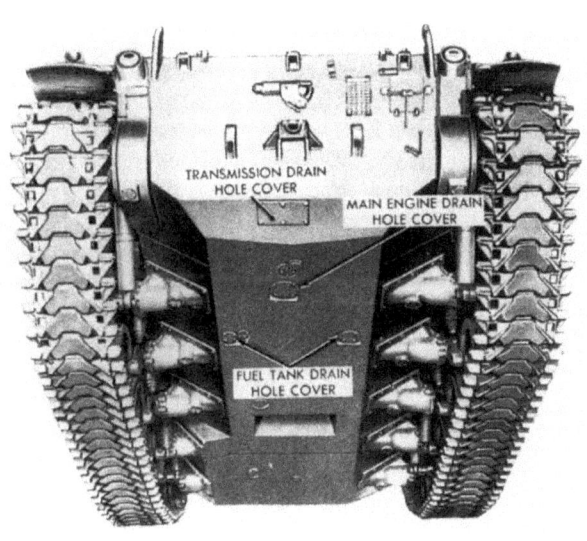

The bottom rear of the M42 hull appears at the right. Below, one of the loaders' seats is shown at the left and the .30 caliber machine gun and mount is at the right.

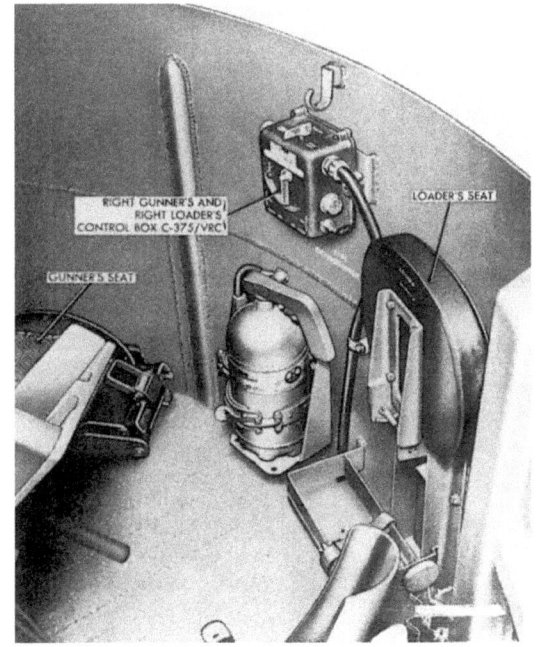

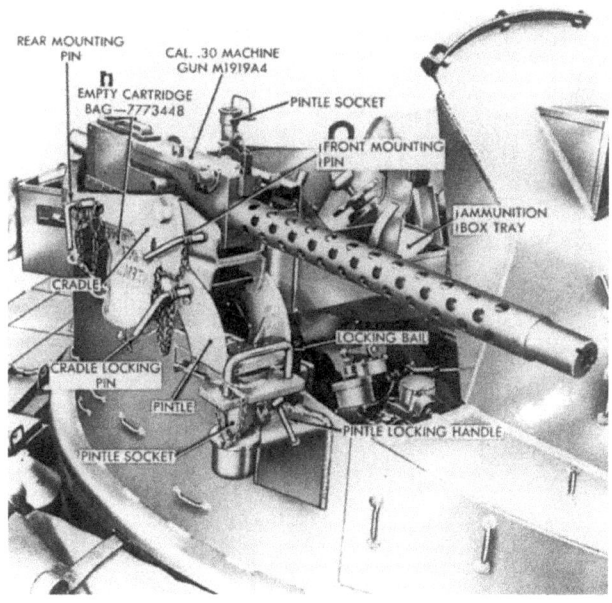

229

The twin 40mm self-propelled gun M42 above at the left still carries the T141 designation, but the 40mm guns are fitted with the later three prong flash suppressors. Above at the right, this M42 has been equipped with an experimental radar fire control system.

Production of the M42 continued at ACF Industries until December 1953 and at the Cleveland Tank Plant until June 1956. Although it was replaced in the antiaircraft role by the M163 Vulcan Air Defense System beginning in 1969, the Duster remained in service as a ground weapon in Vietnam. It also continued to serve in National Guard units and in allied armies. During this service, it was the subject of numerous modification programs. The NAPCO power pack was installed experimentally in 1982 and the Cadillac Gage gun control system was evaluated at Fort Bliss. In Taiwan, the 40mm guns were fitted with 32 round magazines and some vehicles were rearmed with TOW missiles. In Italy,

Breda Meccanica Bresciana developed a new, fully enclosed, two man turret assembled by welding armor steel plate 9mm to 16mm thick. This turret was armed with a new single barrel 40mm L/70 automatic gun with a firing rate of 450 rounds per minute. A total of 90 ready rounds were carried in the magazine and a radar fire control system was offered as an option.

As with the light tanks themselves, many concept studies were in progress on possible replacements for the antiaircraft vehicles then in service. The Questionmark II Conference produced three proposals for new self-propelled antiaircraft weapons. One of these, intended to replace the

Below is a sectional drawing of the M42 chassis fitted with the Breda Meccanica Bresciana turret armed with the single 40mm L/70 automatic gun.

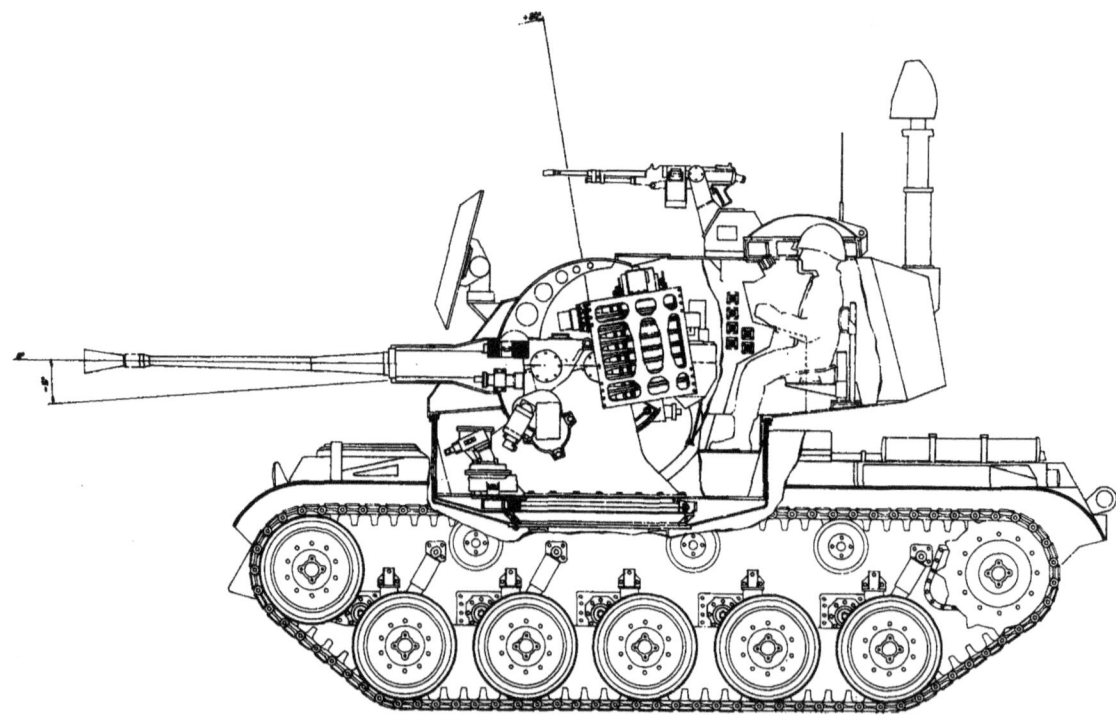

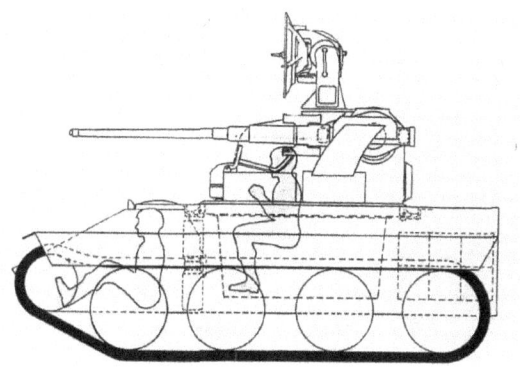

At the left above is the twin 40mm gun mount on the new light tank chassis proposed at the Questionmark II Conference. At the top right is the TS-11 concept from Questionmark III mounting twin 37mm guns on a lightweight tracked chassis.

T141, mounted the same dual 40mm gun on the chassis of the proposed new light tank. The other two were armed with the 75mm gun T83E1 Skysweeper. One of these installed the Skysweeper on a tracked platform attached to a tracked prime mover. The other mounted the T83E1 on the chassis of the 76mm gun tank T41E1.

At the Questionmark III Conference, five self-propelled antiaircraft vehicles were proposed. Two of these were intended to replace the twin 40mm self-propelled gun M42 and three were proposed to succeed the multiple .50 caliber machine gun motor carriage M16A1. One of the M42 replacements mounted a twin 40mm gun on a wheeled chassis and the other installed two radar controlled T37E2 37mm guns on a lightweight tracked chassis. As a replacement for the half-track M16A1, one proposal was armed with the M45 quad .50 caliber machine gun mount on a wheeled chassis using ¼ ton truck components and another installed four .50 caliber machine guns on a modified Ontos chassis. The third M16A1 replacement featured a six barrel 20mm T171 cannon on a modified Ontos chassis.

Questionmark IV presented seven proposals for self-propelled antiaircraft vehicles. Three were replacements for the twin 40mm gun M42 and four were intended to supercede the M16A1. Two of the M42 replacements used a 70mm rocket boosted gun with a firing rate of 650 rounds per minute. One was mounted on a wheeled chassis and the other was installed in a lightweight, armored, tracked vehicle. The third M42 replacement was armed with two 30mm automatic cannon and two .50 caliber machine guns in a coaxial mount on a lightweight, armored, tracked chassis. One M16A1 replacement carried the quad .50 caliber machine gun mount on an eight wheeled chassis using swivel frame construction between the front and rear parts. Two other proposals used four or six .50 caliber machine guns on an armored tracked chassis. The fourth concept was armed with the six barrel 20mm cannon T171E1 turret mounted on a lightweight, armored, tracked vehicle.

The Roland missile system also was installed on the M109 chassis. This short range, all weather, air defense system was evaluated under the designation XM975.

Below, the XM975 appears at the left consisting of the Roland missile system mounted on the chassis of the M109 self-propelled howitzer. An artist's concept of the weapon system in action is shown at the right.

RECOVERY, ENGINEER, AND ARTILLERY SUPPORT VEHICLES

Following the precedent set by the T6 tank recovery vehicle based upon the light tank M24, a new light recovery vehicle was designed using the chassis of the 76mm gun tank M41. It was designated as the recovery vehicle T50, but it was never standardized or placed in production.

As mentioned previously, the chassis for the T235 and T236 self-propelled gun and howitzer with the lockout system for its suspension provided an excellent platform for a lightweight wrecker or recovery vehicle. Pilots for two versions of such a vehicle were assembled. Both utilized the same basic chassis with the armored front and armor protection around the drivers compartment. Each vehicle had a crane base with a 360 degree traverse on which was installed a box boom with associated components. Both were fitted with a 30,000 pound capacity boom line winch and a 60,000 pound capacity tow winch. The vehicles were manned by a crew of three consisting of a driver, a crane operator, and a rigger. The first vehicle, designated as the T119, had no additional armor, but a light fiber glass reinforced plastic cab could be provided for installation over the controls on the right side of the boom when protection was needed from the weather. On the second vehicle, designated as the T120, the crane operator and rigger rode in a cab protected by ½ inch thick steel armor. Each of the two men had an armored cupola with six M17 periscopes. A two section door was in the rear of the cab and an additional door was in each side wall.

Above is the mock-up of the T50 tank recovery vehicle based upon the chassis of the 76mm gun tank M41.

Engineering tests of the T119 were completed at Aberdeen Proving Ground in March 1959 and the service tests were finished at Fort Sill in December. The T120 was shipped to Fort Knox for user tests on 2 March 1959. With the introduction of the General Motors 8V71T diesel engine, OCM 37167, dated 17 September 1959, assigned the designations T119E1 and T120E1 to the two vehicles when powered by the diesel engine. However, the new engine was not installed in the T119. The T120 was retrofitted with the 8V71T engine thus becoming the T120E1. After further evaluation, it was standardized as the light, full tracked, armored recovery vehicle M578.

Below, the light recovery vehicle T119 appears at the left and the T120 with the armored cab is at the right.

The light recovery vehicle M578 is shown above at Detroit Arsenal on 10 May 1965. The various features of the M578 are identified in the photographs below.

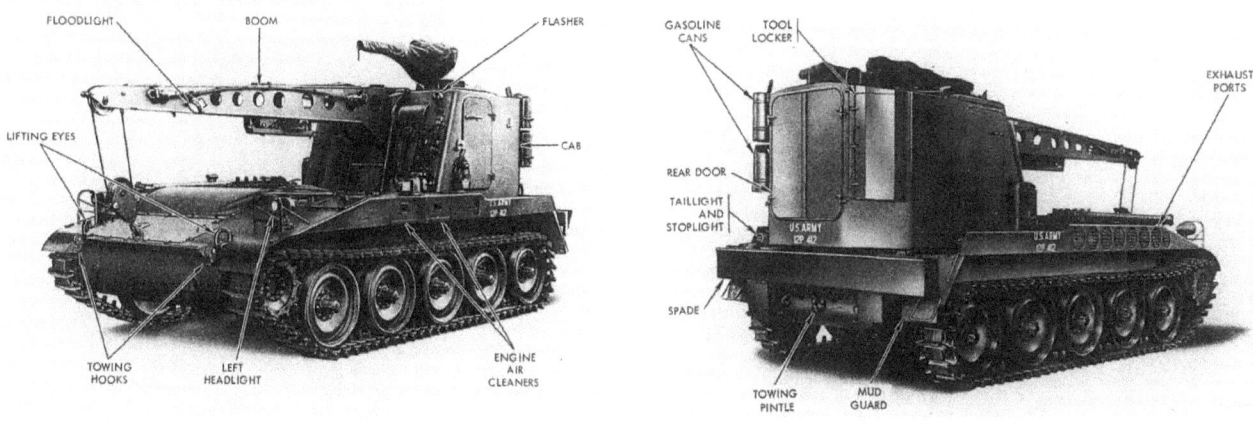

Below are two views of the M578 during its evaluation program. The only armament on the vehicle was the .50 caliber machine gun shown in the left view.

The lifting capacity of the crane on the M578 is illustrated above by the removal of the turret from an M48A2 tank. Below, the external components and stowage are identified on the M578.

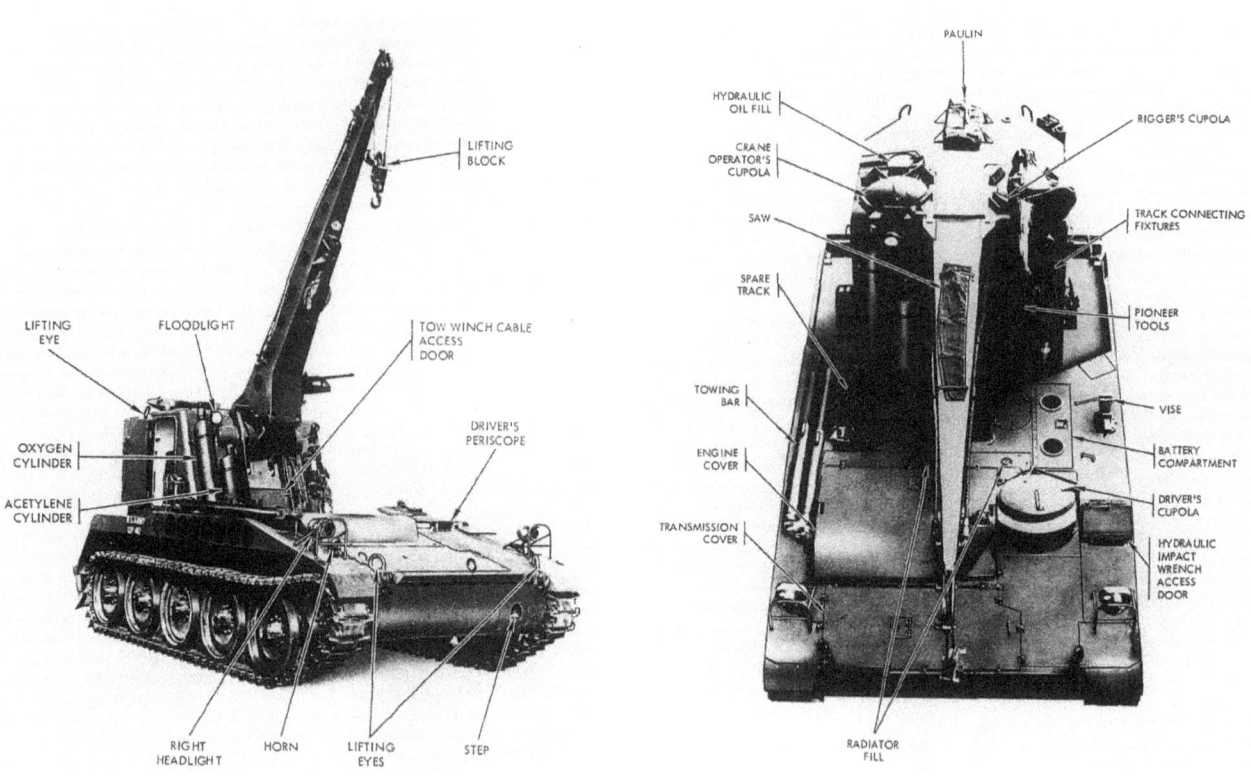

LIFTING BLOCK

LIFTING EYE

FLOODLIGHT

TOW WINCH CABLE ACCESS DOOR

DRIVER'S PERISCOPE

OXYGEN CYLINDER

ACETYLENE CYLINDER

RIGHT HEADLIGHT

HORN

LIFTING EYES

STEP

PAULIN

HYDRAULIC OIL FILL

CRANE OPERATOR'S CUPOLA

RIGGER'S CUPOLA

SAW

TRACK CONNECTING FIXTURES

SPARE TRACK

PIONEER TOOLS

TOWING BAR

VISE

ENGINE COVER

BATTERY COMPARTMENT

DRIVER'S CUPOLA

TRANSMISSION COVER

HYDRAULIC IMPACT WRENCH ACCESS DOOR

RADIATOR FILL

Above is the production version of the M9 armored combat earthmover (ACE) as manufactured by BMY. The dimensions of the ACE are shown in the drawings at the bottom of the page.

The development effort for an armored combat earthmover (ACE) extended over thirty years and was carried out under several different names. An earlier nomenclature was the universal engineer tractor (UET). The vehicle was type classified as the M9 ACE in early 1977 and Pacific Car and Foundry Company received a contract to build 15 prototypes which were completed in the early 1980s. After some additional improvements, a competitive production contract was awarded to BMY for 566 vehicles to be delivered over a period extending from fiscal year 1986 through fiscal year 1991.

The M9 was intended to provide a highly mobile, armored, earth moving capability for combat engineers supporting armored, mechanized, and light infantry units. With a weight of 36,000 pounds in the travel mode, it could be transported by air and was amphibious. The Cummins V903C diesel engine and the Clark 13.5 HR 3610-2 transmission were located in the rear along with the operator's station. The latter was protected by aluminum armor plus steel and aramid laminate plates. The operator was provided with an armored cupola with eight vision blocks. For heavy dozing operations, the eight cubic yard scraper bowl could be filled with earth increasing the weight to about 54,000 pounds. The M9's hydropneumatic suspension had four road wheel stations per side and the vehicle rode on 18 inch wide, rubber padded, tracks. A 25,000 pound winch was located in the lower rear hull. The apron and dozer blade were positioned by use of the hydropneumatic suspension system. The M9 had a maximum road speed of 30 miles per hour and a cruising range of about 200 miles. The tracks provided propulsion in water at a maximum speed of about three miles per hour.

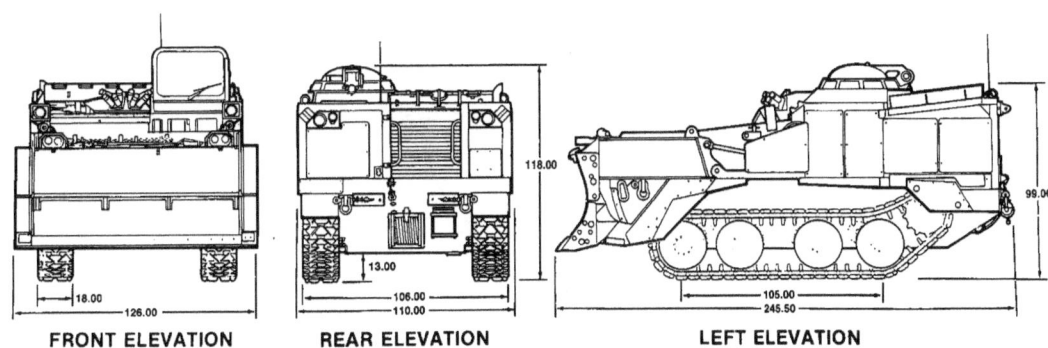

FRONT ELEVATION   REAR ELEVATION   LEFT ELEVATION

The dozer blade on the M9 ACE is shown in two raised positions above. Below, the ACE is engaged in earth moving operations.

The drawings below show the ACE in both the sprung and unsprung conditions as required for travel and earth moving operations.

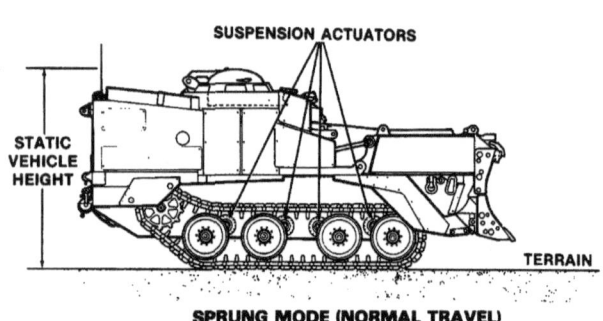

SPRUNG MODE (NORMAL TRAVEL)

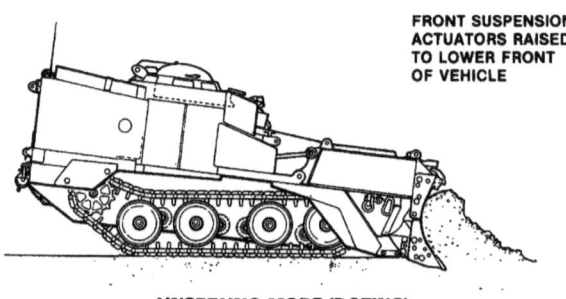

UNSPRUNG MODE (DOZING)

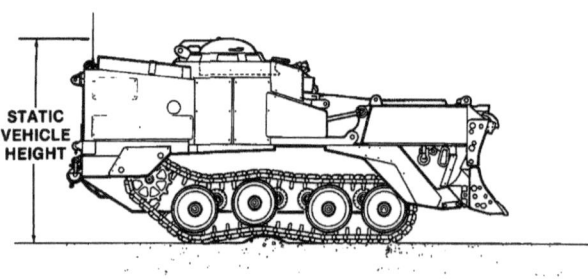

SPRUNG MODE (OBSTACLE ENCOUNTERED)

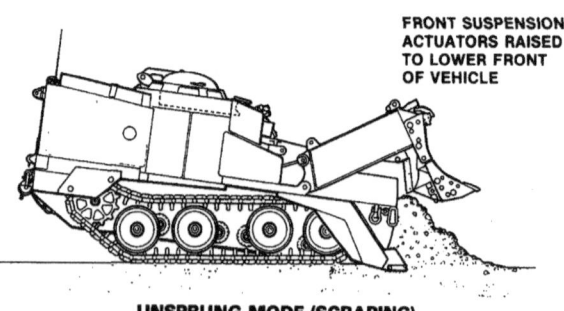

UNSPRUNG MODE (SCRAPING)

The M992 field artillery ammunition support vehicle (FAASV) appears above based upon the same chassis as the M109A2 self-propelled howitzer. The optional ammunition handling crane is not included on the front of this vehicle.

Operations with the self-propelled artillery revealed the need for an ammunition resupply vehicle with armor protection. The five ton trucks and the M548 cargo carriers then in use to support the M109 and M110 howitzers had no such protection. An armored vehicle would permit the rapid resupply of the self-propelled howitzers in the forward area allowing them to maintain continuous fire in support of the maneuver forces. After some experimentation, two versions were designed as the field artillery ammunition support vehicle (FAASV). The designations XM992 and XM1050 were assigned for vehicles to support the M109 and M110 howitzers respectively, but the XM1050 was dropped at a later date.

Based upon the M109A2 howitzer chassis, the FAASV extended the rear hull and replaced the turret with an armored cab superstructure fitted with NBC protection and a Halon automatic fire extinguisher system. It was equipped with on-board ammunition handling equipment consisting of a stacker and a conveyor. A separate crane on the front of the vehicle for loading ammunition into the FAASV was offered as an option. The mechanical stacker assisted in removing the projectiles from the honeycomb racks. The rounds were then assembled, fuzed, and transferred into the supported howitzer by the hydraulic conveyor. The armored rear door on the FAASV hinged upward providing

overhead protection between the FAASV and the howitzer during the transfer operation. The vehicle could carry a total of 90 155mm projectiles, fuzes and propellant charges.

Standardized as the M992 FAASV, 675 vehicles were produced at BMY beginning in 1984. The M992 was fielded in Europe starting in March 1986 and 50 M992s arrived in South Korea in November 1989. Initially, it was intended to provide an FAASV for each self-propelled howitzer, but this was not possible because of funding limitations.

An additional 125 FAASVs were built configured to support the M109A6 Paladin. Designated as the M992A1, they featured changes to the propellant stowage as well as the rear door and conveyor. Later, 664 M992s were upgraded to the M992A1 standard plus additional improvements to the auxiliary power unit, the radiator, and other components. These vehicles were designated as the M992A2.

The cutaway drawing at the right show the interior arrangement of the M992 FAASV. Note the ammunition conveyor in the extended position. Below are additional views of the FAASV with the various hatches open. The overhead protection provided to the conveyor by the open rear door is obvious.

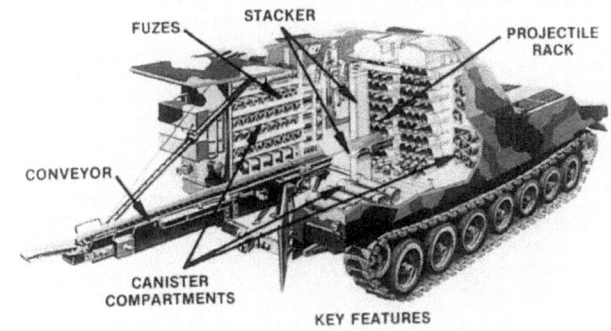

In the artist's concept above and the photograph below, the FAASV can be seen with the optional ammunition handling crane installed on the front hull.

These three photographs by Michael Green show the 155mm self-propelled howitzer M109A2 operating with the FAASV.

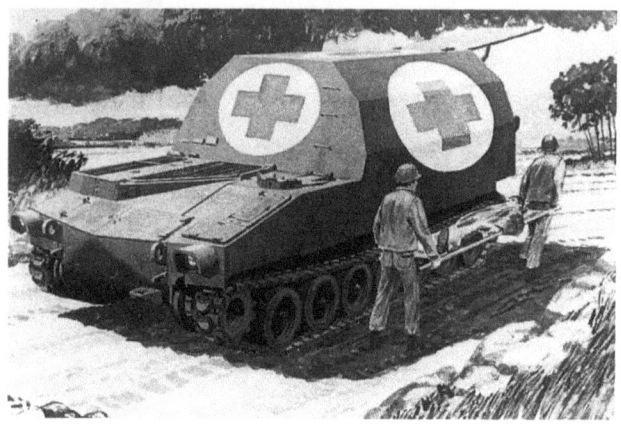

These artist's concept drawings depict the proposed medical evacuation vehicle (above), the maintenance assistance vehicle (top right), and the armored forward area rearm vehicle (right).

BMY also proposed the use of the M992 chassis as the basis for a variety of other support vehicles. These included an armored forward area rearm vehicle (AFARV) to resupply the tanks in the forward area. A medical evacuation vehicle (MEV) and a maintenance assistance vehicle (MAV) also were proposed. Other variations of the FAASV chassis under consideration were a fire direction center/command post vehicle, a fuel supply vehicle, and a forward observer vehicle.

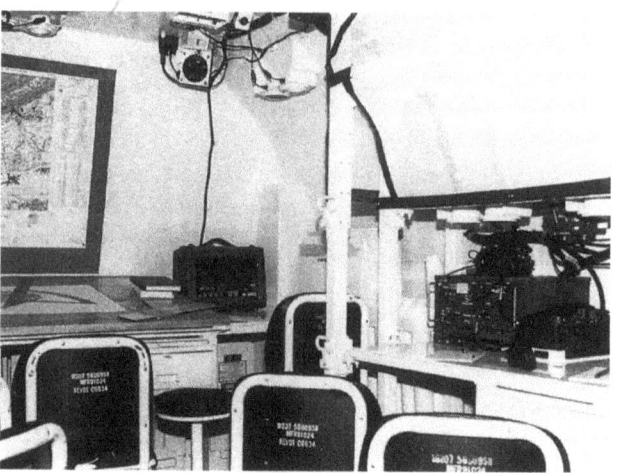

The fire direction center vehicle/command post vehicle is shown in these two photographs and the drawing below.

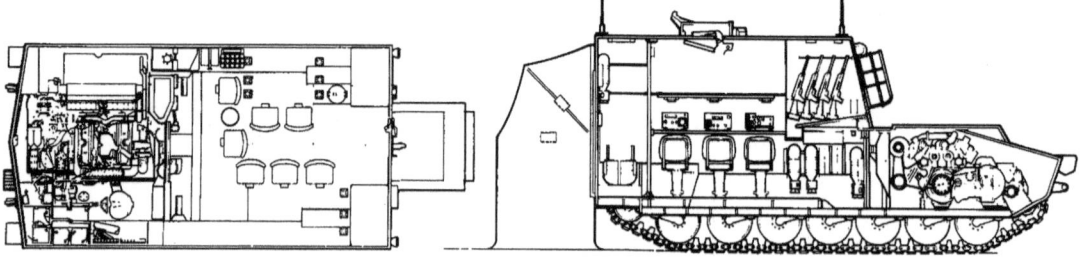

PART V

LIGHT COMBAT VEHICLES ON ACTIVE DUTY

The 76mm gun tanks M41 above are part of C Company, 82nd Reconnaissance Battalion, 2nd Armored Division in Germany during September 1953.

WORLDWIDE SERVICE

By 1953, most of the problems with the early production 76mm gun tanks M41 had been resolved and they began to replace the light tank M24 in the division reconnaissance battalions. Although most of the M41 production was allocated to the Seventh Army in Europe, some were shipped to South Korea for evaluation. By this time, the fighting there had ended and the truce was in effect.

The troops greatly appreciated the high power to weight ratio of the M41 and the resulting high speed and agility of the vehicle. For reconnaissance use, it was somewhat hampered by the loud whine of the supercharged engine and its short cruising range. The latter problem was reduced by the later introduction of the fuel injection engine in the M41A2 and M41A3. The Walker Bulldog was relatively easy

to operate and service. As one veteran armor officer remembered, it was the last tank that the troops in the operating unit could maintain without outside support.

Although it was never used in combat by American troops, the M41s provided valuable service during two decades of the Cold War. It also served in many National Guard and Reserve units. By the early 1970s, the M41 had been replaced by the Sheridan in U.S. Army units, but it continued to serve for many years with allied armies.

The M41 at the right is supporting troops of the 7th Infantry Division during maneuvers north of the Imjin River in Korea. This photograph was dated 1962.

Above, an M41 of the 759th Tank Battalion is firing on the range at the Seventh Army Training Center, Grafenwoehr, Germany on 22 July 1955. Below, this M41 is participating in training exercises in northern Italy on 22 May 1957. Note the gas masks on the troops.

The markings on the M41 above indicate that it belongs to the 107th Armored Cavalry. Below is a lineup of National Guard M41s at Camp Roberts, California.

One method of compensating for the short cruising range of the M41 is illustrated above. These fuel drums could be jettisoned in an emergency. At the left below is the Spanish M41E modification.

The M41A3 was selected to replace the light tank M24 in the Army of the Republic of Vietnam (ARVN). In addition to South Vietnam, the M41 series also was supplied to the Japanese Ground Self-Defense Force. In Spain, the M41 was modified by Talbot of Spain to use the General Motors 8V71T diesel engine with the original CD-500-3 transmission. These vehicles were referred to as the M41E. Some M41s also were rebuilt as antitank vehicles with missile armament.

Brazilian Army M41s were upgraded by the Bernardini Company in Sao Paulo, Brazil. The modifications included the installation of the Saab-Scania DS-14A04 diesel engine with the CD-500-3 transmission. Originally, the modified tank, designated as the M41B, retained the M32 76mm gun. Later versions were armed with the M32 gun bored out to accept the 90mm ammunition for the Cockerill Mk III gun. This weapon was shorter than the original M32 and was designated as the Ca 76/90 M32 BR1. The Ca 76/90 M32 BR2 was a full length version installed in the later modified tanks. These vehicles also were fitted with side skirts and spaced armor on the forward parts of the hull and turret. With an improved fire control system and the Ca 76/90 M32 BR3 gun, these tanks were designated as the M41C.

The Brazilian M41C conversion appears at the left.

The rear deck of the M41 with the NAPCO power pack can be seen above. At the right is the Danish modified M41 (upper) and the new ammunition available for the 76mm gun (lower).

NAPCO in the United States developed a power pack using the General Motors 8V71T diesel engine with the original CD-500-3 transmission. The Cummins VTA-903T diesel engine also was used with the original transmission by Levy Autoparts Company in Canada and this combination was installed in M41s modified for the Danish Army. In Denmark, 53 M41s were upgraded and designated as the M41 DK-1. In addition to the new power plant, the tanks were fitted with an NBC system in the turret bustle as well as fire detection and suppression equipment. The .50 caliber machine gun was replaced by a 7.62mm weapon and the improved fire control system included a thermal sight for the gunner and a laser range finder. New electric gun controls were installed and a halogen searchlight was mounted on the gun shield. The M32 76mm gun was retained, but it was provided with new APFSDS-T ammunition developed by AAI Corporation. Four smoke grenade launchers were installed at the front on each side of the turret and the appearance of the vehicle was altered by the new stowage bins and the skirts over the suspension. The modification of the 53 tanks was completed in 1988.

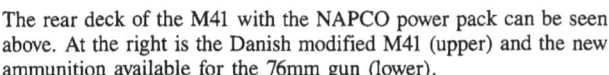

HIGH EXPLOSIVE, M352

HIGH EXPLOSIVE, MK404 IR PROXIMITY FUZE
(AIR DEFENSE)

ARMOR PIERCING, FIN STABILIZED,
DISCARDING SABOT

SMOKE, WP, M361 A1

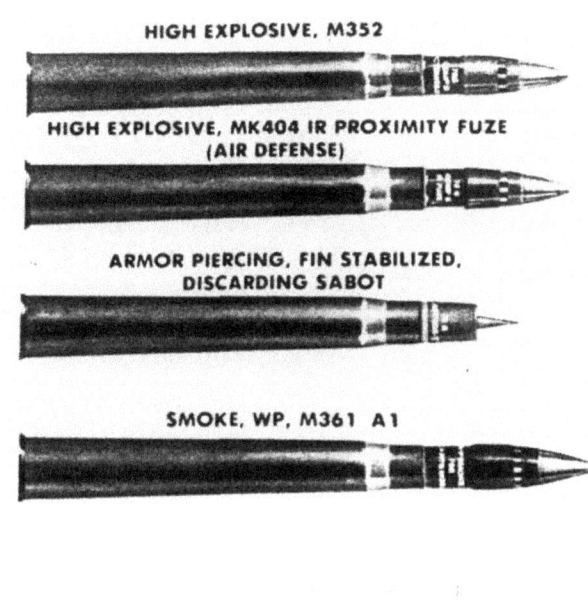

Another view of the Danish modified M41 is at the right. Below, the Belgian Cockerill 90mm Mark IV gun is installed in a new two man turret on the M41 (left) and in the standard turret (right).

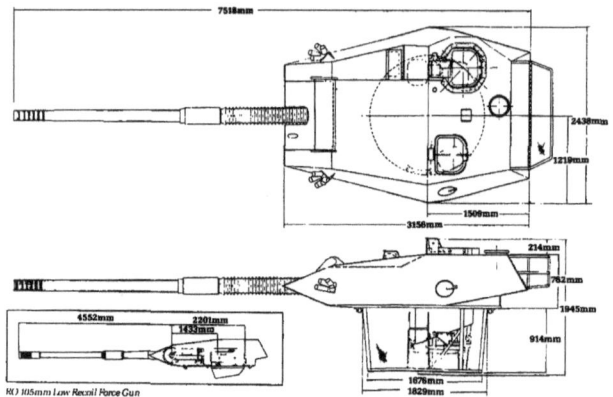

RO 105mm Low Recoil Force Gun

At the left is a drawing of the low recoil force turret developed by Royal Ordnance armed with the 105mm L7 gun. This was the turret installed on the Stingray by Cadillac Gage. Above, the low recoil force turret is mounted on the chassis of the M41 tank.

The M41 German Tank Improvement (GTI) program powered the M41 with the MTU MB 833 Aa-501 diesel engine. It also used the original transmission with an intermediate gear box. The 76mm gun was retained, but 90mm and 105mm guns were offered as options. Other improvements included new torsion bars, shock absorbers, and replaceable pad tracks. The hydraulic turret drive was replaced by an electric system including a stabilizer. A daylight and thermal periscopic sight incorporated a laser range finder and a collective NBC protection system was added to the turret bustle.

These were only a few of the upgrades applied to the M41 series. The wide range of modifications should assure the continued use of the Walker Bulldog for many years to come.

The M41 modified under the German tank improvement program appears in the view above and in the four view drawing below.

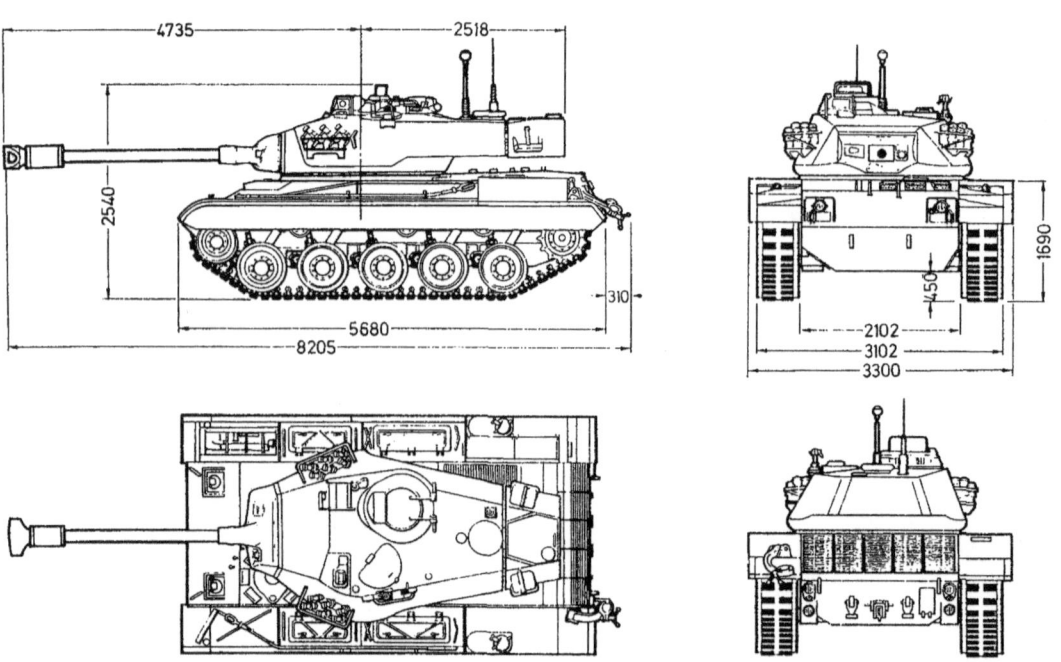

Above, the commander's shield and rear armor enclosure are mounted on the M551 at the left. Details of the .50 caliber machine gun and mount can be seen on the Sheridan at the top right during a Strike Command exercise in January 1969.

Service tests of the Sheridan revealed that serious problems still existed. However, because of its long and expensive development program, there was great pressure to field the M551 at the earliest possible date. As mentioned previously, 54 Sheridans, minus their missile systems, were shipped to Vietnam in January 1969. The operations in Vietnam will be reviewed in the following section. In April 1969, the first unit was equipped with the new vehicle in Europe and in August, the M551 was deployed to South Korea. Also during 1969, the 4th Battalion (Airborne) 68th Armor received its Sheridans becoming the first airborne tank unit in the U.S. Army. The Sheridan was to serve with the 4/68th and its successor, the 3/73rd Armor, in the 82nd Airborne Division for more than 25 years.

Above, an M551 of the 4th Squadron, 9th Cavalry climbs out of Kyle Lake at Fort Campbell, Kentucky on 23 May 1968. Below, a Sheridan swims past a Coast Guard vessel near Fort Meade, Maryland on 4 December 1968.

The Sheridan above is near Schweinfurt, Germany on 17 December 1969. Below, M551s of the 4th Armored Division move out to the range at the Seventh Army Training Center, Grafenwoehr, Germany.

Above, combustible case ammunition is being loaded into this M551 at Grafenwoehr on 15 September 1970. Note the ballistic protective covers placed on the round prior to stowing on board the vehicle. Below is a 4th Armored Division Sheridan during the cold German winter.

At the left, the external telephone is being checked on this Sheridan of the 4th Battalion, 68th Armor, 82nd Airborne Division at Fort Bragg, North Carolina on 5th February 1973. Above, two M551s of the 82nd Airborne are at Fort Stewart, Georgia during January 1972.

At the right above, an M551 of the 11th Armored Cavalry Regiment is on border patrol duty between East and West Germany during May 1979. Below, a Sheridan of the 1st Infantry Division climbs out of the Main River in Germany during Exercise Certain Thrust on 19 October 1970.

These two photographs taken by R. P. Vaughan show Sheridans of E Troop, 2/2 Armored Cavalry Regiment in Germany during 1974. The commander's armor shields are clearly visible without the .50 caliber machine gun installed.

Turret details of the M551 can be seen in these two photographs by R. P. Vaughan. Both Sheridans belong to the 2nd Armored Cavalry Regiment in Germany. The top and bottom views were dated 1974 and 1976 respectively.

Sheridans of the 82nd Airborne Division appear above and below. The top photograph by R. P. Vaughan was dated 1980. The view below was taken during the deployment to Honduras in March 1988.

The Sheridan replaced the M41 in the divisional cavalry units as well as in the armored cavalry regiments. As described earlier, with the installation of the Hughes AN/VVG-1 laser range finder, the Sheridan was redesignated as the AR/AAV M551A1. This range finder was retrofitted to many of the Sheridans in the field as well as those being rebuilt.

Because of its complexity and high maintenance requirements, the Sheridan was never popular with the troops, although these problems were less severe in Europe compared to arctic or tropical service. As a result, the Army in late 1978 directed that the Sheridan be withdrawn from all units in the active army except for the 82nd Airborne Division. Here, the Sheridan filled a requirement for which no other vehicle existed. The light weight permitted rapid deployment by air to any point in the world and its ability to be dropped by parachute or landed by low altitude parachute extraction was unique for a vehicle with its firepower and level of protection.

Below at the left, an M551 is being refueled at Fort Irwin, California prior to its deployment during Exercise Gallant Eagle on 1 April 1982. At the bottom right, a Sheridan has just been offloaded from a C130 Hercules aircraft during Exercise Ocean Venture in April-May 1984.

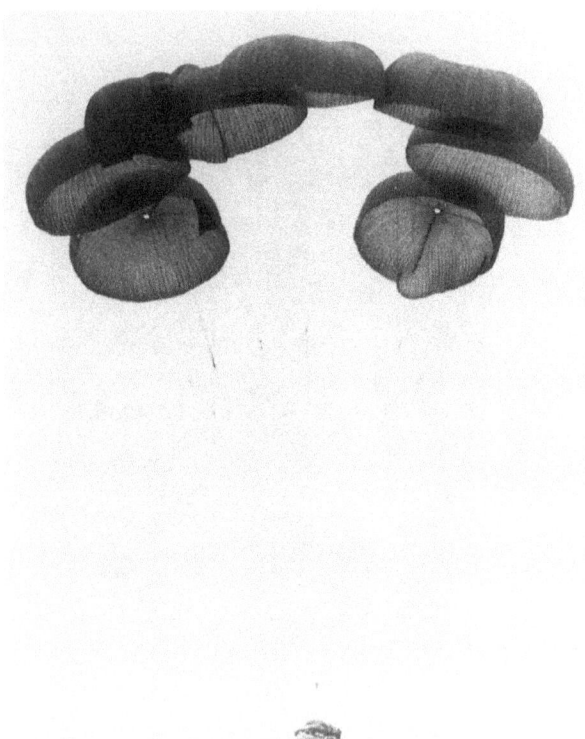

At the left, a Sheridan is suspended from its eight parachutes during a low velocity air drop. Below, the vehicle is on the ground with the parachutes still descending. Both photographs were provided by Colonel Dana Dillon.

The photograph below and those on the following page show a Sheridan of the 3rd Platoon, C Company, 3/73rd Armor after an air drop in Puerto Rico during Exercise Ocean Venture. The crew is preparing the vehicle for action after the drop.

Above is a lineup of M41A3 76mm gun tanks in use by the Army of the Republic of Vietnam (ARVN). These photographs and those on the three following pages were taken by James Loop in August 1968 during his service in Vietnam.

VIETNAM

Although it was not employed in Vietnam by the U.S. Army, the 76mm gun tank M41A3 was for a long period the primary tank in the Army of the Republic of Vietnam (ARVN). It provided excellent service in this role which continued until the ARVN received the M48A3 in the latter days of the war.

Note the armor shield added for the tank commander on the ARVN M41A3 below.

Details of the M41A3 76mm gun tanks supplied to South Vietnam can be seen in these photographs. Below, one of the vehicles is undergoing maintenance while in a defilade firing position.

These ARVN M41A3s are shown in defensive positions. Note the two types of armor shields provided for the tank commander.

Rear turret details (above) and the searchlight installation (below) can be seen on these ARVN M41A3s photographed by James Loop.

Above, B Troop, 3rd Squadron, 4th Cavalry are test firing the weapons on their new Sheridans at Range #3 on the outskirts of their base at Cu Chi, Vietnam. This photograph was dated 1 February 1969.

As early as 1966, there were proposals by the Army Staff in Washington to deploy the Sheridan to Vietnam. However, they were rejected since there was no requirement for the missile armament in Vietnam and satisfactory conventional ammunition for the 152mm gun-launcher was not yet available. By 1968, the problems with the combustible case 152mm ammunition had been partially solved and two cavalry squadrons in Vietnam were scheduled to receive the Sheridan. Originally, the two units selected were the 1st and 3rd Squadrons of the 4th Cavalry which were assigned to

the 1st and 25th Infantry Divisions respectively. However, after a visit to Vietnam by General Creighton Abrams in late 1968, the plan was modified to equip a divisional cavalry squadron and a squadron from an armored cavalry regiment with the new vehicle. Thus, the 3rd Squadron, 4th Cavalry and the 1st Squadron, 11th Armored Cavalry Regiment were selected. In the 3/4th Cavalry, the platoons received Sheridans in exchange for their M48A3 tanks on a one for one basis, but in the 11th ACR, three Sheridans replaced every two armored cavalry assault vehicles (ACAVs). The

Below, James Loop photographed this Sheridan of the 11th Armored Cavalry Regiment in Vietnam during February 1969.

At the right is an M551 of the 3/4th Cavalry in Vietnam on 12 February 1969. Note the wire mesh on the front of the vehicle used to intercept rocket propelled grenades.

ACAVs had less armor protection than the Sheridan and only carried machine gun armament. Needless to say, the 11th ACR viewed the change in a much more favorable light than the 3/4th Cavalry. The latter unit's opinion of the Sheridan received a further blow during its first action on 15 February 1969. One vehicle detonated a 25 pound pressure sensitive mine which ruptured the hull and ignited the combustible case 152mm ammunition. The driver was killed in the explosion. The fact that a similar explosion under an M48A3 would have resulted only in the loss of a road wheel or two did not help the morale of the Sheridan crews. Fortunately, subsequent fighting showed the effectiveness of the M551. Descriptions of two of these actions are quoted here from the Department of the Army Study "Mounted Combat in Vietnam" by General Donn A. Starry.

"After the mine incident, the effectiveness of the Sheridan was continually suspect in the 4th Cavalry. Then, on 10 March 1969, at a night bivouac at a road junction east of Tay Ninh City, a Troop A listening post reported enemy movement and the troop went to full alert. Sheridan crews used night observation devices to scan the battlefield. Observing a large group of advancing North Vietnamese, the Sheridans fired canister into the enemy ranks. Confused by the overwhelming volume of fire, the North Vietnamese broke and ran. The next morning more than forty enemy dead, including a battalion commander and a company commander, were found on the battlefield. Reports of this action quickly spread through the squadron, restoring some measures of confidence in the Sheridan.

In contrast, the 11th Armored Cavalry's first combat with the Sheridan was successful. In early February 1969, anticipating an enemy offensive, the regiment's 1st Squadron moved to Bien Hoa as a reaction force. Task Force Privette, commanded by Major William C. Privette, the squadron executive officer, included Troops A and B of the 1st Squadron. After an enemy mortar and rocket attack on 23 February, Task Force Privette moved out on an armored sweep and immediately encountered an enemy force. Placing the Sheridans on line, the two cavalry troops moved forward, firing canister into the enemy ranks. In the face of this firepower, the Viet Cong panicked and fled, leaving behind over eighty dead. This fight demonstrated the devastating effect of the 152mm canister round. The troops were impressed with the Sheridan's firepower as compared with that of the armored cavalry assault vehicle."

Below, Sheridans of C Troop, 3/4th Cavalry are operating with the 25th Infantry Division in Vietnam on 22 February 1969.

Above, the 3rd Platoon, C Troop, 3/4th Cavalry secures their Sheridans for the night. Note the fencing installed to intercept any rocket propelled grenades. Below at the right, A Troop, 3/4th Cavalry moves across a rice paddy during a search and destroy mission in Vietnam on 8 July 1969.

After the evaluation period, additional Sheridans were shipped to Vietnam and the various cavalry units had received over 200 by late 1970.

The vulnerability of the Sheridan to mine blast was greatly reduced with the installation of the belly armor kit provided for the later vehicles. However, the danger remained from the rocket propelled grenades (RPGs) which also could easily ignite the combustible case ammunition. As a result, the Sheridan crews tended to quickly abandon the vehicle when it was hit. Even though the Vietnam Sheridans were not armed with missiles, the electrical fire control system often failed in the wet, high humidity, conditions that prevailed in Vietnam. Despite these problems, the Sheridan served effectively until the withdrawal of the American troops.

Below, an M551 from C Troop, 3/4th Cavalry enters a heavily mined area near Boi Loi Woods south of Fire Support Base Hampton during operations in 1969.

The Sheridans on this page are operating near Quan Loi, Vietnam on 28 December 1969. Note the gun shield from an armored cavalry assault vehicle (ACAV) installed on the M551 above.

Above, the 185th Maintenance Battalion is at work on these M551s near Quan Loi, Vietnam on 22 December 1969. Below, C Troop, 1st Squadron, 11th Armored Cavalry Regiment is operating out of Fire Support Base Dennis north of Loc Ninh, Vietnam on 22 January 1970.

Details of the turret can be seen on the M551s above. Below is a Sheridan from the 11th Armored Cavalry Regiment which has been fitted with the ACAV gun shield in addition to the standard armor protection for the tank commander.

Above is a Sheridan from A Troop 1/1st Cavalry near Tam Ky, Vietnam during March 1971. The M551 below belongs to E Troop, 17th Cavalry, 173rd Airborne Brigade. Note the exterior stowage on both of these vehicles.

Above, a Marine Corps M50A1 Ontos rolls off of a landing craft at Da Nang, Vietnam on 8 March 1965. Below at the right, this M50A1 is operating with the 1st Marine Division on 8 April 1966.

Among the first combat vehicles deployed to Vietnam by the U. S. Marine Corps in March 1965 was the M50A1 Ontos. All of the vehicles of the 3rd Antitank Battalion were in Vietnam by July and the 1st Antitank Battalion arrived in March 1966. The Ontos was effectively employed in the battle for Hue City and two platoons were engaged during the siege of Khe Sanh. However, maintenance problems and the occasional accidental discharge of the recoilless rifles resulted in the decision by the Marine Corps to phase it out of service. As early as December 1967, the 3rd Anti-tank Battalion was deactivated and the 1st Antitank Battalion was reduced to one company attached to the 1st Tank Battalion.

The Ontos below is in position to support the 1st Battalion, 12th Marines on Hill 41. This photograph was dated 30 September 1966.

Above, three M50A1s are supporting an infantry unit during Operation Mobile in the Chu Lai area of Vietnam on 28 May 1966. The Ontos at the right has its rifles fully elevated for long range fire. Below at the left, a searchlight has been installed on this M50A1 for night perimeter guard duty.

Above at the right, one of the 106mm recoilless rifles is being reloaded. Note that this had to be done from outside the vehicle. Below, these M50A1s are operating with B Company, 1st Battalion, 4th Marines during 1966.

Above, an M56 90mm self-propelled gun of D Company, 16th Armor, 173rd Airborne Brigade is in position to fire during training at Bien Hoa, Vietnam on 10 June 1965.

Another light antitank weapon that saw action during this period was the 90mm self-propelled gun M56. With no enemy tanks to engage, the Scorpion provided fire support for the airborne units.

At the right and below, these M56 self-propelled guns of the 173rd Airborne Brigade are in action against the Viet Cong during July 1966.

Above and at the right are photographs of the 105mm self-propelled howitzer M108 in Vietnam. Note the empty cartridge cases in the top view.

Both the 105mm M108 and the 155mm M109 self-propelled howitzers were used during the Vietnam fighting. The same applied to the M107 175mm self-propelled gun and the M110 8 inch self-propelled howitzer.

Below, this photograph taken by James Loop shows a 155mm self-propelled howitzer M109 in Vietnam during June 1968.

Above, a 155mm self-propelled howitzer M109 from C Battery, 2nd Battalion, 138th Artillery is in action on Hill 88 on 19 March 1969. Below, an M109 from the 5th Battalion, 4th Artillery is at Long Vei, Vietnam on 11 February 1971.

Above, a 175mm self-propelled gun M107 from A Battery, 8th Battalion, 4th Artillery awaits a fire mission at Landing Zone Elliot in Vietnam on 20 March 1969. Below, an M107 attached to the 1st Marine Division is preparing to fire in Vietnam during February 1970.

Above, maintenance is being performed on this 8 inch self-propelled howitzer M110 of D Battery, 13th Artillery about 32 kilometers northwest of Saigon, Vietnam on 8 June 1969. Below, an M110 of A Battery, 7th Battalion, 8th Artillery moves into position to support the 199th Light Infantry Brigade in the Binh Chanh district of Vietnam on 8 April 1968.

Above, this twin 40mm self-propelled gun M42 was photographed by Staff Sergeant James W. Burdick in Vietnam about 1/2 mile from the Cambodian border.

Although there was no antiaircraft role for it, the twin 40mm self-propelled gun M42A1 was effective in supporting ground troops. The Duster was extremely popular in this role.

The M42s at the right and below were photographed by James Loop in Vietnam during June 1968. Note that the flash suppressors have been removed from the 40mm guns.

The Duster above was participating in Operation Greely in Vietnam on 15 July 1967. Below, this M42 of B Battery, 60th Artillery is firing near the Cambodian border on 14 August 1967.

This M578 light recovery vehicle was photographed by James Loop in Vietnam during June 1968.

The M578 light recovery vehicle also saw its first action in Vietnam. Although considered inadequate compared to the heavier M88 recovery vehicle, it did provide useful service with the Sheridans and other light vehicles.

At the right and below are additional views of the M578 in Vietnam from James Loop. Note the shield added to the .50 caliber machine gun.

Above, an M578 moves a bale of concertina wire to the perimeter of the 11th Armored Cavalry Regiment base camp south of Xuan Loc, Vietnam on 27 December 1966. Below, James Loop photographed this M578 during February 1969. Note the two .50 caliber machine guns.

Above at the left, maintenance is being performed on the suspension of Sheridan C-31 at Tocumen Air Field, Panama during Operation Just Cause. The effect of the 152mm round on a reinforced concrete wall of the Panama Defense Force Headquarters can be seen at the top right.

PANAMA

The value of the Sheridan's unique capabilities was clearly demonstrated during operation Just Cause in Panama during December 1989. In preparation for this operation, four Sheridans of C Company, 3/73rd Armor, 82nd Airborne Division, were carried to Panama in a C5A transport aircraft arriving on 16 December. The four vehicles were concealed in a motor pool at Howard Air Force Base and prepared for combat. On 20 December, an additional eight Sheridans of C Company, 3/73rd Armor arrived by low velocity air delivery (LVAD or heavy drop). One of these was lost when it ended up suspended by its parachutes in the trees. Some of the combustible case ammunition broke up during the landing dumping loose propellant inside the vehicle. The engineers considered it to be too dangerous to attempt a recovery and it was rigged for demolition and blown in place. This was the first time that tanks had been dropped by parachute directly into combat.

The Sheridan's performance during operation Just Cause was evaluated as "absolutely critical". The 152mm HEAT round easily penetrated reinforced concrete walls six

to ten inches thick providing entry for the infantry. The Sheridans were extremely effective in neutralizing snipers that interfered with the infantry. The vehicles also drove over or eliminated road blocks consisting of rubble, concertina wire, trucks and buses. They completed 100 per cent of their assigned missions and in their final action secured the Vatican Embassy from which General Noriega later surrendered. Although the Sheridans performed successfully during operation Just Cause, the age of the vehicles and their increased maintenance requirements reinforced the need for an early replacement.

At the right is the west end of La Commandancia, headquarters of the Panama Defense Force. All of the photographs on this page were taken by Captain Kevin Hammond and furnished by Captain Scott Womack.

Above, a Sheridan of the 3rd Platoon, C Company, 3/73rd Armor stands guard during Operation Just Cause. Below, Sheridan C-23 takes up a blocking position near the Vatican Embassy prior to the surrender of Manuel Noriega.

Above, a Sheridan from the 3rd platoon, A Company, 3/73rd Armor is alongside one of the C5 aircraft used to transport the Battalion to Saudi Arabia.

WAR IN THE PERSIAN GULF

On 2 August 1990, the Iraqi armed forces drove south invading the small country of Kuwait. President Bush called for an International Force to protect neighboring Saudi Arabia and to force an Iraqi withdrawal from Kuwait. On 8 August 1990, American forces were ordered to Saudi Arabia as part of operation Desert Shield. As usual, the 82nd Airborne Division with its Sheridans were among the first to arrive. The Sheridans were to be the only American armored force on the ground until the arrival of the Marine

Corps M60A1s and the M1s of the 24th Mechanized Infantry Division in late August and early September.

As remembered by Dennis Wilson, the Command Sergeant Major of the 3/73rd Armor, a major problem during the months of August and September was the overheating of the vehicle. As a result, the Sheridans were moved to the training areas on carriers during the day and the training was carried out at night. When the weather cooled off in the Fall, the overheating problem disappeared.

Below, a lineup of Sheridans from C Company, 3/73rd Armor appears at the left and the platoon leader's M551 from the 2nd Platoon, A Company is loaded onto a transporter at the right.

Above at the left, Sheridan B-34 is being loaded onto a C130 transport during the deployment for Desert Storm in January 1991. At the top right, the 3/73rd Armor is preparing to return home after Desert Storm. The markings indicate that the Sheridans in the foreground are from B Company.

Initially during Desert Shield, B Company of the 3/73rd was deployed to protect the port of Al Jubayl until this mission was taken over by the U.S. Marines. B Company then rejoined the battalion at Champion Main (a Saudia Arabian air defense base near the town of Safwah north of Dhahran).

On 12 January 1991, Lieutenant Colonel Charles A. Donnell took command of the 3/73rd and his recollections of the battalion's operations until the end of Desert Storm are quoted below.

"When I arrived, Desert Shield was for all intents and purposes over. At the beginning of Desert Shield, the battalion had deployed almost in total (the two battalion headquarters tanks never made it over, so 56 out of the 58 Sheridans assigned were in the theater). In November and December 1990, the original Sheridans had been swapped for newly modified vehicles. This modification project, begun in concept by Colonel Dana Dillon and pushed first by Colonel Frank Hartline and later by Colonel Jim Grazioplane, resulted in the installation of the tank thermal sight (TTS) from the M60A3 tank in the Sheridan. Substantial modifications were required to both the sights and the vehicles (new "choke sight" ballistic reticles, pinning the automatic lead feature, installing a missile sight system, reinforcing the turret roof, etc.).

The new tanks were a blessing for two reasons. First the TTS was absolutely great. For the first time in the Sheridan's history, night missile fire was possible. Second, the modified tanks had been expertly tuned by the people at the Anniston Army Depot. All of the systems from automotive to missile were in excellent working order. This resulted in an exceptionally high operational readiness rate during the course of Desert Storm.

When I got to the battalion it was located with the rest of the 82nd at Champion Main. Once the air war started, we became part of the massive movement of the XVIII Airborne Corps to the west. The decision was made to air deploy 100% of our combat vehicles to the forward assembly area near Rafah, Saudi Arabia. This saved wear and tear on the vehicles, released heavy trucks for other missions, and conserved road space. We were essentially assigned six C130H aircraft that flew three crews around the clock for a total of 72 sorties from King Faad airport to Rafah airport. All in all we took 56 tanks, 14 LAV-25s, 1 LAV-R (recovery), 1 M113A2 and 1 HMMWV. The last two were my vehicles and both fit in one aircraft together. To my knowledge, this was the largest ever intratheater airlift of combat vehicles.

We arrived at Rafah and waited for the battalion's wheels. The XO and S3 led the ground deployment from Champion Main to Rafah. We linked up, sorted out personnel and equipment and began the task organization of our units. First Brigade took Charlie Company (their habitually associated "slice") and Second Brigade, who had been detached from the 82nd and attached to the French Sixth Light Armored Division, took Bravo Company. I retained Alpha and Delta Companies.

The battalion's primary mission for the initial part of the ground war was Division Reserve. We were to travel behind the First Brigade and be prepared to execute several contingency plans. Since I had been involved with Sheridans quite a lot and knew their maintenance history, I requested and received heavy equipment transports (HETs) to haul them during the attack. We backed the tanks onto the trailers and secured the tracks with sandbags. That way, if we needed to dismount quickly, we could drop the ramps and drive off. We practiced this procedure and found we could establish a tactical formation in less than five minutes. There were 32 tractors and 30 trailers (A Co. – 14 tanks/D Co. – 14 tanks/ the S2 & S3 M577/ the S4 M577 & my M113A2). The battalion Scout Platoon, mounted in LAV-25s, was to conduct a moving right flank screen.

As I understood the overall plan, the French were to begin the ground war in our sector by breaching defenses along the escarpment, then attacking and securing Main Supply Route (MSR) Texas all the way to the town of As Salman, a distance of about

At the left is the Platoon Leader's M551A1(TTS) of the 3rd Platoon, D Company, 3/73rd Armor during Desert Storm. All of the tanks have names on the cannon barrel starting with the letter D. This one is Damn Yankee. The housings for the tank thermal sight and the laser range finder are both visible with their covers closed.

60 miles, by the end of G-Day (24 February 1991). The Second Brigade was chosen to lead the attack on the escarpment and my Bravo Company was in the lead echelon. The brigade attacked a full day before G-Day, as they were supposed to, and secured lanes for the French to come through. Dust was so thick that visibility was reduced to about 100 meters at best. The rest of the 82nd followed close on the heels of the French with my Charlie Company forming the advance guard of the First Brigade who led the Division.

Bravo Company was involved in several actions on the first and second days of the attack, G-1 and G-Day. The 2nd platoon led by 2nd Lieutenant Clint Sanders was accompanied by the Company Commander Captain Tony Emmy. They were under sporadic artillery fire from the moment they crossed the border, as the learned after noticing holes in their water and oil cans (with the dust as bad as it was, Captain Emmy told me that he couldn't tell the difference between exploding artillery and dust swirls). They got into a brief firefight with Iraqi infantry and Lieutenant Sanders launched a Shillelagh missile at a bunker equipped with an anti-tank gun. To my knowledge, this is the only use of the Shillelagh in combat. He was carrying the

missile in the gun tube and didn't have time to unload and reload a more appropriate type of ammunition. Sergeant First Class Blackwood told me that they captured in excess of 400 prisoners at this location.

During this attack, one of the tanks of Bravo Company, commanded by Staff Sergeant Alvie Benskin, broke down before it reached the escarpment. There were several missed communications and, with the attack proceeding as quickly as it was, Staff Sergeant Benskin was left behind. While waiting for help that would not arrive, he diagnosed the problem as a disintegrated generator gear drive. With only on board tools and no wrecker or other lift support, Benskin and his crew repaired the gear drive and reinstalled it through the small inspection holes in the bottom of the hull. They then set out on their own across the escarpment into Iraq to rejoin their platoon.

The battalion itself (called Task Force 3/73) departed its assembly area on time on G-Day and began the trip up MSR Texas into Iraq. As it turned out, the French did not capture As Salman the first night and Task Force 3/73 was forced to spend the night, still mounted on the HETs, on the side of MSR Texas. After we had set up for the night, we saw, and the scout platoon reported, a Sheridan driving alongside the road heading north. Command Sergeant Major Wilson and I took a quick inventory of our tanks in the Task Force and found that all were accounted for. As it turned out, it was Staff Sergeant Benskin. We grabbed him and told him to stay with us until we could link him up with his company. I did not want him driving by the French at night with the unfamiliar Sheridan.

The MSR remained quiet the first night. It was during this period that the Task Force captured its first prisoners. Lieutenant Hyatt, the Scout Platoon leader, had been leapfrogging his LAVs forward to screen the right flank of the Task Force and the main body of the engineer battalion that followed in the convoy. He saw several personnel running through the dust and captured four Iraqi soldiers. Military police support was available to take these prisoners to the rear and Command Sergeant Major Wilson and I were amused at the surprise on the faces of the Iraqis when they were handed over to a pretty female MP with 'Barbie of Arabia' stenciled on her helmet camouflage band.

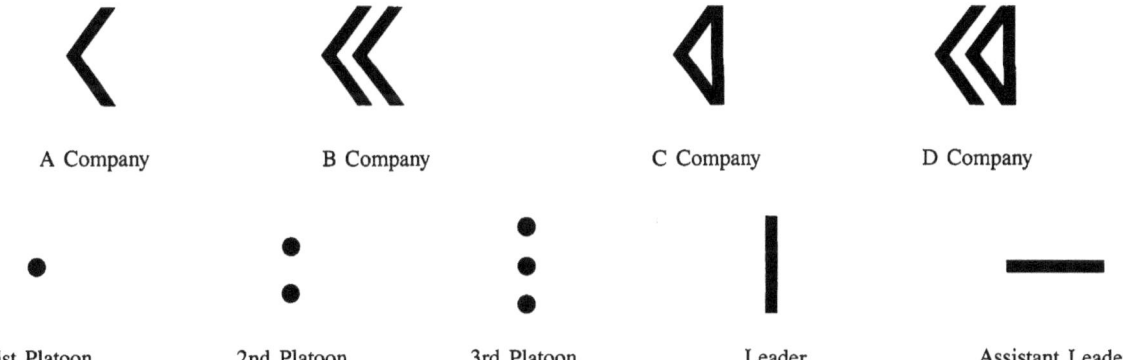

A Company B Company C Company D Company

1st Platoon 2nd Platoon 3rd Platoon Leader Assistant Leader

The symbols shown above were combined to identify the various vehicles in the 3/73rd Armor during Desert Storm. The battalion headquarters tanks did not carry the company chevrons. They were marked only with the vertical or horizontal bar to identify the Battalion Commander or the Operations Officer (S3) respectively. However, the battalion headquarters Sheridans were not deployed to the Persian Gulf. This system of markings differed slightly from an earlier version used by the 3/73rd. Examples of the various combined symbols are illustrated below. These are, from left to right: Platoon Sergeant, 2nd Platoon, A Company; tank D12 or D13, 1st Platoon, D Company; Battalion Operations Officer; Company Commander, B Company.

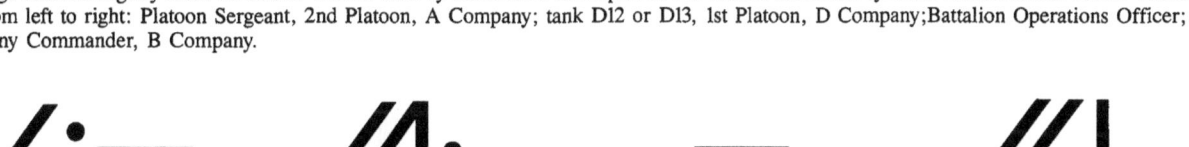

At the right is a lineup of Sheridans from D Company, 3/73rd Armor. The markings indicate that the end vehicle, named Deathstalker, belongs to the Platoon Sergeant of the 3rd Platoon.

Bravo Company was reattached on the second night. At this time the Division was directed to move east down MSR Virginia and then north to take up blocking positions along the Baghdad to Basra highway. Just after we turned off of MSR Virginia and onto MSR Yankee North, I dismounted the tanks. From this point until we were redeployed to Saudi Arabia, the tanks were on their tracks.

The only action involving any battalion element after we got to the highway was the Third Brigade attack on the Talil Air Force Base. The base had been overrun by the 24th Mechanized Infantry Division, but there were some holdouts. Alpha Company was requested to show them that we meant business. It worked and the air base was captured.

The tanks had held up very well during the whole attack. Other than Staff Sergeant Benskin's tank, which he repaired himself, all had made it to the Euphrates River without difficulty."

Lieutenant Colonel Donnell was fortunate in having excellent crew maintenance for his Sheridans. When the 3/73rd deployed to Rafah, the vehicles were exposed to very fine dust that clogged the air filters and could penetrate the lubricant seals. However, frequent cleaning of the engine air filters and lubrication of the suspension system components prevented any serious problems.

The 3rd Platoon Leader's tank, Damn Yankee, from D Company appears at the right.

Above is a closeup view of an M551A1 during Desert Shield. At the top right is an M551A1(TTS) during Desert Storm showing the tank thermal sight with the housing cover open. The tranceiver cover on the laser range finder is open in both views.

The M578 light recovery vehicle also served in the desert. It was among the first vehicles to arrive by sea with the 24th Mechanized Infantry Division. The late version of the 155mm self-propelled howitzer series was deployed in support of the maneuver forces and were in action during Desert Storm. The M9 ACE small engineer vehicle also served making its combat debut in the Persian Gulf.

At the right, the M9 ACE is operating during Desert Shield. Below are two views of the 155mm self-propelled howitzer M109A2 during the same operation.

The 155mm self-propelled howitzer M109A2 above and the M578 light recovery vehicle below were photographed during the preparations for Desert Storm. The M578 is towing a cargo carrier.

The Sheridan's successor, the XM8 armored gun system, is shown above during its evaluation program. In the view at the right, the 105mm gun is firing.

THE FUTURE

The critical contribution of the Sheridan in Panama and its subsequent deployment to the Persian Gulf emphasized the urgent need for an early replacement for the vehicle which had been in service for a quarter century. As described before, numerous attempts had been made to develop such a replacement vehicle. Several versions, ranging from the rapid deployment force light tank to the armored gun system, had been studied over a period of years, but none had been placed in production. The operations of the Sheridan in Panama and the Gulf apparently supplied the impetus required to fund the procurement of a replacement vehicle. The development of the XM8 by FMC Corporation (now United Defense) will provide a vehicle to meet the future requirements of the airborne forces as well as other light units. Although not officially designated as a light tank, the XM8 is exactly that to the airborne forces which require mobile protected firepower that can be deployed by air to any point in the world in a few hours time. The new vehicle also could serve as the basis of a new family of light-weight combat vehicles.

Below, the armored gun system provides the chassis for the prototype of another weapon system. These photographs show the line of sight anti-tank (LOSAT) weapon system built by Loral Vought Systems Corporation mounted on the chassis of the CCVL. The LOSAT carries 12 kinetic energy missiles in two six packs. The long rod penetrators in these missiles reach a velocity of approximately 5000 feet per second to penetrate heavy armor.

PART VI

REFERENCE DATA

The markings on the M551A1(TTS) above indicate that it belongs to the Platoon Sergeant, 3rd Platoon, D Company, 3/73rd Armor during Operation Desert Storm.

This Third Army Staff Sergeant stands in front of his Walker Bulldog 76mm gun tank.

Above is a lineup of late production M41 series 76mm gun tanks.

The 76mm gun tank T92 appears below in its original configuration.

These M551A1 Sheridans of the 3/73rd Armor were photographed in Saudi Arabia in 1990 during Operation Desert Shield. The original green camouflage has been smeared with mud to make the vehicles less obvious in the desert terrain.

The newly arrived M551A1(TTS) Sheridans on this page have the desert sand color scheme in preparation for Operation Desert Storm. The unit markings of the 3/73rd Armor have not yet been added. Ammunition and other stowage for the Sheridan are illustrated below.

Above is a lineup of M109A2/3 155mm self-propelled howitzers preparing for action in the Persian Gulf. Below, a Marine Corps M50 Ontos displays its armament of six 106mm recoilless rifles.

The lightest version of the XM8 105mm armored gun system can be seen above with level 1 protection. Additional armor has been added to the vehicle below to provide level 2 protection.

Passive armor boxes have been installed on the hull and turret of the 105mm armored gun system XM8 above to give it level 3 protection.

298

VEHICLE DATA SHEETS

All of the production light tanks in the U.S. Army since World War II are described in the data sheets of this section. In addition, data sheets are included for some of the experimental tanks and many of the self-propelled artillery vehicles based on lightweight chassis.

Whenever they were available, the original arsenal drawings provided the vehicle dimensions. Other source documents were the characteristic sheets, notes on materiel, and technical manuals for the appropriate vehicles. In the case of many experimental vehicles, information was obtained from the test reports issued at Fort Knox or Aberdeen Proving Ground. Some dimensions such as ground clearance or fire height would vary with the suspension spring compression resulting from the load on the vehicle. In this case, the design reference values are quoted to permit comparison between the various vehicles.

Some of the terms may require clarification. The fire height is defined as the distance from the ground to the centerline of the main weapon bore at zero elevation. The ground contact length at zero penetration is the distance between the centers of the front and rear road wheels. This value is used to calculate the ground contact area and the ground pressure of the vehicle. The combat weight of the vehicle is used in the latter calculation. This combat weight includes the crew with a full load of fuel and ammunition. If available, the exact weight of an experimental vehicle is listed. However, in some cases only approximate weights could be obtained. For production vehicles, the average weight is often rounded off to the nearest 1000 pounds. When available, the maximum values are quoted for the gross and net engine horsepower and torque. The gross horsepower and torque are the values obtained with only those accessories essential to engine operation without the effect of items such as air cleaners or generators. The net values reflect the operation of the engine as installed in the vehicle with all of its accessories. The power to weight ratios were calculated using the combat weight. The terms left and right are from the perspective of someone seated in the vehicle driver's seat.

During the operational life of the vehicle, the stowage arrangements were frequently changed. In that case, the stowage specified when the vehicle was new or during its period of greatest use is listed. Some items also may have been omitted because of security restrictions.

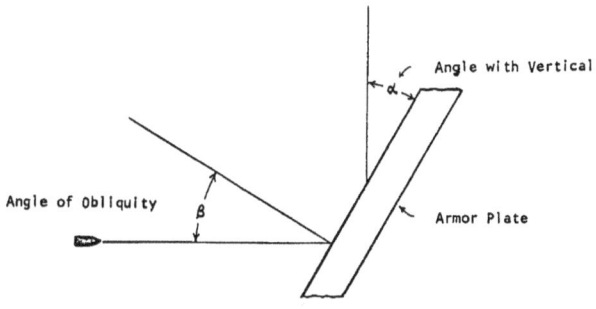

Security considerations also limit the information available on certain vehicles. This particularly applies to the use of composite special armor. On the early vehicles, the armor is specified by type, thickness, and angle with the vertical. This angle is measured between a vertical plane and the armor plate surface as indicated by the angle alpha in the sketch. Note also in this two dimensional drawing that the angle beta is the angle of obliquity. The latter is defined as the angle between a line perpendicular to the armor plate and the path of a projectile impacting the plate.

GENERAL DATA

Crew:	4	men
Length: Gun forward	292.1	inches
Length: Gun in travel position	249.3	inches
Length: Without gun	222.1	inches
Gun Overhang: Gun forward	70	inches
Width: Over fenders	127	inches
Height: Over cupola	102	inches
Tread:	101.75	inches
Ground Clearance:	17.5	inches
Fire Height:	approx. 75	inches
Turret Ring Diameter: (inside)	69	inches
Weight, Combat Loaded:	48,280	pounds
Weight, Unstowed:	42,680	pounds
Power to Weight Ratio: Net	16.2	hp/ton
Gross	20.7	hp/ton
Ground Pressure: Zero penetration	9.4	psi

ARMOR

Type: Turret, rolled and cast homogeneous steel; Hull, rolled and cast homogeneous steel; Welded assembly

Hull Thickness:	Actual	Angle w/Vertical
Front, Upper	1.0 inches (25mm)	60 degrees
Lower	1.25 inches (32mm)	45 degrees
Sides, Front	1.0 inches (25mm)	12 degrees
Rear	0.75 inches (19mm)	12 degrees
Rear, Upper	0.75 inches (19mm)	55 degrees
Lower	0.75 inches (19mm)	40 degrees
Top	0.5 inches (13mm)	90 degrees
Floor, Front	1.25 inches (32mm)	90 degrees
Rear	0.375 inches (10mm)	90 degrees

Turret Thickness:		
Gun Shield	1.25-1.0 inches (32-25mm)	60 degrees
Front	1.25 inches (32mm)	55 degrees
Sides	1.0 inches (25mm)	10 degrees
Rear	1.0 inches (25mm)	0 degrees
Top	0.5 inches (13mm)	90 degrees

ARMAMENT

Primary: 76mm Gun T94 in Mount T137 in turret

Traverse: Electric-hydraulic and manual	360 degrees
Traverse Rate: (max) 12 seconds/360 degrees	
Elevation: Electric-hydraulic and manual	+20 to −9 degrees
Elevation Rate: (max)	6 degrees/second
Firing Rate: (max)	12 rounds/minute
Loading System:	Manual
Stabilizer System:	None

Secondary:
(1) .50 caliber MG HB M2 flexible AA mount on turret
(1) .50 caliber MG HB M2 coaxial w/76mm gun in turret
(2) .30 caliber MG M1919A4 in turret blisters
Provision for (4) .45 caliber SMG M3

AMMUNITION

60 rounds 76mm
1980 rounds .50 caliber
900 rounds .45 caliber
3750 rounds .30 caliber

FIRE CONTROL AND VISION EQUIPMENT

Primary Weapon:	Direct	Indirect
	Range Finder T37 (stereo)	Azimuth Indicator
		Elevation Quadrant M9
		Gunner's Quadrant M1A1
Blister Machine Guns:	Periscope T32	
Vision Devices:	Direct	Indirect
Driver	Hatch	Periscope M17 (4)
Commander	Vision blocks (6)	Periscope M15 (1)
	in cupola, hatch	
Gunner	None	Periscope T32 (1)
Loader	Hatch and pistol port	Periscope T32 (1)

Total Periscopes: M15 (1), M17 (4), T32 (2)
Total Pistol Ports: Turret (1)
Total Vision Blocks: (6) in cupola on turret top

ENGINE

Make and Model: Continental AOS-895-1	
Type: 6 cylinder, 4 cycle, opposed, supercharged	
Cooling System: Air Ignition: Magneto	
Displacement:	895.9 cubic inches
Bore and Stroke:	5.75 x 5.75 inches
Compression Ratio:	5.5:1
Net Horsepower: (max)	390 hp at 2800 rpm
Gross Horsepower: (max)	500 hp at 2800 rpm
Net Torque: (max)	800 ft-lb at 2100 rpm
Gross Torque: (max)	945 ft-lb at 2400 rpm
Weight:	1660 pounds, dry
Fuel: 80 octane gasoline	143 gallons
Engine Oil:	58 quarts

POWER TRAIN

Transmission: Cross-drive CD-500-1, 2 ranges forward, 1 reverse
    Single stage hydraulic torque converter
    Stall multiplication: 4:1
    Overall Usable Ratios: low 14.7:1   reverse 14.7:1
                              high 3.9:1
Steering Control: Mechanical, wobble stick
    Steering Rate: 6.8 rpm
Brakes: Multiple disc
Final Drive: Spur gear   Gear Ratio: 3.769:1
Drive Sprocket: At rear of vehicle with 12 teeth
    Pitch Diameter: 23.182 inches

RUNNING GEAR

Suspension: Torsion bar
    10 individually sprung dual road wheels (5/track)
    Tire Size: 25.5 x 4.5 inches
    6 dual track return rollers (3/track)
    Dual compensating idler at front of each track
    Idler Size: 22.5 x 4.5 inches, steel, no tire
    Shock absorbers fitted on first 2 and last 2 road wheels on each side
    Track tension idler installed between last road wheel and sprocket
Tracks: Center guide T91
    Type: (T91) Single pin, 21 inch width, steel
    Pitch: 6 inches
    Shoes per Vehicle: 150 (75/track)
    Ground Contact Length: 122 inches

ELECTRICAL SYSTEM

Nominal Voltage: 24 volts DC
Main Generator: (2) 24 volts, 150 amperes, in parallel driven by main engine
Auxiliary Generator: None
Battery: (4) 12 volts, 2 sets of 2 in series, 1 set per generator

COMMUNICATIONS

Radio: SCR 508, SCR 528, AN/GRC-3, or AN/GRC-4 in turret bustle
Interphone: 4 stations plus external extension kit AN/VIA-1

FIRE PROTECTION

(2) 10 pound carbon dioxide, fixed
(1) 5 pound carbon dioxide, portable

PERFORMANCE

Maximum Speed: Level road	41 miles/hour
Maximum Tractive Effort: TE at stall	45,000 pounds
Per Cent of Vehicle Weight: TE/W	93 per cent
Maximum Grade:	60 per cent
Maximum Trench:	8 feet
Maximum Vertical Wall:	26 inches
Maximum Fording Depth:	44 inches
Minimum Turning Circle: (diameter)	pivot
Cruising Range: Roads	150 miles

GENERAL DATA

Crew:	4	men
Length: Gun forward	317.1	inches
Length: Gun in travel position	273.2	inches
Length: Without gun	222.1	inches
Gun Overhang: Gun forward	95	inches
Width: Over fenders	127	inches
Height: Over cupola	107.9	inches
Tread:	101.75	inches
Ground Clearance:	17.5	inches
Fire Height:	approx. 75	inches
Turret Ring Diameter: (inside)	69	inches
Weight, Combat Loaded:	51,600	pounds
Weight, Unstowed:	45,980	pounds
Power to Weight Ratio: Net	15.1	hp/ton
Gross	19.4	hp/ton
Ground Pressure: Zero penetration	10.1	psi

ARMOR

Type: Turret, rolled and cast homogeneous steel; Hull, rolled and cast homogeneous steel; Welded assembly

Hull Thickness:

	Actual	Angle w/Vertical
Front, Upper	1.0 inches (25mm)	60 degrees
Lower	1.25 inches (32mm)	45 degrees
Sides, Front	1.0 inches (25mm)	12 degrees
Rear	0.75 inches (19mm)	12 degrees
Rear, Upper	0.75 inches (19mm)	55 degrees
Lower	0.75 inches (19mm)	40 degrees
Top	0.5 inches (13mm)	90 degrees
Floor, Front	1.25 inches (32mm)	90 degrees
Rear	0.375 inches (10mm)	90 degrees

Turret Thickness:

Gun Shield	1.25-1.0 inches (32-25mm)	60 degrees
Front	1.25 inches (32mm)	56 degrees
Sides	1.0 inches (25mm)	10 degrees
Rear	1.0 inches (25mm)	0 degrees
Top	0.5 inches (13mm)	90 degrees

ARMAMENT

Primary: 76mm Gun T91 in Mount T138 in turret

Traverse: Electric-hydraulic and manual	360 degrees
Traverse Rate: (max)	11 seconds/360 degrees
Elevation: Electric-hydraulic and manual	+20 to −9 degrees
Elevation Rate: (max)	6 degrees/second
Firing Rate: (max)	12 rounds/minute
Loading System:	Manual
Stabilizer System:	Azimuth and elevation

Secondary:
(1) .50 caliber MG HB M2 flexible AA mount on turret
(1) .50 caliber MG HB M2 coaxial w/76mm gun in turret
(2) .30 caliber MG M1919A4 in turret blisters
Provision for (4) .45 caliber SMG M3

AMMUNITION
40 rounds 76mm
1540 rounds .50 caliber
900 rounds .45 caliber
3500 rounds .30 caliber

FIRE CONTROL AND VISION EQUIPMENT

	Direct	Indirect
Primary Weapon:	Range Finder (color coincidence)	Azimuth Indicator
		Elevation Quadrant M9
	Lead Computer	Gunner's Quadrant M1A1
Blister Machine Guns:	Periscope T32	

Vision Devices:

	Direct	Indirect
Driver	Hatch	Periscope M17 (4)
Commander	Vision blocks (6) in cupola, hatch	Periscope M15 (1)
Gunner	None	Periscope T32 (1)
Loader	Hatch and pistol port	Periscope T32 (1)

Total Periscopes: M15 (1), M17 (4), T32 (2)
Total Pistol Ports: Turret (1)
Total Vision Blocks: (6) in cupola on turret top

ENGINE

Make and Model: Continental AOS-895-1	
Type: 6 cylinder, 4 cycle, opposed, supercharged	
Cooling System: Air Ignition: Magneto	
Displacement:	895.9 cubic inches
Bore and Stroke:	5.75 x 5.75 inches
Compression Ratio:	5.5:1
Net Horsepower: (max)	390 hp at 2800 rpm
Gross Horsepower: (max)	500 hp at 2800 rpm
Net Torque: (max)	800 ft-lb at 2100 rpm
Gross Torque: (max)	945 ft-lb at 2400 rpm
Weight:	1660 pounds, dry
Fuel: 80 octane gasoline	143 gallons
Engine Oil:	58 quarts

POWER TRAIN

Transmission: Cross-drive CD-500-1, 2 ranges forward, 1 reverse
Single stage hydraulic torque converter
Stall multiplication: 4:1
Overall Usable Ratios: low 14.7:1   reverse 14.7:1
high 3.9:1

Steering Control: Mechanical, wobble stick
Steering Rate: 6.8 rpm
Brakes: Multiple disc
Final Drive: Spur gear   Gear Ratio: 3.769:1
Drive Sprocket: At rear of vehicle with 12 teeth
Pitch Diameter: 23.182 inches

RUNNING GEAR

Suspension: Torsion bar
10 individually sprung dual road wheels (5/track)
Tire Size: 25.5 x 4.5 inches
6 dual track return rollers (3/track)
Dual compensating idler at front of each track
Idler Size: 22.5 x 4.5 inches, steel, no tire
Shock absorbers fitted on first 2 and last 2 road wheels on each side

Tracks: Center guide T91
Type: (T91) Single pin, 21 inch width, steel
Pitch: 6 inches
Shoes per Vehicle: 150 (75/track)
Ground Contact Length: 122 inches

ELECTRICAL SYSTEM

Nominal Voltage: 24 volts DC
Main Generator: (2) 24 volts, 150 amperes, in parallel driven by main engine
Auxiliary Generator: None
Battery: (4) 12 volts, 2 sets of 2 in series, 1 set per generator

COMMUNICATIONS

Radio: SCR 508, SCR 528, AN/GRC-3, or AN/GRC-4 in turret bustle
Interphone: 4 stations plus external extension kit AN/VIA-1

FIRE PROTECTION

(2) 10 pound carbon dioxide, fixed
(1) 5 pound carbon dioxide, portable

PERFORMANCE

Maximum Speed: Level road	41 miles/hour
Maximum Tractive Effort: TE at stall	45,000 pounds
Per Cent of Vehicle Weight: TE/W	87 per cent
Maximum Grade:	60 per cent
Maximum Trench:	8 feet
Maximum Vertical Wall:	26 inches
Maximum Fording Depth:	44 inches
Minimum Turning Circle: (diameter)	pivot
Cruising Range: Roads	150 miles

76mm GUN TANKS M41 (T41E1) AND M41A1 (T41E2)

GENERAL DATA

Crew:	4 men
Length: Gun forward, M41 w/early muzzle brake	318.6 inches
M41A1 w/late muzzle brake	319.8 inches
Length: Gun in travel position, M41 w/early muzzle brake	274.5 inches
M41A1 w/late muzzle brake	276.8 inches
Length: Without gun	229.1 inches
Gun Overhang: Gun forward, M41 w/early muzzle brake	89.5 inches
M41A1 w/late muzzle brake	90.7 inches
Width: Over fenders	125.9 inches
Height: Over AA MG	118.8 inches
Tread:	102.5 inches
Ground Clearance:	17.5 inches
Fire Height:	approx. 75 inches
Turret Ring Diameter: (inside)	73 inches
Weight, Combat Loaded: M41	51,200 pounds
M41A1	51,800 pounds
Weight, Unstowed: M41 and M41A1	44,700 pounds
Power to Weight Ratio: Net, M41	17.4 hp/ton
M41A1	17.2 hp/ton
Gross, M41	19.5 hp/ton
M41A1	19.3 hp/ton
Ground Pressure: Zero penetration, M41	9.6 psi
M41A1	9.7 psi

ARMOR

Type: Turret, rolled and cast homogeneous steel; Hull, rolled and cast homogeneous steel; Welded assembly

Hull Thickness:	Actual	Angle w/Vertical
Front, Upper	1.0 inches (25mm)	60 degrees
Lower	1.25 inches (32mm)	45 degrees
Sides, Upper Front	1.0 inches (25mm)	0 degrees
Upper Rear	0.75 inches (19mm)	0 degrees
Lower by driver	1.0 inches (25mm)	45 degrees
Lower not by driver	0.5 inches (13mm)	60 degrees
Rear, Upper (doors)	0.5 inches (13mm)	56 degrees
Lower	0.75 inches (19mm)	40 degrees
Top	0.75 inches (19mm)	90 degrees
Floor, Front	1.5 inches (38mm)	90 degrees
Rear	0.375 inches (10mm)	90 degrees
Turret Thickness:		
Gun Shield	1.25 inches (32mm)	50 degrees
Front	1.0 inches (25mm)	18 degrees
Sides	1.0 inches (25mm)	10 and 30 degrees
Rear	1.0 inches (25mm)	0 degrees
Top, Front	0.75 inches (19mm)	73 degrees
Rear	0.5 inches (13mm)	90 degrees

ARMAMENT

Primary: 76mm Gun M32 (T91E3) in Mount M76 (T138E1) in turret (M41)
76mm Gun M32 (T91E3) in Mount M76A1 (T138E2) in turret (M41A1)

Traverse: Electric-hydraulic and manual	360 degrees
Traverse Rate: (max)	10 seconds/360 degrees
Elevation: Manual (M41)	+20 to −10 degrees
Electric-hydraulic and manual (M41A1)	+20 to −10 degrees
Elevation Rate: (max) (M41A1)	4 degrees/second
Firing Rate: (max)	12 rounds/minute
Loading System:	Manual
Stabilizer System:	None

Secondary:
(1) .50 caliber MG HB M2 flexible AA mount on turret
(1) .50 caliber MG HB M2E1 coaxial w/76mm gun in turret
or
(1) .30 caliber MG M1919A4E1 coaxial w/76mm gun in turret
Provision for (1) .45 caliber SMG M3A1
Provision for (1) .30 caliber Carbine M2

AMMUNITION

57 rounds 76mm (M41)	180 rounds .45 caliber
65 rounds 76mm (M41A1)	90 rounds .30 caliber (carbine)
600 rounds .50 caliber AA	8 hand grenades
2175 rounds .50 caliber coaxial	
or	
5225 rounds .30 caliber coaxial (M41)	
4900 rounds .30 caliber coaxial (M41A1)	

FIRE CONTROL AND VISION EQUIPMENT

Primary Weapon:	Direct	Indirect
	Periscope M20 (T35) or M20A1	Azimuth Indicator M31 (T24)
	Telescope M97 (T156)	Elevation Quadrant M9
	Ballistic Drive M4 (T23)	Gunner's Quadrant M1 or M1A1

Vision Devices:	Direct	Indirect
Driver	Hatch	Periscope M17 (4) and Periscope M19 (infrared) (1)
Commander	Vision blocks (5) in cupola, hatch	Periscope M20 (T35) or M20A1 (1)
Gunner	None	Periscope M20 (T35) or M20A1 (1)
Loader	Hatch	Periscope M13 or M13B1 (1)

Total Periscopes: M13 or M13B1 (1), M17 (4), M19 (infrared) (1), M20 (T35) or M20A1 (2)
Total Vision Blocks: (5) in cupola on turret top

ENGINE

Make and Model: Continental AOS-895-3	
Type: 6 cylinder, 4 cycle, opposed, supercharged	
Cooling System: Air Ignition: Magneto	
Displacement:	895.9 cubic inches
Bore and Stroke:	5.75 x 5.75 inches
Compression Ratio:	5.5:1
Net Horsepower: (max)	440 hp at 2400 rpm
Gross Horsepower: (max)	500 hp at 2800 rpm
Net Torque: (max)	900 ft-lb at 2100 rpm
Gross Torque: (max)	960 ft-lb at 2400 rpm
Weight:	approx. 1900 pounds, dry
Fuel: 80 octane gasoline	140 gallons
Engine Oil:	44 quarts

POWER TRAIN

Transmission: Cross-drive CD-500-3, 2 ranges forward, 1 reverse w/automatic lock-up in high
Single stage hydraulic torque converter
Stall Multiplication: 4:1

Overall Usable Ratios:	low 14.7:1	direct	1:1
	high 3.9:1	reverse	14.7:1

Steering Control: Mechanical, T-bar
Steering Rate: 6.8 rpm
Brakes: Multiple disc
Final Drive: Spur gear   Gear Ratio: 4.25:1
Drive Sprocket: At rear of vehicle with 12 teeth
Pitch Diameter: 23.422 inches

RUNNING GEAR

Suspension: Torsion bar
10 individually sprung dual road wheels (5/track)
Tire Size: 25.5 x 4.5 inches
6 dual track return rollers (3/track)
Dual Compensating idler at front of each track
Idler Size: 22.5 x 4.5 inches, steel, no tire (early)
Idler Tire Size: 25.5 x 4.5 inches (late)
Shock absorbers fitted on first 2 and last road wheels on each side
Tracks: Center guide T91E3
Type: (T91E3) Single pin, 21 inch width, steel w/detachable rubber pad
Pitch: 6 inches
Shoes per Vehicle: 150 (75/track)
Ground Contact Length: 127 inches

ELECTRICAL SYSTEM

Nominal Voltage: 24 volts DC
Main Generator: (1) 24 volts, 150 amperes, driven by main engine
Auxiliary Generator: (1) 24 volts, 300 amperes, driven by auxiliary engine
Battery: (4) 12 volts, 2 sets of 2 in series connected in parallel

COMMUNICATIONS

Radio: AN/GRC-3 thru 8 series in turret bustle
Interphone: 4 stations plus external extension kit AN/VIA-1

FIRE PROTECTION

(2) 10 pound carbon dioxide, fixed
(1) 5 pound carbon dioxide, portable

PERFORMANCE

Maximum Speed: Level road	45 miles/hour
Maximum Tractive Effort: TE at stall	44,000 pounds
Per Cent of Vehicle Weight: TE/W (M41)	86 per cent
(M41A1)	85 per cent
Maximum Grade:	60 per cent
Maximum Trench:	6 feet
Maximum Vertical Wall:	28 inches
Maximum Fording Depth:	48 inches
Minimum Turning Circle: (diameter)	pivot
Cruising Range: Roads	approx. 100 miles

76mm GUN TANKS M41A2 AND M41A3

GENERAL DATA

Crew:	4 men
Length: Gun forward	319.8 inches
Length: Gun in travel position	276.8 inches
Length: Without gun	229.1 inches
Gun Overhang: Gun forward	90.7 inches
Width: Over fenders	125.9 inches
Height: Over AA MG	118.8 inches
Tread:	102.5 inches
Ground Clearance:	17.5 inches
Fire Height:	approx. 75 inches
Turret Ring Diameter: (inside)	73 inches
Weight, Combat Loaded: M41A2	51,200 pounds
M41A3	51,800 pounds
Weight, Unstowed: M41A2 and M41A3	44,700 pounds
Power to Weight Ratio: Net, M41A2	16.3 hp/ton
M41A3	16.1 hp/ton
Gross, M41A2	20.5 hp/ton
M41A3	20.3 hp/ton
Ground Pressure: Zero penetration, M41A2	9.6 psi
M41A3	9.7 psi

ARMOR

Type: Turret, rolled and cast homogeneous steel; Hull, rolled and cast homogeneous steel; Welded assembly

Hull Thickness:	Actual	Angle w/Vertical
Front, Upper	1.0 inches (25mm)	60 degrees
Lower	1.25 inches (32mm)	45 degrees
Sides, Upper Front	1.0 inches (25mm)	0 degrees
Upper Rear	0.75 inches (19mm)	0 degrees
Lower by driver	1.0 inches (25mm)	45 degrees
Lower not by driver	0.5 inches (13mm)	60 degrees
Rear, Upper (doors)	0.5 inches (13mm)	56 degrees
Lower	0.75 inches (19mm)	40 degrees
Top	0.75 inches (19mm)	90 degrees
Floor, Front	1.5 inches (38mm)	90 degrees
Rear	0.375 inches (10mm)	90 degrees
Turret Thickness:		
Gun Shield	1.25 inches (32mm)	50 degrees
Front	1.0 inches (25mm)	18 degrees
Sides	1.0 inches (25mm)	10 and 30 degrees
Rear	1.0 inches (25mm)	0 degrees
Top, Front	0.75 inches (19mm)	73 degrees
Rear	0.5 inches (13mm)	90 degrees

ARMAMENT

Primary: 76mm Gun M32 in Mount M76 in turret (M41A2)
76mm Gun M32 in Mount M76A1 in turret (M41A3)

Traverse: Electric-hydraulic and manual	360 degrees
Traverse Rate: (max)	10 seconds/360 degrees
Elevation: Manual (M41A2)	+20 to −10 degrees
Electric-hydraulic and manual (M41A3)	+20 to −10 degrees
Elevation Rate: (max) (M41A3)	4 degrees/second
Firing Rate: (max)	12 rounds/minute
Loading System:	Manual
Stabilizer System:	None

Secondary:
(1) .50 caliber MG HB M2 flexible AA mount on turret
(1) .30 caliber MG M1919A4E1 coaxial w/76mm gun in turret
Provision for (1) .45 caliber SMG M3A1
Provision for (1) .30 caliber Carbine M2

AMMUNITION

57 rounds 76mm (M41A2)	180 rounds .45 caliber
65 rounds 76mm (M41A3)	90 rounds .30 caliber (carbine)
600 rounds .50 caliber	8 hand grenades
5225 rounds .30 caliber (M41A2)	
4900 rounds .30 caliber (M41A3)	

FIRE CONTROL AND VISION EQUIPMENT

		Direct	Indirect
Primary Weapon:		Periscope M20 or M20A1	Azimuth Indicator M31
		Telescope M97	Elevation Quadrant M9
		Ballistic Drive M4	Gunner's Quadrant M1A1
Vision Devices:		Direct	Indirect
Driver		Hatch	Periscope M17 (4) and Periscope M19 (infrared) (1)
Commander		Vision blocks (5) in cupola, hatch	Periscope M20 or M20A1 (1)
Gunner		None	Periscope M20 or M20A1 (1)
Loader		Hatch	Periscope M13 or M13B1 (1)

Total Periscopes: M13 or M13B1 (1), M17 (4), M19 (infrared) (1), M20 or M20A1 (2)
Total Vision Blocks: (5) in cupola on turret top

ENGINE

Make and Model: Continental AOSI-895-5
Type: 6 cylinder, 4 cycle, opposed, supercharged, fuel injection
Cooling System: Air   Ignition: Magneto

Displacement:	895.9 cubic inches
Bore and Stroke:	5.75 x 5.75 inches
Compression Ratio:	5.5:1
Net Horsepower: (max)	440 hp at 2400 rpm
Gross Horsepower: (max)	500 hp at 2800 rpm
Net Torque: (max)	900 ft-lb at 2100 rpm
Gross Torque: (max)	960 ft-lb at 2400 rpm
Weight:	approx. 1900 pounds, dry
Fuel: 80 octane gasoline	140 gallons
Engine Oil:	44 quarts

POWER TRAIN

Transmission: Cross-drive CD-500-3, 2 ranges forward, 1 reverse w/automatic lock-up in high
Single stage hydraulic torque converter
Stall Multiplication: 4:1
Overall Usable Ratios: low 14.7:1   direct 1:1
high 3.9:1   reverse 14.7:1
Steering Control: Mechanical, T-bar
Steering Rate: 6.8 rpm
Brakes: Multiple disc
Final Drive: Spur gear   Gear Ratio: 4.25:1
Drive Sprocket: At rear of vehicle with 12 teeth
Pitch Diameter: 23.422 inches

RUNNING GEAR

Suspension: Torsion bar
10 individually sprung dual road wheels (5/track)
Tire Size: 25.5 x 4.5 inches
6 dual track return rollers (3/track)
Dual compensating idler at front of each track
Idler Size: 22.5 x 4.5 inches, steel, no tire (early)
Idler Tire Size: 25.5 x 4.5 inches (late)
Shock absorbers fitted on first 2 and last road wheels on each side
Tracks: Center guide T91E3
Type: (T91E3) Single pin, 21 inch width, steel w/detachable rubber pad
Pitch: 6 inches
Shoes per Vehicle: 150 (75/track)
Ground Contact Length: 127 inches

ELECTRICAL SYSTEM

Nominal Voltage: 24 volts DC
Main Generator: (1) 24 volts, 150 amperes, driven by main engine
Auxiliary Generator: (1) 24 volts, 300 amperes, driven by auxiliary engine
Battery: (4) 12 volts, 2 sets of 2 in series connected in parallel

COMMUNICATIONS

Radio: AN/GRC-3 thru 8 series in turret bustle
Interphone: 4 stations plus external extension kit AN/VIA-1

FIRE PROTECTION

(2) 10 pound carbon dioxide, fixed
(1) 5 pound carbon dioxide, portable

PERFORMANCE

Maximum Speed: Level road	45 miles/hour
Maximum Tractive Effort: TE at stall	44,000 pounds
Per Cent of Vehicle Weight: TE/W (M41A2)	86 per cent
(M41A3)	85 per cent
Maximum Grade:	60 per cent
Maximum Trench:	6 feet
Maximum Vertical Wall:	28 inches
Maximum Fording Depth:	48 inches
Minimum Turning Circle: (diameter)	pivot
Cruising Range: Roads	approx. 110 miles
w/jettison tanks	approx. 280 miles

90mm GUN TANK T49

GENERAL DATA

Crew:	4 men
Length: Gun forward	313.0 inches
Length: Gun in travel position	274.5 inches
Length: Without gun	223.4 inches
Gun Overhang: Gun forward	89.6 inches
Width: Over fenders	128.9 inches
Height: Over AA MG	127.3 inches
Tread:	102.5 inches
Ground Clearance:	17.5 inches
Fire Height:	approx. 75 inches
Turret Ring Diameter: (inside)	73 inches
Weight, Combat Loaded:	53,200 pounds
Weight, Unstowed:	46,650 pounds
Power to Weight Ratio: Net	16.8 hp/ton
Gross	18.8 hp/ton
Ground Pressure: Zero penetration	10.0 psi

ARMOR

Type: Turret, rolled and cast homogeneous steel; Hull, rolled and cast homogeneous steel; Welded assembly

Hull Thickness:	Actual	Angle w/Vertical
Front, Upper	1.0 inches (25mm)	60 degrees
Lower	1.25 inches (32mm)	45 degrees
Sides, Upper Front	1.0 inches (25mm)	0 degrees
Upper Rear	0.75 inches (19mm)	0 degrees
Lower by driver	1.0 inches (25mm)	45 degrees
Lower not by driver	0.5 inches (13mm)	60 degrees
Rear, Upper (doors)	0.5 inches (13mm)	56 degrees
Lower	0.75 inches (19mm)	40 degrees
Top	0.75 inches (19mm)	90 degrees
Floor, Front	1.5 inches (38mm)	90 degrees
Rear	0.375 inches (10mm)	90 degrees
Turret Thickness:		
Gun Shield	1.25 inches (32mm)	50 degrees
Front	1.0 inches (25mm)	18 degrees
Sides, Upper 6½ inches	1.0 inches (25mm)	0 degrees
Lower	1.0 inches (25mm)	10 and 30 degrees
Rear	1.0 inches (25mm)	0 degrees
Top, Front	0.75 inches (19mm)	73 degrees
Rear	0.5 inches (13mm)	90 degrees

ARMAMENT

Primary: 90mm Gun T132E3 in Mount T145 in turret

Traverse: Amplidyne and manual	360 degrees
Traverse Rate: (max)	13 seconds/360 degrees
Elevation: Amplidyne and manual	+19.5 to −9.5 degrees
Elevation Rate: (max)	4 degrees/second
Firing Rate: (max)	10 rounds/minute
Loading System:	Manual
Stabilizer System:	None

Secondary:
(1) .50 caliber MG HB M2 flexible AA mount on turret
(1) .30 caliber MG M1919A4E1 coaxial w/90mm gun in turret
Provision for (1) .45 caliber SMG M3A1
Provision for (1) .30 caliber Carbine M2

AMMUNITION

46 rounds 90mm	180 rounds .45 caliber
600 rounds .50 caliber	90 rounds .30 caliber (carbine)
6225 rounds .30 caliber	8 hand grenades

FIRE CONTROL AND VISION EQUIPMENT

Primary Weapon:	Direct	Indirect
	Range Finder T41E3	Azimuth Indicator M31
	Periscope M20	Elevation Quadrant M13
	Telescope T156E1	Gunner's Quadrant M1A1
	Ballistic Computer T23E3	

Vision Devices:	Direct	Indirect
Driver	Hatch	Periscope M17 (4) and Periscope M19 (infrared) (1)
Commander	Vision blocks (5) in cupola, hatch	Periscope M20 (1)
Gunner	None	Periscope M20 (1)
Loader	Hatch	Periscope M13 (1)

Total Periscopes: M13 (1), M17 (4), M19 (infrared) (1), M20 (2)
Total Vision Blocks: (5) in cupola on turret top

ENGINE

Make and Model: Continental AOS-895-3	
Type: 6 cylinder, 4 cycle, opposed, supercharged	
Cooling System: Air Ignition: Magneto	
Displacement:	895.9 cubic inches
Bore and Stroke:	5.75 x 5.75 inches
Compression Ratio:	5.5:1
Net Horsepower: (max)	440 hp at 2400 rpm
Gross Horsepower: (max)	500 hp at 2800 rpm
Net Torque: (max)	900 ft-lb at 2100 rpm
Gross Torque: (max)	960 ft-lb at 2400 rpm
Weight:	approx. 1900 pounds, dry
Fuel: 80 octane gasoline	140 gallons
Engine Oil:	44 quarts

POWER TRAIN

Transmission: Cross-drive CD-500-3, 2 ranges forward, 1 reverse w/automatic lock-up in high
Single stage hydraulic torque converter
Stall Multiplication: 4:1

Overall Usable Ratios:	low 14.7:1	direct 1:1
	high 3.9:1	reverse 14.7:1

Steering Control: Mechanical, T-bar
Steering Rate: 6.8 rpm
Brakes: Multiple disc
Final Drive: Spur gear   Gear Ratio: 4.25:1
Drive Sprocket: At rear of vehicle with 12 teeth
Pitch Diameter: 23.422 inches

RUNNING GEAR

Suspension: Torsion bar
10 individually sprung dual road wheels (5/track)
Tire Size: 25.5 x 4.5 inches
6 dual track return rollers (3/track)
Dual compensating idler at front of each track
Idler Size: 22.5 x 4.5 inches, steel, no tire
Shock absorbers fitted on first 2 and last road wheels on each side
Tracks: Center guide T91E3
Type: (T91E3) Single pin, 21 inch width, steel w/detachable rubber pad
Pitch: 6 inches
Shoes per Vehicle: 150 (75/track)
Ground Contact Length: 127 inches

ELECTRICAL SYSTEM

Nominal Voltage: 24 volts DC
Main Generator: (1) 24 volts, 150 amperes, driven by main engine
Auxiliary Generator: (1) 24 volts, 300 amperes, driven by auxiliary engine
Battery: (4) 12 volts, 2 sets of 2 in series connected in parallel

COMMUNICATIONS

Radio: AN/GRC-3 thru 8 series in turret bustle
Interphone: 4 stations plus external extension kit AN/VIA-1

FIRE PROTECTION

(2) 10 pound carbon dioxide, fixed
(1) 5 pound carbon dioxide, portable

PERFORMANCE

Maximum Speed: Level road	45 miles/hour
Maximum Tractive Effort: TE at stall	44,000 pounds
Per Cent of Vehicle Weight: TE/W	83 per cent
Maximum Grade:	60 per cent
Maximum Trench:	6 feet
Maximum Vertical Wall:	28 inches
Maximum Fording Depth:	48 inches
Minimum Turning Circle: (diameter)	pivot
Cruising Range: Roads	approx. 100 miles

GENERAL DATA

Crew:	4	men
Length: Gun forward	271.0	inches
Length: Gun to rear	283.5	inches
Length: Without gun	182.5	inches
Gun Overhang: Gun forward	88.5	inches
Width: Over tracks	109.75	inches
Height: Over cupola	98.75	inches
Tread:	85.75	inches
Ground Clearance:	17.5	inches
Fire Height:	approx. 69	inches
Turret Ring Diameter: (inside)	73.75	inches
Weight, Combat Loaded:	37,400	pounds
Weight, Unstowed:	33,150	pounds
Power to Weight Ratio: Net	15.8	hp/ton
Gross	18.2	hp/ton
Ground Pressure: Zero penetration	11.7	psi

ARMOR

Type: Turret, rolled and cast homogeneous steel; Hull, rolled and cast homogeneous steel; Welded assembly

Hull Thickness:	Actual	Angle w/Vertical
Front, Upper	1.0 inches (25mm)	60 degrees
Lower	1.0 inches (25mm)	40 degrees
Sides	0.875 inches (22mm)	0 degrees
Rear, Upper	0.75 inches (19mm)	15 degrees
Lower	0.75 inches (19mm)	45 degrees
Top	0.5 inches (13mm)	90 degrees
Floor, Front	1.0 inches (25mm)	90 degrees
Rear	0.375 inches (10mm)	90 degrees
Turret Thickness:		
Gun Shield	1.0 inches (25mm)	60 degrees
Front	0.875 inches (22mm)	60 degrees
Sides	0.875 inches (22mm)	7 and 28 degrees
Rear	0.875 inches (22mm)	15 degrees
Top, Front	0.5 inches (13mm)	75 degrees
Rear	0.5 inches (13mm)	90 degrees

ARMAMENT

Primary: 76mm Gun T185 in Mount T138E2 in turret

Traverse: Electric-hydraulic and manual	360 degrees
Traverse Rate: (max)	15 seconds/360 degrees
Elevation: Manual	+20 to −10 degrees
Firing Rate: (max)	12 rounds/minute
Loading System:	Manual
Stabilizer System:	None

Secondary:
(1) .50 caliber MG HB M2 in cupola on turret
(1) .30 caliber MG M1919A4E1 coaxial w/76mm gun in turret
(1) .30 caliber MG M1919A4 on turret roof
Provision for (1) .45 caliber SMG M3A1
Provision for (1) .30 caliber Carbine M2

AMMUNITION

60 rounds 76mm	180 rounds .45 caliber
600 rounds .50 caliber	90 rounds .30 caliber (carbine)
5000 rounds .30 caliber	8 hand grenades

FIRE CONTROL AND VISION EQUIPMENT

Primary Weapon:	Direct	Indirect
	Periscopic sight	Azimuth Indicator
	Telescope	Elevation Quadrant M9
		Gunner's Quadrant M1A1
Vision Devices:	Direct	Indirect
Driver	Hatch	Periscope M17 (4) and
		Periscope M19 (infrared) (1)
Commander	Vision blocks (6)	None
	in cupola, hatch	
Gunner	None	Periscopic sight (1)
Loader	Vision blocks (1)	None
	hatch	

Total Periscopes: M17 (4), M19 (infrared) (1), periscopic sight (1)
Total Vision Blocks: (7)

ENGINE

Make and Model: Continental AOI-628-1
Type: 8 cylinder, 4 cycle, opposed, fuel injection
Cooling System: Air   Ignition: Magneto

Displacement:	628.3 cubic inches
Bore and Stroke:	5 x 4 inches
Compression Ratio:	6.7:1
Net Horsepower: (max)	280 hp at 3200 rpm
Gross Horsepower: (max)	340 hp at 3200 rpm
Net Torque: (max)	503 ft-lb at 2500 rpm
Gross Torque: (max)	587 ft-lb at 2500 rpm
Weight:	1098 pounds, dry
Fuel: 80-86 octane gasoline	150 gallons
Engine Oil:	16 quarts

POWER TRAIN

Transmission: XT-300, 3 ranges forward, 1 reverse
    Single stage hydraulic torque converter
    Stall Multiplication: 3.8:1

Overall Usable Ratios:	low 19.65:1	direct 1.21:1
	high 4.71:1	reverse 22:1

Steering Control: Clutch-brake, control handles
Brakes: Multiple disc
Final Drive: Spur gear   Gear Ratio: 5.08:1
Drive Sprocket: At front of vehicle with 12 teeth
    Pitch Diameter: 21.492 inches

RUNNING GEAR

Suspension: Flat track, torsion bar
    8 individually sprung dual road wheels (4/track)
    Tire Size: 34 x 5 inches
    Rear road wheel serves as trailing idler
    Shock absorbers fitted on first and last road wheels on each side
Tracks: Center guide
    Type: Single pin, 14 inch width
    Pitch: 5.5 inches
    Shoes per Vehicle: 130 (65/track)
    Ground Contact Length: 114 inches

ELECTRICAL SYSTEM

Nominal Voltage: 24 volts DC
Main Generator: (1) 24 volts, 100 amperes, driven by main engine
Auxiliary Generator: None
Battery: (2) 12 volts, in series

COMMUNICATIONS

Radio: AN/GRC-3 thru 8 series on turret floor
Interphone: AN/UIC-1, 3 stations plus external head set w/plug connection

FIRE PROTECTION

(2) 10 pound carbon dioxide, fixed
(1) 5 pound carbon dioxide, portable

PERFORMANCE

Maximum Speed: Level road	35 miles/hour
Maximum Tractive Effort: TE at stall	42,300 pounds
Per Cent of Vehicle Weight: TE/W	113 per cent
Maximum Grade:	60 per cent
Maximum Trench:	6 feet
Maximum Vertical Wall:	36 inches
Maximum Fording Depth:	48 inches
Minimum Turning Circle: (diameter)	pivot
Cruising Range: Roads	approx. 165 miles

76mm GUN TANK T92

GENERAL DATA

Crew:	4 men
Length: Gun forward	247.50 inches
Length: Gun to rear	300.75 inches
Length: Without gun	189.75 inches
Gun Overhang: Gun forward	57.75 inches
Width: Over tracks	124.0 inches
Height: Over periscopes	89.1 inches
Tread:	108.0 inches
Ground Clearance:	17.0 inches
Fire Height:	67.4 inches
Turret Ring Diameter: (inside)	89 inches
Weight, Combat Loaded:	37,160 pounds
Weight, Unstowed:	33,204 pounds
Power to Weight Ratio: Net	15.1 hp/ton
Gross	18.3 hp/ton
Ground Pressure: Zero penetration, T110 track	9.6 psi
T85E1 track	10.9 psi

ARMOR

Type: Turret, rolled and cast homogeneous steel and cast aluminum; Hull, rolled and cast homogeneous steel and rolled aluminum; Welded assembly

Hull Thickness:	Actual	Angle w/Vertical
Front, Upper	0.5 inches (13mm)	65 to 83 degrees
Doors (aluminum)	1.0 inches (25mm)	65 degrees
Lower (inner)	0.5 inches (13mm)	35 degrees
Lower (outer)	0.5 inches (13mm)	50 degrees
Right Side, by final drive	0.375 inches (10mm)	0 degrees
by engine	0.75 inches 19mm)	0 degrees
by turret	0.675-1.0 inches (17-25mm)	0 degrees
by fuel	0.375-0.75 inches (10-19mm)	0 degrees
Left Side, by final drive	0.375 inches (10mm)	0 degrees
by driver	1.0 inches (25mm)	0 degrees
by turret	1.0 inches (25mm)	0 degrees
by fuel	0.75 inches (19mm)	0 degrees
Top	0.5 inches (13mm)	90 degrees
Floor, under driver	1.0 inches (25mm)	90 degrees
remainder	0.375 inches (10mm)	90 degrees

Turret Thickness:		
Front, Cradle	1.25 inches (32mm)	0 degrees
Sides	0.5 inches (13mm)	45 degrees
Sides, Cradle	0.75 inches (19mm)	0 degrees
Rear	1.25 inches (32mm)	0 degrees
Rear, Cradle	0.75 inches (19mm)	0 degrees
Top	0.5 inches (13mm)	90 degrees
Cupola Sides	1.125 inches (29mm)	0 degrees

ARMAMENT

Primary: 76mm Gun T185E1 in cradle mount in turret

Traverse: Hydraulic and manual	360 degrees
Traverse Rate: (max)	15 seconds/360 degrees
Elevation: Hydraulic and manual	+20 to −10 degrees
Elevation Rate: (max)	4 degrees/second
Firing Rate: (max)	12 rounds/minute
Loading System:	Semiautomatic loader
Stabilizer System:	None

Secondary:
- (1) .50 caliber MG HB M2 in right cupola
- (1) .30 caliber MG M37 in left cupola
- (1) .30 caliber MG M37 coaxial w/76mm gun in cradle
- Provision for (1) .45 caliber SMG M3A1
- Provision for (1) .30 caliber Carbine M2

AMMUNITION

60 rounds 76mm	180 rounds .45 caliber
700 rounds .50 caliber	90 rounds .30 caliber (carbine)
5000 rounds .30 caliber	8 hand grenades

FIRE CONTROL AND VISION EQUIPMENT

Primary Weapon:	Direct	Indirect
	Periscope M16E2	Azimuth Indicator T24 mod.
	Elbow Telescope	Elevation Quadrant M9
	Ballistic drive	Gunner's Quadrant M1A1

Vision Devices:	Direct	Indirect
Driver	Hatch	Periscope M17 (4) and Periscope M19 (infrared) (1)
Commander	Vision blocks (10) in turret and cupola, hatch	Periscope M16E2 (1) Periscope T42 mod. (1) over cradle periscope (2)
Gunner	Vision blocks (10) in turret and cupola hatch	Periscope M16E2 (1) Periscope T42 mod. (1) over cradle periscope (2)
Loader	Vision blocks (2)	None

Total Periscopes: M16E2 (2), M17 (4), M19 (infrared) (1), T42 mod. (2) over cradle periscopes (4)

Total Vision Blocks: Turret (20), Hull (2)

ENGINE

Make and Model: Continental AOI-628-1	
Type: 8 cylinder, 4 cycle, opposed, fuel injection	
Cooling System: Air Ignition: Magneto	
Displacement:	628.3 cubic inches
Bore and Stroke:	5 x 4 inches
Compression Ratio:	6.7:1
Net Horsepower: (max)	280 hp at 3200 rpm
Gross Horsepower: (max)	340 hp at 3200 rpm
Net Torque: (max)	503 ft-lb at 2500 rpm
Gross Torque: (max)	587 ft-lb at 2500 rpm
Weight:	1098 pounds, dry
Fuel: 80-86 octane gasoline	150 gallons
Engine Oil:	16 quarts

POWER TRAIN

Transmission: XT-300, 3 ranges forward, 1 reverse
Single stage hydraulic torque converter
Stall Multiplication: 3.8:1

Overall Usable Ratios:	low 19.65:1	direct 1.21:1
	high 4.71:1	reverse 22:1

Steering Control: Clutch-brake, control handles
Brakes: Multiple disc
Final Drive: Spur gear   Gear Ratio: 6.17:1
Drive Sprocket: At front of vehicle with 11 teeth (T110 tracks)
At front of vehicle with 13 teeth (T85E1 tracks)
Pitch Diameter: 21.80 inches (T110 tracks), 22.979 inches (T85E1 tracks)

RUNNING GEAR

Suspension: Torsilastic
8 individually sprung dual road wheels (4/track)
Tire Size: 21.75 x 4.25 inches
4 dual track return rollers (2/track)
Rear road wheel serves as trailing idler
Double shock absorbers fitted on first and last road wheels on each side
Tracks: Center guide T110 and T85E1
Type: (T110) Band type, 16 inch width, each section 43.365 inches long
(T85E1) Double pin, 14 inch width, rubber chevron
Pitch: Cross bar (T110) 6.195 inches
(T85E1) 5.5 inches
Track Sections: (T110) 18 (9/track)
Cross Bars: (T110) 126 (63/track)
Shoes per Vehicle: (T85E1) 134 (67/track)
Ground Contact Length: 121.5 inches

ELECTRICAL SYSTEM

Nominal Voltage: 24 volts DC
Main Generator: (1) 24 volts, 300 amperes, driven by main engine
Auxiliary Generator: (1) 24 volts, 72 amperes, driven by auxiliary engine
Battery: (4) 12 volts, 2 sets of 2 in series connected in parallel

COMMUNICATIONS

Radio: AN/GRC-3, AN/VRC-24 in turret
Interphone: 4 stations plus external extension kit AN/VIA-4

FIRE PROTECTION

(3) 10 pound carbon dioxide, fixed
(1) 5 pound carbon dioxide, portable

PERFORMANCE

Maximum Speed: Level road	35 miles/hour
Maximum Tractive Effort: TE at stall	60,000 pounds
Per Cent of Vehicle Weight: TE/W	161 per cent
Maximum Grade:	60 per cent
Maximum Trench:	6 feet
Maximum Vertical Wall:	30 inches
Maximum Fording Depth:	40 inches
Minimum Turning Circle: (diameter)	pivot
Cruising Range: Roads	approx. 210 miles

GENERAL DATA

Crew:	4 men
Length:	258 inches
Width: Over hull	110 inches
Height:	108 inches
Tread:	92.5 inches
Ground Clearance:	19 inches
Fire Height:	approx. 76 inches
Turret Ring Diameter: (inside)	76 inches
Weight, Combat Loaded:	33,247 pounds
Weight, Unstowed:	28,632 pounds
Power to Weight Ratio: Net	12.6 hp/ton
Gross	17.1 hp/ton
Ground Pressure: Zero penetration	6.7 psi

ARMOR

Type: Turret, rolled and cast homogeneous steel; Hull, rolled 7039 aluminum alloy; Welded assembly. Highly sloped hull armor surrounded by lightweight flotation cells filled with polystyrene foam.

ARMAMENT

Primary: 152mm Gun-Launcher XM81E3 in turret mount

Traverse: Electric and manual	360 degrees
Traverse Rate: (max)	15 seconds/360 degrees
Elevation: Electric and manual	+20 to −10 degrees
Elevation Rate: (max)	4 degrees/second
Firing Rate: (max)	4 rounds/minute
Loading System:	Manual
Stabilizer System:	Azimuth and elevation

Secondary:

(1) .50 caliber MG HB M2 flexible AA mount on turret
(1) .50 caliber spotting rifle XM121 coaxial w/152mm gun-launcher
(1) 7.62mm MG M73 coaxial w/152mm gun-launcher in turret
Provision for (2) .45 caliber SMG M3A1

AMMUNITION

20 rounds 152mm	360 rounds .45 caliber
1000 rounds .50 caliber	8 hand grenades
100 rounds .50 caliber (spotting rifle)	
3000 rounds 7.62mm	

FIRE CONTROL AND VISION EQUIPMENT

Primary Weapon:	Direct	Indirect
	Periscope (infrared) XM38	Azimuth Indicator
	Telescope XM112	Elevation Quadrant
	Spotting Rifle XM121	Gunner's Quadrant M1A1
Vision Devices:	Direct	Indirect
Driver	Hatch	Periscope M27 (3) and Periscope M24 (infrared) (1)
Commander	Vision blocks (10), hatch	None
Gunner	None	Periscope XM38 (infrared) (1)
Loader	Hatch	Periscope M13 (1)

Total Periscopes: M13 (1), M24 (infrared) (1), M27 (3), XM38 (infrared) (1)
Total Vision Blocks: (10) around commander's hatch

ENGINE

Make and Model: General Motors 6V53T	
Type: 6 cylinder, 2 cycle, vee, supercharged	
Cooling System: Liquid Ignition: Compression	
Displacement:	318.4 cubic inches
Bore and Stroke:	63.875 x 4.5 inches
Compression Ratio:	17:1
Net Horsepower: (max)	210 hp at 2800 rpm
Gross Horsepower: (max)	285 hp at 2800 rpm
Gross Torque: (max)	435 ft-lb at 1900 rpm
Weight: (aluminum block)	1092 pounds
Fuel: 40 cetane diesel oil	148 gallons
Engine Oil:	24 quarts

POWER TRAIN

Transmission: XTG-250, 4 ranges forward, 1 reverse with power take-off for water propulsion unit
Single stage hydraulic torque converter
Stall Multiplication: 2.5:1

Overall Usable Ratios:	1st 8.92:1	4th	1.44:1
	2nd 6.04:1	reverse 1	12.60:1
	3rd 3.24:1	reverse 2	5.75:1

Steering Control: Mechanical, T-bar (land), steering levers (water)
Brakes: Multiple disc
Final Drive: Integral w/XTG-250   Gear Ratio: 2.22:1
Drive Sprocket: At rear of vehicle with 8 teeth
Pitch Diameter: approx. 10.5 inches

RUNNING GEAR

Suspension: Flat track, torsion bar
10 individually sprung dual road wheels (5/track)
Tire Size: 24 x 2.75 inches
Dual adjustable idler at front of each track
Idler Size: 14 x 2.75 inches
Shock absorbers on first and last road wheels on each side
Tracks: Double center guide, band type
Type: Band type 19 inch width
Pitch: Crossbar, 4 inches
Sections per Vehicle: 24 (12/track)
Ground Contact Length: 130 inches, estimated

ELECTRICAL SYSTEM

Nominal Voltage: 24 volts DC
Main Generator: (1) 24 volts, 300 amperes, driven by main engine
Auxiliary Generator: None
Battery: (2) 12 volts, in series

COMMUNICATIONS

Radio: RT-246-VRC in turret
Interphone: 4 stations plus external extension

FIRE PROTECTION

(1) 2.5 pound Halon, fixed
(1) 5 pound Halon, portable

PERFORMANCE

Maximum Speed: Level road	35 miles/hour
Maximum Grade:	60 per cent
Maximum Trench:	5 feet
Maximum Vertical Wall:	18 inches
Maximum Fording Depth:	floats
Minimum Turning Circle: (diameter)	pivot
Cruising Range: Roads	approx. 300 miles

152mm GUN-LAUNCHER AR/AAV M551

GENERAL DATA

Crew:	4 men
Length:	248.3 inches
Width: Over tracks	110 inches
Height: Over AA MG	116 inches
Tread:	92.5 inches
Ground Clearance:	19 inches
Fire Height:	approx. 76 inches
Turret Ring Diameter: (inside)	76 inches
Weight, Combat Loaded:	33,460 pounds
Weight, Unstowed:	28,525 pounds
Power to Weight Ratio: Net	15.2 hp/ton
Gross	17.9 hp/ton
Ground Pressure: Zero penetration	6.8 psi

ARMOR

Type: Turret, rolled and cast homogeneous steel; Hull, rolled 7039 aluminum alloy; Welded assembly. Highly sloped hull armor surrounded by lightweight flotation cells filled with polystyrene foam.

ARMAMENT

Primary: 152mm Gun-Launcher M81 (XM81E12) Modified or M81E1 in turret mount

Traverse: Electric and manual	360 degrees
Traverse Rate: (max) w/o stabilizer	10 seconds/360 degrees
Elevation: Electric and manual	+19.5 to −8 degrees
Elevation Rate: (max)	4 degrees/second
Firing Rate: (max)	4 rounds/minute
Loading System:	Manual
Stabilizer System:	Azimuth and elevation

Secondary:
(1) .50 caliber MG HB M2 flexible AA mount on turret
(1) 7.62mm MG M73 or M219 coaxial w/152mm gun-launcher in turret
(8) grenade launchers (smoke)
Provision for (2) .45 caliber SMG M3A1

AMMUNITION

*10 missiles MGM-51A, MGM-51B, or MGM-51C	360 rounds .45 caliber
*20 rounds 152mm	8 XM19 smoke grenades
1000 rounds .50 caliber	8 hand grenades
3000 rounds 7.62mm	

*Total of missiles plus 152mm later reduced to 29 when CBSS installed

FIRE CONTROL AND VISION EQUIPMENT

Primary Weapon:	Direct	Indirect
	Telescope M119 or M127	Azimuth Indicator M31A1
	Periscope M44	Elevation Quadrant M13A1C
	(passive night vision)	Gunner's Quadrant M1A1

Vision Devices:	Direct	Indirect
Driver	Hatch	Periscope M47 (3) and Periscope M48 (infrared) (1)
Commander	Vision blocks (10) in cupola, hatch	Night vision sight (1)
Gunner	None	Periscope M44 (passive night vision)
Loader	Hatch	Periscope M37 (1)

Total Periscopes: M37 (1), M44 (passive night vision) (1), M47 (3), M48 (infrared) (1)
Total Vision Blocks: (10) around cupola

ENGINE

Make and Model: General Motors 6V53T	
Type: 6 cylinder, 2 cycle, vee, supercharged	
Cooling System: Liquid Ignition: Compression	
Displacement:	318.4 cubic inches
Bore and Stroke:	3.875 x 4.5 inches
Compression Ratio:	17:1
Net Horsepower: (max)	255 hp at 2800 rpm
Gross Horsepower: (max)	300 hp at 2800 rpm
Net Torque: (max)	520 ft-lb at 2100 rpm
Gross Torque: (max)	615 ft-lb at 2100 rpm
Weight: (aluminum block)	1092 pounds, dry
Fuel: 40 cetane diesel oil	158 gallons
Engine Oil:	21 quarts

POWER TRAIN

Transmission: XTG-250-1A, 4 ranges forward, 2 reverse
Single stage hydraulic torque converter
Stall Multiplication: 2.5:1

Overall Usable Ratios:	1st	8.92:1	4th	1.44:1
	2nd	6.04:1	reverse 1	12.60:1
	3rd	3.24:1	reverse 2	5.75:1

Steering Controls: Mechanical, T-bar
Brakes: Multiple disc
Final Drive: Integral w/XTG-250-1A   Gear Ratio: 2.22:1
Drive Sprocket: At rear of vehicle with 11 teeth
Pitch Diameter: 16.732 inches

RUNNING GEAR

Suspension: Flat track, torsion bar
10 individually sprung dual road wheels (5/track)
Tire Size: 28 x 2.75 inches
Dual adjustable idler at front of each track
Idler Size: 14.5 x 2.75 inches
Shock absorbers fitted on first and last road wheels on each side
Tracks: Double center guide T138
Type: (T138) Single pin, 17.5 inch width, cast steel w/rubber pads
Pitch: 4.7 inches
Shoes per Vehicle: 204 (102/track)
Ground Contact Length: 140 inches

ELECTRICAL SYSTEM

Nominal Voltage: 24 volts DC
Main Generator: (1) 24 volts, 300 amperes, driven by main engine
Auxiliary Generator: None
Battery: (4) 12 volts, 2 sets of 2 in series connected in parallel

COMMUNICATIONS

Radio: AN/VRC-12, 46, 47, or 53 in turret bustle
Interphone: 4 stations plus external extension C2296/VRC

FIRE PROTECTION

(1) 3.25 pound Halon, fixed
(1) 2.75 pound Halon, portable

PERFORMANCE

Maximum Speed: Level road	43 miles/hour
Maximum Speed: Water	3.6 miles/hour
Maximum Tractive Effort: TE at stall	19,150 pounds
Per Cent of Vehicle Weight: TE/W	57 per cent
Maximum Grade:	60 per cent
Maximum Trench:	8 feet
Maximum Vertical Wall:	33 inches
Maximum Fording Depth:	floats
Minimum Turning Circle: (diameter)	pivot
Cruising Range: Roads	approx. 350 miles

152mm GUN-LAUNCHER AR/AAV M551A1(TTS)

GENERAL DATA

Crew:	4 men
Length:	248.3 inches
Width: Over tracks	110 inches
Height: Over AA MG	116 inches
Tread:	92.5 inches
Ground Clearance:	19 inches
Fire Height:	approx. 76 inches
Turret Ring Diameter: (inside)	76 inches
Weight, Combat Loaded:	33,600 pounds
Weight, Unstowed:	28,970 pounds
Power to Weight Ratio: Net	15.2 hp/ton
Gross	17.9 hp/ton
Ground Pressure: Zero penetration	6.9 psi

ARMOR

Type: Turret, rolled and cast homogeneous steel; Hull, rolled 7039 aluminum alloy; Welded assembly. Highly sloped hull armor surrounded by lightweight flotation cells filled with polystyrene foam.

ARMAMENT

Primary: 152mm Gun-Launcher M81E1 in turret mount

Traverse: Electric and manual	360 degrees
Traverse Rate: (max) w/o stabilizer	10 seconds/360 degrees
Elevation: Electric and manual	+19.5 to −8 degrees
Elevation Rate: (max)	4 degrees/second
Firing Rate: (max)	4 rounds/minute
Loading System:	Manual
Stabilizer System:	Azimuth and elevation

Secondary:
- (1) .50 caliber MG HB M2 flexible AA mount on turret
- (1) 7.62mm MG M240 coaxial w/152mm gun-launcher in turret
- (8) M176 grenade launchers (smoke)
- Provision for (1) .45 caliber SMG M3A1
- Provision for (1) 40mm M79 grenade launcher

AMMUNITION

9 missiles MGM-51A, MGM-51B, or MGM-51C	180 rounds .45 caliber
20 rounds 152mm	8 smoke grenades
1000 rounds .50 caliber	12 grenades 40mm
3000 rounds 7.62mm	

FIRE CONTROL AND VISION EQUIPMENT

Primary Weapon:	Direct	Indirect
	Telescope M127A1	Azimuth Indicator M31A1C
	Tank thermal sight	Elevation Quadrant M13A1C
	Laser range finder AN/VVG-1	Gunner's Quadrant M1A1

Vision Devices:	Direct	Indirect
Driver	Hatch	Periscope M47 (3) and Periscope M48 (infrared) (1)
Commander	Vision blocks (10) in cupola, hatch	Tank thermal sight extension
Gunner	None	Tank thermal sight
Loader	Hatch	Periscope M37 (1)

Total Periscopes: M37 (1), M47 (3), M48 (infrared) (1), tank thermal sight (1)
Total Vision Blocks: (10) around cupola

ENGINE

Make and Model: General Motors 6V53T	
Type: 6 cylinder, 2 cycle, vee, supercharged	
Cooling System: Liquid Ignition: Compression	
Displacement:	318.4 cubic inches
Bore and Stroke:	3.875 x 4.5 inches
Compression Ratio:	17:1
Net Horsepower: (max)	255 hp at 2800 rpm
Gross Horsepower: (max)	300 hp at 2800 rpm
Net Torque: (max)	520 ft-lb at 2100 rpm
Gross Torque: (max)	615 ft-lb at 2100 rpm
Weight: (cast iron block)	1325 pounds, dry
Fuel: 40 cetane diesel oil	158 gallons
Engine Oil:	21 quarts

POWER TRAIN

Transmission: XTG-250-1A, 4 ranges forward, 2 reverse
Single stage hydraulic torque converter
Stall Multiplication: 2.5:1

Overall Usable Ratios:	1st 8.92:1	4th	1.44:1
	2nd 6.04:1	reverse 1	12.60:1
	3rd 3.24:1	reverse 2	5.75:1

Steering Controls: Mechanical, T-bar
Brakes: Multiple disc
Final Drive: Integral w/XTG-250-1A   Gear Ratio: 2.22:1
Drive Sprocket: At rear of vehicle with 11 teeth
Pitch Diameter: 16.732 inches

RUNNING GEAR

Suspension: Flat track, torsion bar
10 individually sprung dual road wheels (5/track)
Tire Size: 28 x 2.75 inches
Dual adjustable idler at front of each track
Idler Size: 14.5 x 2.75 inches
Shock absorbers fitted on first and last road wheels on each side
Tracks: Double center guide T138
Type: (T138) Single pin, 17.5 inch width, cast steel w/rubber pads
Pitch: 4.7 inches
Shoes per Vehicle: 204 (102/track)
Ground Contact Length: 140 inches

ELECTRICAL SYSTEM

Nominal Voltage: 24 volts DC
Main Generator: (1) 24 volts, 300 amperes, driven by main engine
Auxiliary Generator: None
Battery: (4) 12 volts, 2 sets of 2 in series connected in parallel

COMMUNICATIONS

Radio: AN/VRC-12, 46, 47, or 53 in turret bustle
Interphone: 4 stations plus external extension C2296/VRC

FIRE PROTECTION

(1) 3.25 pound Halon, fixed
(1) 2.75 pound Halon, portable

PERFORMANCE

Maximum Speed: Level road	43 miles/hour
Maximum Speed: Water	3.6 miles hour
Maximum Tractive Effort: TE at stall	19,150 pounds
Per Cent of Vehicle Weight: TE/W	57 per cent
Maximum Grade:	60 per cent
Maximum Trench:	8 feet
Maximum Vertical Wall:	33 inches
Maximum Fording Depth:	floats
Minimum Turning Circle: (diameter)	pivot
Cruising Range: Roads	approx. 350 miles

105mm ARMORED GUN SYSTEM XM8

GENERAL DATA

Crew:	3	men
Length: Gun forward, level 1 armor	361.4	inches
level 2 and 3 armor	365.2	inches
Length: Gun to rear level 1, 2, and 3 armor	354.6	inches
Length: Without gun, level 1 armor	241.9	inches
level 2 and 3 armor	246.6	inches
Gun Overhang: Gun forward, level 1, 2, and 3 armor	121.4	inches
Width: Over fenders, level 1 armor	104.0	inches
Over tracks	100.0	inches
Height: Over cupola, level 1 armor	100.6	inches
level 2 armor	99.6	inches
level 3 armor	98.6	inches
Tread:	85	inches
Ground Clearance: level 1 armor	17.0	inches
level 2 armor	16.0	inches
level 3 armor	15.0	inches
Fire Height: level 1 armor	75.9	inches
level 2 armor	74.9	inches
level 3 armor	73.9	inches
Turret Ring Diameter: (inside)	78.0	inches
Weight: Air drop, level 1	36,900	pounds
Combat, level 1	38,800	pounds
Roll on-roll off, level 2	44,000	pounds
Combat, level 3	52,000	pounds
Power to Weight Ratio: Combat, level 1, gross	28.3	hp/ton
Combat, level 2, gross	25.0	hp/ton
Combat, level 3, gross	21.2	hp/ton
Ground Pressure: Zero penetration, Combat, level 1	9.1	psi
Combat, level 2	10.3	psi
Combat, level 3	12.2	psi

ARMOR

The welded 5083 aluminum alloy structure of the hull and turret is reinforced with ceramic and applique armor to achieve three levels of protection.

ARMAMENT

Primary: 105mm Gun XM35 in soft recoil mount in turret

Traverse: Hydraulic and manual	360 degrees
Traverse Rate: (max)	8.5 seconds/360 degrees
Elevation: Hydraulic and manual	+20 to −10 degrees
Elevation Rate: (max)	11 degrees/second
Firing Rate: (max)	12 rounds/minute
Loading System:	Automatic
Stabilizer System:	Azimuth and elevation

Secondary:
- (1) .50 caliber MG HB M2 or (1) 40mm Mark 19 automatic grenade launcher or (1) 7.62mm M240 machine gun flexible mount on turret roof
- (1) 7.62mm M240 machine gun coaxial w/105mm gun in turret
- (16) smoke grenade launchers on turret

AMMUNITION

- 30 rounds 105mm (21 in automatic loader)
- 600 rounds .50 caliber
- 4500 rounds 7.62mm
- 32 smoke grenades (16 in launchers)

FIRE CONTROL AND VISION EQUIPMENT

Primary Weapon: Laser range finder
Day/night tank thermal sight
Telescope w/fiber optics
Digital fire control computer

Vision Devices:	Direct	Indirect
Driver	Hatch	5 wide angle periscopes
Commander	Hatch	7 wide angle periscopes
Gunner	Hatch	Day/night sight

Total Periscopes: (12)

ENGINE

Make and Model: Detroit Diesel 6V92TA	
Type: 6 cylinder, 2 cycle, vee, supercharged	
Cooling System: Liquid Ignition: Compression	
Displacement:	552 cubic inches
Bore and Stroke:	4.84 x 5 inches
Compression Ratio:	17:1
Gross Horsepower: (max)	550 hp at 2400 rpm
Gross Torque: (max)	1446 ft-lb at 1500 rpm
Weight:	1900 pounds, dry
Fuel: Diesel or JP-8 150	gallons
Engine Oil:	20 quarts, 16 at refill

POWER TRAIN

Transmission: General Electric HMPT 500-3EC, 3 ranges forward, 1 reverse
Hydromechanical, infinitely variable ratio
Hydrostatic steering
Steering Control: T-bar
Brakes: Multiple disc
Final Drive: Spur gear   Gear Ratio: 4.4:1
Drive Sprocket: At rear of vehicle with 11 teeth
Pitch Diameter: 21.29 inches

RUNNING GEAR

Suspension: Flat track, torsion bar
12 individually sprung dual road wheels (6/track)
Tire Size: 3.38 x 24 inches
Dual adjustable idler at front of each track
Idler Size: 2.41 x 17.25 inches
Shock absorbers on 1st, 2nd, 3rd, 5th, and 6th road wheels on each side
Tracks: Center guide, T150 modified
Type: Double pin, 15 inch width, steel w/detachable rubber pads
Pitch: 6 inches
Shoes per Vehicle: 154 (77/track), new 156 (78/track)
Ground Contact Length: 142.2 inches

ELECTRICAL SYSTEM

Nominal Voltage: 24 volts DC
Main Generator: (1) 24 volts, 300 amperes, driven by main engine
Auxiliary Generator: None
Battery: (4) 12 volts, 2 sets of 2 in series connected in parallel

COMMUNICATIONS

Radio: SINCGARS, AN/VCR-87A, 89A, or 92A
Interphone: 3 stations AN/VIC-1

FIRE AND NBC PROTECTION

Automatic Halon fire extinguisher system in crew compartment
Dry powder fire extinguisher system in engine compartment
Individual masks for NBC protection

PERFORMANCE

Maximum Speed: Level road	45 miles/hour
Maximum Tractive Effort: TE at stall	54,500 pounds
Per Cent of Vehicle Weight: TE/W, level 1 armor	140 per cent
Maximum Grade:	60 per cent
Maximum Trench:	7 feet
Maximum Vertical Wall:	32 inches
Maximum Fording Depth:	40 inches
Minimum Turning Circle: (diameter)	pivot
Cruising Range: Roads	approx. 300 miles

106mm MULTIPLE SELF-PROPELLED RIFLES M50 (T165E2) AND M50A1

GENERAL DATA
Crew:	3 men
Length:	150.8 inches
Width: Over fenders	102.3 inches
Height:	83.9 inches
Tread:	73.0 inches
Ground Clearance:	14.6 inches
Fire Height: Top rifles only	approx. 78 inches
Turret Ring Diameter: (inside)	31.7 inches
Weight, Combat Loaded: M50	19,050 pounds
Weight, Unstowed: M50	16,450 pounds
Power to Weight Ratio: Net, M50	13.0 hp/ton
Gross, M50	15.2 hp/ton
M50A1	18.9 hp/ton
Ground Pressure: Zero penetration	5.1 psi

ARMOR
Type: Turret, rolled homogeneous steel; Hull, rolled homogeneous steel; Welded assembly

Hull Thickness:	Actual	Angle w/Vertical
Front, Upper	0.5 inches (13mm)	71 degrees
Lower	0.5 inches (13mm)	45 degrees
Sides, Upper	0.5 inches (13mm)	42 degrees
Lower	0.5 inches (13mm)	0 degrees
Rear, Upper	0.5 inches (13mm)	27 degrees
Lower	0.5 inches (13mm)	0 degrees
Top	0.5 inches (13mm)	90 degrees
Floor	0.25 inches (6mm)	90 degrees
Turret Thickness:		
Front	0.5 inches (13mm)	30 degrees
Sides	0.5 inches (13mm)	30 degrees
Rear	0.5 inches (13mm)	30 degrees
Top	0.5 inches (13mm)	90 degrees

ARMAMENT
Primary: (6) 106mm Recoilless Rifle M40A1C on turret Mount T149E5

Traverse: Manual	80 degrees
	(40 degrees left or right)
Elevation: Manual	+20 to −10 degrees
Firing Rate: (max)	6 round salvo
Loading System:	Manual
Stabilizer System:	None

Secondary:
- (4) .50 caliber Spotting Rifle M8C on top of 106mm rifles
- (1) .30 caliber MG M1919A4 flexible or fixed on turret
- Provision for (1) .45 caliber SMG M3A1

AMMUNITION
- 18 rounds 106mm
- 80 rounds .50 caliber
- 180 rounds .45 caliber
- 1000 rounds .30 caliber

FIRE CONTROL AND VISION EQUIPMENT
Primary Weapon:	Direct	Indirect
	(M50) Periscopic Sight	(M50A1) Elevation Quadrant
	M20A3C (M50A1)	M13A1C
	(M50A1) Periscopic Sight	(M50) Gunner's Quadrant
	M20A3G	M1A1
Vision Devices:	Direct	Indirect
Driver	Hatch	Periscope M13 (1)
Gunner	Hatch	Periscopic Sight
Loader	None	None
Total Periscopes: M13 (1)		

ENGINE
Make and Model: M50, General Motors Model 302	
M50A1, Chrysler HT-361-318	
Type: M50, 6 cylinder, 4 cycle , in-line	
M50A1, 8 cylinder, 4 cycle, vee	
Cooling System: Liquid Ignition: Battery	
Displacement: M50	301.6 cubic inches
M50A1	360.8 cubic inches
Bore and Stroke: M50	4.0 x 4.0 inches
M50A1	4.125 x 3.375 inches
Compression Ratio: M50	7.5:1
M50A1	7.8:1
Net Horsepower: (max) M50	124 hp at 2400 rpm
Gross Horsepower: (max) M50	145 hp at 3400 rpm
M50A1	180 hp at 3450 rpm
Net Torque: (max) M50	252 ft-lb at 1400 rpm
Gross Torque: (max) M50	255 ft-lb at 2000 rpm
M50A1	283 ft-lb at 2400 rpm
Weight: M50	630 pounds, dry
M50A1	710 pounds, dry
Fuel: 80 octane gasoline	47 gallons
Engine Oil: M50	11 quarts
M50A1	7 quarts

POWER TRAIN
Transmission: M50, Allison X-drive XT-90-2, 3 ranges forward, 1 reverse
M50A1, Allison X-drive XT-90-5, 3 ranges forward, 1 reverse
Single stage hydraulic torque converter w/lock-up clutch
Stall Multiplication: 3.8:1

Overall Usable Ratios:	low	27.0:1	high	6.4:1
	intermediate	14.4:1	reverse	19.3:1

Steering Controls: Mechanical, steering levers
Steering System: Clutch-brake
Steering Rate: Variable
Brakes: Multiple disc
Final Drive: Integral w/transmission   Gear Ratio: 5.075:1
Drive Sprocket: At front of vehicle with 15 teeth
Pitch Diameter: 19.223 inches

RUNNING GEAR
Suspension: Torsilastic
- 8 individually sprung dual road wheels (4/track)
- Tire Size: 20 x 4.5 inches
- 8 track skid bumpers (4/track)
- The rear road wheels serve as adjustable trailing idlers
- Shock absorbers fitted on road wheels 1, 2, and 3 on each side

Tracks: Center and outside guide, T123
- Type: (T123) Band type, 20 inch width, each section 60 inches long
- Pitch: Cross bar, 4 inches
- Cross Bars: 150 (75/track)
- Track Sections: 10 (5/track)
- Ground Contact Length: 94.2 inches

ELECTRICAL SYSTEM
Nominal Voltage: 24 volts DC
Main Generator: M50, (1) 24 volts, 25 amperes, driven by main engine
M50A1, (alternator), (1) 28 volts, 60 amperes, driven by main engine
Auxiliary Generator: None
Battery: (2) 12 volts, in series

COMMUNICATIONS
Radio: AN/PRC-10 in left sponson
Interphone: M50, none
M50A1, AN/UIC-1

FIRE PROTECTION
(1) 5 pound carbon dioxide, portable

PERFORMANCE
Maximum Speed: Level road	30 miles/hour
Maximum Tractive Effort: TE at stall, M50	15,000 pounds
Per Cent of Vehicle Weight: TE/W, M50	79 per cent
Maximum Grade:	60 per cent
Maximum Trench:	4.5 feet
Maximum Vertical Wall:	28 inches
Maximum Fording Depth: w/o fording kit	24 inches
w/fording kit	60 inches
Minimum Turning Circle: (diameter)	18 feet
Cruising Range: Roads, M50	approx. 115 miles
M50A1	approx. 100 miles

90mm SELF-PROPELLED GUN M56 (T101)

GENERAL DATA

Crew:	4 men
Length: Gun in travel position	229.8 inches
Length: w/o gun	179.4 inches
Gun Overhang: Gun in travel position	50.4 inches
Width: Over fenders	101.3 inches
Over tracks	98.0 inches
Height: Over blast shield	78.9 inches
Tread:	78.0 inches
Ground Clearance:	12.8 inches
Fire Height:	66.0 inches
Weight, Combat Loaded:	15,750 pounds
Weight, Unstowed:	12,500 pounds
Power to Weight Ratio: Net	21.0 hp/ton
Gross	25.4 hp/ton
Ground Pressure: Zero penetration	4.2 psi

ARMOR

None

ARMAMENT

Primary: 90 mm Gun M54 (T125) in Mount M88 (T170E1)

Traverse: Manual	60 degrees
	(30 degrees left or right)
Elevation: Manual	+15 to −10 degrees
Firing Rate: (max)	10 rounds/minute
Loading System:	Manual
Stabilizer System:	None

Secondary:
Provision for (4) .30 caliber Carbine M2

AMMUNITION

29 rounds 90mm
240 rounds .30 caliber (carbine)
8 hand grenades

FIRE CONTROL AND VISION EQUIPMENT

Primary Weapon:	Direct	Indirect
	Telescope T186	Gunner's Quadrant M1A1
		Fuze Setter M27

Vision Devices:	Direct	Indirect
Driver	Open vehicle	None
Commander	Open vehicle	None
Gunner	Open vehicle	None
Loader	Open vehicle	None

ENGINE

Make and Model: Continental AOI-402-5	
Type: 6 cylinder, 4 cycle, opposed, fuel injection	
Cooling System: Air Ignition: Magneto	
Displacement:	403.2 cubic inches
Bore and Stroke:	4.625 x 4.0 inches
Compression Ratio:	6.9:1
Net Horsepower: (max)	165 hp at 3000 rpm
Gross Horsepower: (max)	200 hp at 3000 rpm
Net Torque: (max)	325 ft-lb at 2200 rpm
Gross Torque: (max)	347 ft-lb at 2800 rpm
Weight:	746 pounds, dry
Fuel: 80 octane gasoline	55 gallons
Engine Oil:	13 quarts, 12 quarts at refill

POWER TRAIN

Transmission: Cross-drive, CD-150-4, 2 ranges forward, 1 reverse
    Single stage hydraulic torque converter w/lock-up clutch
    Stall Multiplication: 4.0:1
    Overall Usable Ratios: low 13.95:1   reverse 13.95:1
                        high 6.08:1

Steering Controls: Mechanical, steering wheel
Brakes: Multiple disc
Final Drive: Planetary gear   Gear Ratio: 4.8:1
Drive Sprocket: At front of vehicle with 15 teeth
    Pitch Diameter: 19.09 inches

RUNNING GEAR

Suspension: Flat track, torsion tube over bar at stations 1 and 4, torsion bar
    at stations 2 and 3
    8 individually sprung road wheels (4/track) w/pneumatic tires
    Rim Size: 12 x 6 inches
    Tire Size: 7.50 x 12.00
    Tire Air Pressure: 75 psi
    Compensating idler at rear of each track
    Idler Tire Size: 15.25 x 8.0 inches
    Shock absorbers fitted on first and last road wheels on each side
Tracks: Outside guide
    Type: Band type, 20 inch width, each section 44 inches long
    Pitch: Cross bar, 4 inches
    Cross Bars: 176 (88/track)
    Ground Contact Length: 94 inches

ELECTRICAL SYSTEM

Nominal Voltage: 24 volts DC
Main Generator: (1) 24 volts, 25 amperes, driven by main engine
Auxiliary Generator: None
Battery: (2) 12 volts, in series

COMMUNICATIONS

Radio: AN/PRC-8, 9, or 10 on left side behind driver
Interphone: None

FIRE PROTECTION

(1) 5 pound carbon dioxide, portable

PERFORMANCE

Maximum Speed: Level road	28 miles/hour
Maximum Tractive Effort: TE at stall	14,500 pounds
Per Cent of Vehicle Weight: TE/W	92 per cent
Maximum Grade:	60 per cent
Maximum Trench:	4 feet
Maximum Vertical Wall:	30 inches
Maximum Fording Depth: w/o fording kit	42 inches
w/fording kit	60 inches
Minimum Turning Circle: (diameter)	pivot
Cruising Range: Roads	approx. 140 miles

TWIN 40mm SELF-PROPELLED GUNS M42 AND M42A1

GENERAL DATA

Crew:	6 men
Length: Gun forward	250.3 inches
Length: Gun to rear	229.1 inches
Length: Without gun	229.1 inches
Gun Overhang: Gun forward	21.2 inches
Width: Over fenders	126.9 inches
Height: Over gun shield	112.1 inches
Tread:	102.5 inches
Ground Clearance:	17.3 inches
Fire Height:	82.5 inches
Weight, Combat Loaded:	49,800 pounds
Weight, Unstowed:	44,300 pounds
Power to Weight Ratio: Net	17.9 hp/ton
Gross	20.1 hp/ton
Ground Pressure: Zero penetration	9.3 psi

ARMOR

Type: Turret, rolled homogeneous steel; Hull, rolled homogeneous steel; Welded assembly

Hull Thickness:	Actual	Angle w/Vertical
Front, Upper	0.5 inches (13mm)	33 degrees
Lower	1.0 inches (25mm)	39 degrees
Sides, Upper	0.5 inches (13mm)	0 degrees
Lower left front	0.5 inches (13mm)	45 degrees
Lower, remainder	0.5 inches (13mm)	60 degrees
Rear, Upper	0.5 inches (13mm)	56 degrees
Lower	0.75 inches (19mm)	41 degrees
Top	0.5 inches (13mm)	90 degrees
Floor, Front	1.25 inches (32mm)	90 degrees
Rear	0.375 inches (10mm)	90 degrees

Turret Thickness:		
Gun Shields:	0.5 inches (13mm)	0 to 47 degrees
Sides 0.30 inches (8mm)	0 degrees	
Rear	0.30 inches (8mm)	0 degrees
Top	Open	

ARMAMENT

Primary: 40mm Dual Automatic Gun M2A1 in Mount M4E1 in center of chassis

Traverse: Hydraulic and manual	360 degrees
Traverse Rate: (max)	9 seconds/360 degrees
Elevation: Hydraulic	+85 to −3 degrees
Manual	+87 to −5 degrees
Elevation Rate: (hydraulic max)	25 degrees/second
Firing Rate: (max)	240 rounds/minute
	(120 rounds/gun)
Loading System:	Automatic
Stabilizer System:	None

Secondary:
- (1) .30 caliber MG M1919A4 on front or rear of gun mount
- Provision for (1) .45 caliber SMG M3A1
- Provision for (5) .30 caliber Carbine M2
- Provision for (1) 3.5 inch Rocket Launcher M20

AMMUNITION

480 rounds 40mm	8 hand grenades
180 rounds .45 caliber	4 3.5 inch rockets
1750 rounds .30 caliber	
900 rounds .30 caliber (carbine)	

FIRE CONTROL AND VISION EQUIPMENT

Primary Weapon:	Direct	Indirect
	Computing Sight M38	Azimuth Indicator M27
	w/Reflex Sight M24C	Gunner's Quadrant M1A1
	Ring Sight	

Vision Devices:	Direct	Indirect
Driver	Hatch	Periscope M13 or M13B1 (1) and Periscope M19 (infrared) (1)
Commander	Hatch	Periscope M13 or M13B1 (1)
Gunner	Open top	None
Sight Setter	Open top	None
Loaders	Open top	None

Total Periscopes: M13 or M13B1 (2), M19 (infrared) (1)

ENGINE

Make and Model:	Continental AOS-895-3 (M42)
	Continental AOSI-895-5 (M42A1)
Type:	6 cylinder, 4 cycle, opposed, supercharged (M42)
	6 cylinder, 4 cycle, opposed, supercharged, fuel injection (M42A1)
Cooling System: Air Ignition: Magneto	
Displacement:	895.9 cubic inches
Bore and Stroke:	5.75 x 5.75 inches
Compression Ratio:	5.5:1
Net Horsepower: (max)	446 hp at 2400 rpm
Gross Horsepower: (max)	500 hp at 2800 rpm
Net Torque: (max)	890 ft-lb at 2200 rpm
Gross Torque: (max)	955 ft-lb at 2400 rpm
Weight:	approx. 1900 pounds, dry
Fuel: 80 octane gasoline	140 gallons
Engine Oil:	44 quarts

POWER TRAIN

Transmission: Cross-drive CD-500-3, 2 ranges forward, 1 reverse w/automatic lock-up in high
Single stage hydraulic torque converter
Stall Multiplication: 4:1

Overall Usable Ratios:	low 14.7:1	direct	1:1
	high 3.9:1	reverse	14.7:1

Steering Control: Mechanical, T-bar
Steering Rate: 6.8 rpm
Brakes: Multiple disc
Final Drive: Spur gear   Gear Ratio: 4.25:1
Drive Sprocket: At rear of vehicle with 12 teeth
Pitch Diameter: 23.422 inches

RUNNING GEAR

Suspension: Torsion bar
- 10 individually sprung dual road wheels (5/track)
- Tire Size: 25.5 x 4.5 inches
- 6 dual track return rollers (3/track)
- Dual compensating idler at front of each track
- Idler Size: 22.5 x 4.5 inches, steel, no tire (early vehicles)
- Idler Tire Size: 25.5 x 4.5 inches (late vehicles)
- Shock absorbers fitted on first 2 and last road wheels on each side

Tracks: Center guide T91E3
- Type: (T91E3) Single pin, 21 inch width, steel w/detachable rubber pad
- Pitch: 6 inches
- Shoes per Vehicle: 150 (75/track)
- Ground Contact Length: 127 inches

ELECTRICAL SYSTEM

Main Generator: (1) 24 volts, 150 amperes, driven by main engine
Auxiliary Generator: (1) 24 volts, 300 amperes, driven by auxiliary engine
Battery: (4) 12 volts, 2 sets of 2 in series connected in parallel

COMMUNICATIONS

Radio: AN/VRC-7 thru 10 series in right front hull
AN/GRR-5 in right front hull
Interphone: AN/UIC-1, 3 stations plus external outlet C981-U

FIRE PROTECTION

- (2) 10 pound carbon dioxide, fixed
- (1) 5 pound carbon dioxide, portable

PERFORMANCE

Maximum Speed: Level road	45 miles/hour
Maximum Tractive Effort: TE at stall	44,000 pounds
Per Cent of Vehicle Weight: TE/W	88 per cent
Maximum Grade:	60 per cent
Maximum Trench:	6 feet
Maximum Vertical Wall:	28 inches
Maximum Fording Depth:	48 inches
Minimum Turning Circle: (diameter)	pivot
Cruising Range: Roads, M42	approx. 100 miles
M42A1	approx. 120 miles

105mm SELF-PROPELLED HOWITZERS M52 AND M52A1

GENERAL DATA

Crew:	5 men
Length:	228.4 inches
Width:	123.9 inches
Height: Over MG	130.6 inches
Tread:	102.5 inches
Ground Clearance:	19.3 inches
Fire Height:	83.5 inches
Turret Ring Diameter:	73 inches
Weight, Combat Loaded:	54,100 pounds
Weight, Unstowed:	49,800 pounds
Power to Weight Ratio: Net	16.3 hp/ton
Gross	18.5 hp/ton
Ground Pressure: Zero penetration	8.6 psi

ARMOR

Type: Turret, rolled homogeneous steel; Hull, rolled homogeneous steel; Welded assembly

Hull Thickness:

	Actual	Angle w/Vertical
Front, Upper	0.5 inches (13mm)	52 and 81 degrees
Lower	0.5 inches (13mm)	40 and 66 degrees
Sides	0.5 inches (13mm)	0 degrees
Rear	0.5 inches (13mm)	0 degrees
Top	0.5 inches (13mm)	90 degrees
Floor	0.375 inches (10mm)	90 degrees

Turret Thickness:

Howitzer Shield (casting)	0.5 inches (13mm)	0 to 90 degrees
Front	0.5 inches (13mm)	30 degrees
Sides	0.5 inches (13mm)	0 degrees
Rear	0.5 inches (13mm)	0 degrees
Top	0.5 inches (13mm)	90 degrees

ARMAMENT

Primary: 105mm Howitzer M49 (T96E1) in Mount M85 (T67E1) in turret

Traverse: Manual	120 degrees
	(60 degrees left or right)
Elevation: Manual	+65 to −10 degrees
Firing Rate: (max)	3 rounds/minute
Loading System:	Manual
Stabilizer System:	None

Secondary:
- (1) .50 caliber MG HB M2 flexible AA mount on turret cupola
- Provision for (1) .45 caliber SMG M3A1
- Provision for (4) .30 caliber Carbine M2

AMMUNITION

102 rounds 105mm	8 hand grenades
900 rounds .50 caliber	
180 rounds .45 caliber	
720 rounds .30 caliber (carbine)	

FIRE CONTROL AND VISION EQUIPMENT

Primary Weapon:	Direct	Indirect
	Telescope M101 (T150E1)	Panoramic Telescope M100 (T149E1)
	w/Periscope M23 (T38)	Azimuth Indicator T24E1 (early M52)
	Panoramic Telescope M100	Gunner's Quadrant M1A1

Vision Devices:	Direct	Indirect
Driver	Hatch	Periscope M17 (4)
Commander	Vision blocks (6) in cupola, hatch	Periscope M15A1 (1)
Gunner	None	Periscope M13 (1)
Loaders	None	None

Total Periscopes: M13 (1), M15A1 (1), M17 (4)
Total Vision Blocks: 6 in cupola on turret roof

ENGINE

Make and Model:	Continental AOS-895-3 (M52)
	Continental AOSI-895-5 (M52A1)
Type:	6 cylinder, 4 cycle, opposed, supercharged (M52)
	6 cylinder, 4 cycle, opposed, supercharged, fuel injection (M52A1)
Cooling System: Air Ignition: Magneto	
Displacement:	895.9 cubic inches
Bore and Stroke:	5.75 x 5.75 inches
Compression Ratio:	5.5:1
Net Horsepower: (max)	446 hp at 2400 rpm
Gross Horsepower: (max)	500 hp at 2800 rpm
Net Torque: (max)	890 ft-lb at 2200 rpm
Gross Torque: (max)	955 ft-lb at 2400 rpm
Weight:	approx. 1900 pounds, dry
Fuel: 80 octane gasoline	174 gallons
Engine Oil:	44 quarts

POWER TRAIN

Transmission: Cross-drive CD-500-3, 2 ranges forward, 1 reverse w/automatic lock-up in high
   Single stage hydraulic torque converter
   Stall Multiplication: 3.9:1

Overall usable Ratios:	low 14.9:1	direct	1:1
	high 3.9:1	reverse	14.9:1

Steering Control: Mechanical T-bar
   Steering Rate: 13.2 rpm
Brakes: Multiple disc
Final Drive: Spur gear   Gear Ratio: 4.75:1
Drive Sprocket: At front of vehicle with 12 teeth
   Pitch Diameter: 23.422 inches

RUNNING GEAR

Suspension: Torsion bar
   10 individually sprung dual road wheels (5/track)
   Tire Size: 25.5 x 4.5 inches
   8 dual track return rollers (4/track)
   Trailing idler at rear of each track
   Idler Tire Size: 28 x 4.5 inches
   Shock absorbers fitted on road wheels 1, 2, and 5 on each side
Tracks: Center guide T91E3
   Type: (T91E3) Single pin, 21 inch width, steel w/detachable rubber pad
   Pitch: 6 inches
   Shoes per Vehicle: 149 (74 left, 75 right)
   Ground Contact Length: 149.4 inches

ELECTRICAL SYSTEM

Nominal Voltage: 24 volts DC
Main Generator: (1) 24 volts, 150 amperes, driven by main engine
Auxiliary Generator: None, but space provided for installation
Battery: (4) 12 volts, 2 sets of 2 in series connected in parallel

COMMUNICATIONS

Radio: AN/PRC 8, 9, or 10 in turret
Interphone: AN/VIC-1, 3 stations plus external extension kit C-980U

FIRE PROTECTION

(2) 10 pound carbon dioxide, fixed
(1) 5 pound carbon dioxide, portable

PERFORMANCE

Maximum Speed: Level road	35 miles/hour
Maximum Tractive Effort: TE at stall	41,000 pounds
Per Cent of Vehicle Weight: TE/W	76 per cent
Maximum Grade:	60 per cent
Maximum Trench:	6 feet
Maximum Vertical Wall:	30 inches
Maximum Fording Depth:	48 inches
Minimum Turning Circle: (diameter)	pivot
Cruising Range: Roads, M52	approx. 90 miles
M52A1	approx. 100 miles

155mm SELF-PROPELLED HOWITZERS M44 (T194) AND M44A1

GENERAL DATA

Crew:	5 men
Length:	242.5 inches
Width:	127.5 inches
Height: Over canvas top	122.5 inches
Tread:	102.5 inches
Ground Clearance:	18.8 inches
Fire Height:	approx. 84 inches
Weight, Combat Loaded:	64,000 pounds
Weight, Unstowed:	58,000 pounds
Power to Weight Ratio: Net	13.9 hp/ton
Gross	15.6 hp/ton
Ground Pressure: Zero penetration	10.2 psi

ARMOR

Type: Rolled homogeneous steel; Welded assembly

Hull Thickness:	Actual	Angle w/Vertical
Front, Upper	0.5 inches (13mm)	52 and 81 degrees
Lower	0.5 inches (13mm)	40 and 66 degrees
Sides	0.5 inches (13mm)	0 degrees
Rear	0.5 inches (13mm)	0 degrees
Top	Open	
Floor	0.375 inches (10mm)	90 degrees
Gun Shield Thickness:	0.5 inches (13mm)	0 degrees

ARMAMENT

Primary: 155mm Howitzer M45 (T186E1) in Mount M80 (T167)

Traverse: Hydraulic and manual	60 degrees
	(30 degrees left or right)
Traverse Rate: (max)	10 degrees/second
Elevation: Hydraulic and manual	+65 to −5 degrees
Elevation Rate: (max)	15 degrees/second
Firing Rate: (max)	3 rounds/minute
Loading System:	Manual
Stabilizer System:	None

Secondary:
(1) .50 caliber MG HB M2 on ring mount behind driver
Provision for (1) .45 caliber SMG M3A1
Provision for (4) .30 caliber Carbine M2

AMMUNITION

24 rounds 155mm	720 rounds .30 caliber (carbine)
900 rounds .50 caliber	8 hand grenades
180 rounds .45 caliber	

FIRE CONTROL AND VISION EQUIPMENT

Primary Weapon:	Direct	Indirect
	Telescope M93 (T153)	Panoramic Telescope M12A7K (M12A7E4)
		Gunner's Quadrant M1 or M1A1
		Fuze Setter M14, M22, M23, M26, or M27

Vision Devices:	Direct	Indirect
Driver	Open top	None
Commander	Open top	None
Gunner	Open top	None
Loaders	Open top	None

ENGINE

Make and Model:	Continental AOS-895-3 (M44)
	Continental AOSI-895-5 (M44A1)
Type: 6 cylinder, 4 cycle, opposed, supercharged (M44)	
6 cylinder, 4 cycle, opposed, supercharged, fuel injection (M44A1)	
Cooling System: Air Ignition: Magneto	
Displacement:	895.9 cubic inches
Bore and Stroke:	5.75 x 5.75 inches
Compression Ratio:	5.5:1
Net Horsepower: (max)	446 hp at 2400 rpm
Gross Horsepower: (max)	500 hp at 2800 rpm
Net Torque: (max)	890 ft-lb at 2200 rpm
Gross Torque: (max)	955 ft-lb at 2400 rpm
Weight:	approx. 1900 pounds, dry
Fuel: 80 octane gasoline	150 gallons
Engine Oil:	44 quarts

POWER TRAIN

Transmission: Cross-drive CD-500-3, 2 ranges forward, 1 reverse
w/automatic lock-up in high
Single stage hydraulic torque converter
Stall Multiplication: 3.9:1

Overall Usable Ratios:	low 14.9:1	direct	1:1
	high 3.9:1	reverse	14.9:1

Steering Control: Mechanical T-bar
Steering Rate: 13.2 rpm
Brakes: Multiple disc
Final Drive: Spur gear   Gear Ratio: 4.69:1
Drive Sprocket: At front of vehicle with 12 teeth
Pitch Diameter: 23.422 inches

RUNNING GEAR

Suspension: Torsion bar
10 individually sprung dual road wheels (5/track)
Tire Size: 25.5 x 4.5 inches
8 dual track return rollers (4/track)
Trailing idler at rear of each track
Idler Tire Size: 28 x 4.5 inches
Shock absorbers fitted on road wheels 1, 2, and 5 on each side
Tracks: Center guide T91E3
Type: (T91E3) Single pin, 21 inch width, steel w/detachable rubber pad
Pitch: 6 inches
Shoes per Vehicle: 149 (74 left, 75 right)
Ground Contact Length: 149.4 inches

ELECTRICAL SYSTEM

Nominal Voltage: 24 volts DC
Main Generator: (1) 24 volts, 150 amperes, driven by main engine
Auxiliary Generator: (1) 24 volts, 300 amperes, driven by auxiliary engine
Battery: (4) 12 volts, 2 sets of 2 in series connected in parallel

COMMUNICATIONS

Radio: AN/PRC 8, 9, or 10
Interphone: AN/VIC-1, 5 stations plus external extension kit C-980U

FIRE PROTECTION

(2) 10 pound carbon dioxide, fixed
(1) 5 pound carbon dioxide, portable

PERFORMANCE

Maximum Speed: Level road	35 miles/hour
Maximum Tractive Effort: TE at stall	53,000 pounds
Per Cent of Vehicle Weight: TE/W	83 per cent
Maximum Grade:	60 per cent
Maximum Trench:	6 feet
Maximum Vertical Wall:	30 inches
Maximum Fording Depth:	42 inches
Minimum Turning Circle: (diameter)	pivot
Cruising Range: Roads, M44	approx. 75 miles
M44A1	approx. 82 miles

105mm SELF-PROPELLED HOWITZER M108 AND 155mm SELF-PROPELLED HOWITZER M109

GENERAL DATA

Crew: M108	5 men
M109	6 men
Length: M108	240.7 inches
M109	260.4 inches
Length: Without howitzer	240.7 inches
Cannon Overhang: M108	0.0 inches
M109	19.7 inches
Width: w/o fenders	124.0 inches
Height: Over MG	129.1 inches
Tread:	109.0 inches
Ground Clearance:	17.7 inches
Fire Height:	approx. 78 inches
Turret Ring Diameter: (inside)	100 inches
Weight, Combat Loaded: M108	46,221 pounds
M109	52,461 pounds
Weight, Unstowed: M108	36,000 pounds
M109	44,723 pounds
Power to Weight Ratio: Net, M108	14.9 hp/ton
M109	13.2 hp/ton
Gross, M108	17.5 hp/ton
M109	15.5 hp/ton
Ground Pressure: Zero penetration, M108	9.9 psi
M109	11.2 psi

ARMOR

Type: Turret, rolled 5083 aluminum alloy; Hull, rolled 5083 aluminum alloy;
   Welded assembly

Hull Thickness:	Actual	Angle w/Vertical
Front, Upper	1.25 inches (32mm)	75 degrees
Lower	1.25 inches (32mm)	19 and 60 degrees
Sides	1.25 inches (32mm)	0 degrees
Rear	1.25 inches (32mm)	0 degrees
Top	1.25 inches (32mm)	90 degrees
Floor	1.25 inches (32mm)	90 degrees
Turret Thickness:		
Front	1.25 inches (32mm)	22 degrees
Sides	1.25 inches (32mm)	22 degrees
Rear	1.25 inches (32mm)	0 degrees
Top	1.25 inches (32mm)	90 degrees

ARMAMENT

Primary: (M108) 105mm Howitzer M103 (XM103) in Mount M139 (XM139)
   (M109) 155mm Howitzer M126 (T255E3) or M126A1 in Mount
   M127 (XM127)

Traverse: (M108) Manual	360 degrees
(M109) Hydraulic and manual	360 degrees
Traverse Rate: (max M109)	11 degrees/second
Elevation: (M108) Manual	+75 to −6 degrees
(M109) Hydraulic and manual	+75 to −3 degrees
Elevation Rate: (max M109)	7 degrees/second
Firing Rate: (max M108)	10 rounds/minute
(max M109)	4 rounds/minute
Loading System: (M108)	Manual
(M109)	Semiautomatic
Stabilizer System:	None

Secondary:
   (1) .50 caliber MG HB M2 flexible AA mount on turret hatch
   Provision for (5) 7.62mm Rifle M14 (M108)
   Provision for (6) 7.62mm Rifle M14 (M109)
   Provision for (1) 3.5 inch Rocket Launcher M20 series

AMMUNITION

86 rounds 105mm (M108)	6 3.5 inch rockets
28 rounds 155mm (M109)	12 hand grenades
500 rounds .50 caliber	
750 rounds 7.62mm (M108)	
900 rounds 7.62mm (M109)	

FIRE CONTROL AND VISION EQUIPMENT

Primary Weapon:	Direct	Indirect
(M108) Telescope M118	Panoramic Telescope M117
(M109) Telescope M118C	Elevation Quadrant M15
w/ Periscope M42	

Vision Devices:	Direct	Indirect
Driver	Hatch	Periscope M45 (3)
Commander	Hatch	Periscope M27 (1)
Gunner	None	None
Asst. Gunner	None	None
Loaders	None	None

Total Periscopes: M27 (1), M45 (3)

ENGINE

Make and Model: General Motors 8V71T	
Type: 8 cylinder, 2 cycle, vee, supercharged	
Cooling System: Liquid Ignition: Compression	
Displacement:	567.4 cubic inches
Bore and Stroke:	4.25 x 5 inches
Compression Ratio:	17:1
Net Horsepower: (max)	345 hp at 2300 rpm
Gross Horsepower: (max)	405 hp at 2300 rpm
Net Torque: (max)	895 ft-lb at 1600 rpm
Gross Torque: (max)	980 ft-lb at 1700 rpm
Weight:	2442 pounds, dry
Fuel: 40 cetane diesel oil	135 gallons
Engine Oil:	36 quarts, 28 quarts at refill

POWER TRAIN

Transmission: X-drive, XTG-411-2A, 4 ranges forward, 2 reverse
   Single stage hydraulic torque converter w/lock-up clutch
   Stall Multiplication: 3.3:1

Overall Usable Ratios:	1st 4.69:1	4th	0.79:1
	2nd 3.18:1	reverse 1	5.60:1
	3rd 1.58:1	reverse 2	3.79:1

Steering Controls: Mechanical, steering wheel
   Steering System: Clutch-brake (1st, 2nd, and 1st reverse)
      Geared steer (3rd, 4th, and 2nd reverse)
   Steering Ratio: 1.477:1
Brakes: Multiple disc
Final Drive: Spur gear   Gear Ratio: 4.36:1
Drive Sprocket: At front of vehicle with 10 teeth
   Pitch Diameter: 19.624 inches

RUNNING GEAR

Suspension: Flat track, torsion bar
   14 individually sprung dual road wheels (7/track)
   Tire Size: 24 x 4 inches
   Dual adjustable idler at rear of each track
   Idler Size: 18 x 4 inches
   Shock absorbers fitted on first and last road wheels on each side
Tracks: Center guide T136 and T137
   Type: (T136) Double pin, 15 inch width, steel w/detachable rubber pad
      (T137) Single pin, 15 inch width, steel w/detachable rubber pad
   Pitch: 6 inches
   Shoes per Vehicle: 158 (79/track)
   Ground Contact Length: 156 inches

ELECTRICAL SYSTEM

Nominal Voltage: 24 volts DC
Main Generator: (Alternator) (1) 24 volts, 100 amperes, driven by main engine
Auxiliary Generator: None
Battery: (4) 12 volts, 2 sets of 2 in series connected in parallel

COMMUNICATIONS

Radio: None
Interphone: AN/UIC-1, 5 stations w/extension kit C-980/U

FIRE PROTECTION

   (2) 10 pound carbon dioxide, fixed
   (1) 5 pound carbon dioxide, portable

PERFORMANCE

Maximum Speed: Level road	35 miles/hour
Maximum Tractive Effort: TE at stall	53,750 pounds
Per Cent of Vehicle Weight: TE/W, M108	116 per cent
M109	102 per cent
Maximum Grade:	60 per cent
Maximum Trench:	6 feet
Maximum Vertical Wall:	21 inches
Maximum Fording Depth:	Amphibious w/flotation device
Minimum Turning Circle: (diameter)	pivot
Cruising Range: Roads	approx. 220 miles

155mm SELF-PROPELLED HOWITZERS M109A2 AND M109A3

GENERAL DATA

Crew:	6 men
Length: Howitzer forward	359.4 inches
Length: Without howitzer	243.8 inches
Howitzer Overhang: Howitzer forward	115.6 inches
Width:	124.0 inches
Height: Over MG	129.1 inches
Tread:	109.0 inches
Ground Clearance:	17.7 inches
Fire Height:	approx. 78 inches
Turret Ring Diameter: (inside)	100 inches
Weight, Combat Loaded:	55,000 pounds
Weight, Unstowed:	46,500 pounds
Power to Weight Ratio: Net	12.5 hp/ton
Gross	14.7 hp/ton
Ground Pressure: Zero penetration	11.8 psi

ARMOR

Type: Turret, rolled 5083 aluminum alloy; Hull rolled 5083 aluminum alloy; Welded assembly

Hull Thickness:

	Actual	Angle w/Vertical
Front, Upper	1.25 inches (32mm)	75 degrees
Lower	1.25 inches (32mm)	19 and 60 degrees
Sides	1.25 inches (32mm)	0 degrees
Rear	1.25 inches (32mm)	0 degrees
Top	1.25 inches (32mm)	90 degrees
Floor	1.25 inches (32mm)	90 degrees

Turret Thickness:

Front	1.25 inches (32mm)	22 degrees
Sides	1.25 inches (32mm)	22 degrees
Rear	1.25 inches (32mm)	0 degrees
Top	1.25 inches (32mm)	90 degrees

ARMAMENT

Primary: 155mm Howitzer M185 in Mount M178

Traverse: Hydraulic and manual	360 degrees
Traverse Rate: (max)	11 degrees/second
Elevation: Hydraulic and manual	+75 to −3 degrees
Elevation Rate: (max)	7 degrees/second
Firing Rate: (max)	4 rounds/minute
Loading System:	Semiautomatic
Stabilizer System:	None

Secondary:
(1) .50 caliber MG HB M2 flexible AA mount on turret hatch
Provision for (6) 5.56mm Rifle M16A1

AMMUNITION

36 rounds 155mm including 2 CLGP M712 12 hand grenades
(Copperhead)
500 rounds .50 caliber
1200 rounds 5.56mm

FIRE CONTROL AND VISION EQUIPMENT

Primary Weapon:	Direct	Indirect
	Telescope M118CA1	Panoramic Telescope M145
	w/Periscope M42	Elevation Quadrant M15
Vision Devices:	Direct	Indirect
Driver	Hatch	Periscope M45 (3)
Commander	Hatch	Periscope M27 (1)
Gunner	None	None
Asst. Gunner	None	None
Loaders	None	None

Total Periscopes: M27 (1), M45 (3)

ENGINE

Make and Model: General Motors 8V71T	
Type: 8 cylinder, 2 cycle, vee, supercharged	
Cooling System: Liquid Ignition: Compression	
Displacement:	567.4 cubic inches
Bore and Stroke:	4.25 x 5 inches
Compression Ratio:	17:1
Net Horsepower: (max)	345 hp at 2300 rpm
Gross Horsepower: (max)	405 hp at 2300 rpm
Net Torque: (max)	895 ft-lb at 1600 rpm
Gross Torque: (max)	980 ft-lb at 1700 rpm
Weight:	2442 pounds, dry
Fuel: 40 cetane diesel oil	135 gallons
Engine Oil:	36 quarts, 28 quarts at refill

POWER TRAIN

Transmission: X-drive, XTG-411-2A, 4 ranges forward, 2 reverse
Single stage hydraulic torque converter w/lock-up clutch
Stall Multiplication: 3.3:1

Overall Usable Ratios:	1st 4.69:1	4th	0.79:1
	2nd 3.18:1	reverse 1	5.60:1
	3rd 1.58:1	reverse 2	3.79:1

Steering Controls: Mechanical, steering wheel
Steering System: Clutch-brake (1st, 2nd, and 1st reverse)
Geared steer (3rd, 4th, and 2nd reverse)
Steering Ratio: 1.477:1
Brakes: Multiple disc
Final Drive: Spur gear   Gear Ratio: 4.36:1
Drive Sprocket: At front of vehicle with 10 teeth
Pitch Diameter: 19.624 inches

RUNNING GEAR

Suspension: Flat track, torsion bar
14 individually sprung dual road wheels (7/track)
Tire Size: 24 x 4 inches
Dual adjustable idler at rear of each track
Idler Size: 18 x 4 inches
Shock absorbers fitted on first and last road wheels on each side
Tracks: Center guide T136 and T137
Type: (T136) Double pin, 15 inch width, steel w/detachable rubber pad

Pitch: 6 inches
Shoes per Vehicle: 158 (79/track)
Ground Contact Length: 156 inches

ELECTRICAL SYSTEM

Nominal Voltage: 24 volts DC
Main Generator: (Alternator) (1) 24 volts, 100 amperes, driven by main engine
Auxiliary Generator: None
Battery: (4) 12 volts, 2 sets of 2 in series connected in parallel

COMMUNICATIONS

Radio: None
Interphone: AN/VIC-1, 6 stations w/extension kit C-988/U

FIRE PROTECTION

(2) 10 pound carbon dioxide, fixed
(1) 5 pound carbon dioxide, portable

PERFORMANCE

Maximum Speed: Level road	35 miles/hour
Maximum Tractive Effort: TE at stall	53,750 pounds
Per Cent of Vehicle Weight: TE/W	98 per cent
Maximum Grade:	60 per cent
Maximum Trench:	6 feet
Maximum Vertical Wall:	21 inches
Maximum Fording Depth:	42 inches
Minimum Turning Circle: (diameter)	pivot
Cruising Range: Roads	approx. 220 miles

155mm SELF-PROPELLED HOWITZER M109A6

GENERAL DATA

Crew:	4 men
Length:	384.0 inches
Length: Without howitzer	271.4 inches
Howitzer Overhang: Howitzer in travel position	112.6 inches
Width: Over turret stowage racks	154.4 inches
Height: Over MG mount	127.4 inches
Tread:	109.0 inches
Ground Clearance:	17.1 inches
Fire Height:	approx. 78 inches
Turret Ring Diameter: (inside)	100 inches
Weight, Combat Loaded:	63,600 pounds
Weight, Unstowed:	56,400 pounds
Power to Weight Ratio: Net	10.8 hp/ton
Gross	12.7 hp/ton
Ground Pressure: Zero penetration	13.6 psi

ARMOR

Type: Turret, rolled 5083 aluminum alloy; Hull, rolled 5083 aluminum alloy; Welded assembly; Aramid spall liners and steel applique armor on turret bustle

Hull Thickness:	Actual	Angle w/Vertical
Front, Upper	1.25 inches (32mm)	75 degrees
Lower	1.25 inches (32mm)	19 and 60 degrees
Sides	1.25 inches (32mm)	0 degrees
Rear	1.25 inches (32mm)	0 degrees
Top	1.25 inches (32mm)	90 degrees
Floor	1.25 inches (32mm)	90 degrees
Turret Thickness		
Front	1.25 inches (32mm)	22 degrees
Sides plus steel applique armor on bustle sides	1.25 inches (32mm)	22 degrees
Rear	1.25 inches (32mm)	0 degrees
Top	1.25 inches (32mm)	90 degrees

ARMAMENT

Primary: 155mm Howitzer M284 in Mount M182A1

Traverse: Hydraulic and manual	360 degrees
Traverse Rate: (max)	11 degrees/second
Elevation: Hydraulic and manual	+75 to −3 degrees
Elevation Rate: (max)	7 degrees/second
Firing Rate: (max) 3 rounds/15 seconds,	6 rounds/minute
Loading System:	Semiautomatic
Stabilizer System:	None

Secondary:
  (1) .50 caliber MG HB M2 flexible AA mount on turret hatch
  Provision for (4) 5.56mm Rifles M16A1

AMMUNITION

39 rounds 155mm including CLGP M712 (Copperhead)
500 rounds .50 caliber
800 rounds 5.56mm

FIRE CONTROL AND VISION EQUIPMENT

Primary Weapon:	Direct	Indirect
	Elbow Telescope	Panoramic Telescope
		Elevation Quadrant
		Automatic Fire Control
		w/inertial navigation
		and positioning
Vision Devices:	Direct	Indirect
Driver	Hatch	Periscope M45 (3)
Commander	Hatch	Periscope M27 (1)
Gunner	None	None
Asst. Gunner	None	None

Total Priscopes: M27 (1), M45 (3)

ENGINE

Make and Model: General Motors 8V71T	
Type: 8 cylinder, 2 cycle, vee, supercharged	
Cooling System: Liquid Ignition: Compression	
Displacement:	567.4 cubic inches
Bore and Stroke:	4.25 x 5 inches
Compression Ratio:	17:1
Net Horsepower: (max)	345 hp at 2300 rpm
Gross Horsepower: (max)	405 hp at 2300 rpm
Net Torque: (max)	895 ft-lb at 1600 rpm
Gross Torque: (max)	980 ft-lb at 1700 rpm
Weight:	2442 pounds, dry
Fuel: 40 cetane diesel oil	133 gallons
Engine Oil:	36 quarts, 18 quarts at refill

POWER TRAIN

Transmission: X-drive, XTG-411-4, 4 ranges forward, 2 reverse w/quick disconnect for towing
  Single stage hydraulic torque converter w/lock-up clutch
  Stall Multiplication: 3.3:1

Overall Usable Ratios:	1st 4.69:1	4th 0.79:1
	2nd 3.18:1	reverse 1 5.60:1
	3rd 1.58:1	reverse 2 3.79:1

Steering Controls: Mechanical, steering wheel
  Steering System: Clutch-brake (1st, 2nd, and 1st reverse)
         Geared steer (3rd, 4th, and 2nd reverse)
Steering Ratio: 1.477:1
Brakes: Multiple disc
Final Drive: Spur gear   Gear Ratio: 4.36:1
Drive Sprocket: At front of vehicle with 10 teeth
  Pitch Diameter: 19.624 inches

RUNNING GEAR

Suspension: Flat track, high strength torsion bar
  14 individually sprung dual road wheels (7/track)
  Tire Size: 24 x 4 inches
  Dual adjustable idler at rear of each track
  Idler Size: 18 x 4 inches
  High capacity shock absorbers on first and last road wheels on each side
Tracks: Center guide
  Type: Double pin, 15 inch width, steel w/detachable rubber pad
  Pitch: 6 inches
  Shoes per Vehicle: 158 (79/track)
  Ground Contact Length: 156 inches

ELECTRICAL SYSTEM

Nominal Voltage: 24 volts DC
Main Generator: (Alternator) 24 volts, 650 amperes, driven by main engine
Auxiliary Generator: None
Battery: (4) 12 volts, 2 sets of 2 in series connected in parallel

COMMUNICATIONS

Radio: AN/VIC-1, AN/VRC-89 or SINCGARS
Interphone: 4 stations

FIRE AND NBC PROTECTION

  Automatic Halon fire extinguisher system
  NBC system w/climate control

PERFORMANCE

Maximum Speed: Level road	35 miles/hour
Maximum Tractive Effort: TE at stall	53,750 pounds
Per Cent of Vehicle Weight: TE/W	85 per cent
Maximum Grade:	60 per cent
Maximum Trench:	6 feet
Maximum Vertical Wall:	21 inches
Maximum Fording Depth:	42 inches
Minimum Turning Circle: (diameter)	pivot
Cruising Range: Roads	approx. 215 miles

175mm SELF-PROPELLED GUN M107 AND 8 inch SELF-PROPELLED HOWITZER M110

GENERAL DATA

Crew:		5 men
Length: Cannon in travel position, M107		444.8 inches
	M110	294.4 inches
Length: Without cannon		254.3 inches
Cannon Overhang: M107		190.5 inches
	M110	40.1 inches
Width:		124.0 inches
Height: Cannon in travel position, M107		136.8 inches
	M110	115.6 inches
Tread:		106.0 inches
Ground Clearance:		17.4 inches
Fire Height:		approx. 80 inches
Weight, Combat Loaded: M107		62,100 pounds
	M110	58,500 pounds
Weight, Unstowed: M107		57,600 pounds
	M110	53,500 pounds
Power to Weight Ratio: Net, M107		13.5 hp/ton
	M110	14.4 hp/ton
	Gross, M107	14.5 hp/ton
	M110	15.4 hp/ton
Ground Pressure: Zero penetration, M107		11.7 psi
	M110	11.0 psi

ARMOR
Type: Rolled homogeneous steel; Welded assembly

Hull Thickness:	Actual	Angle w/Vertical
Front	0.5 inches (13mm)	0 degrees
Driver's Compartment		
Sides and Rear	0.5 inches (13mm)	0 degrees
Top	0.5 inches (13mm)	90 degrees
Remainder of Hull	Unarmored	

ARMAMENT
Primary: (M107) 175mm Gun M113 (T256E3) in Mount M158 (T185)

(M110) 8 inch Howitzer M2A2 (M2A1E1) in Mount M158 (T185)

Traverse: Hydraulic and manual	60 degrees (30 degrees left or right)
Traverse Rate: (max)	8 degrees/second
Elevation: Hydraulic and manual	+65 to −2 degrees
Elevation Rate: (max)	6 degrees/second
Firing Rate: (max)	1.5 rounds/minute
Loading System:	Semiautomatic
Stabilizer System:	None

Secondary:

Provision for (1) .45 caliber SMG M3A1

Provision for (4) 7.62mm Rifle M14

AMMUNITION
2 rounds 175mm (M107) 8 hand grenades

2 rounds 8 inch (M110)

180 rounds .45 caliber

720 rounds 7.62mm

FIRE CONTROL AND VISION EQUIPMENT

Primary Weapon:	Direct	Indirect
	(M107) Telescope M116C	Panoramic Telescope M115
	(M110) Telescope M116	Elevation Quadrant M15
		Gunner's Quadrant M1A1
Vision Devices:	Direct	Indirect
Driver	Hatch	Periscope M17 (3)
Commander	Open vehicle	None
Gunner	Open vehicle	None
Loaders	Open vehicle	None
Total Periscopes: M17 (3)		

ENGINE

Make and Model: General Motors 8V71T	
Type: 8 cylinder, 2 cycle, vee, supercharged	
Cooling System: Liquid Ignition: Compression	
Displacement:	567.4 cubic inches
Bore and Stroke:	4.25 x 5 inches
Compression Ratio:	17:1
Net Horsepower: (max)	345 hp at 2300 rpm
Gross Horsepower: (max)	405 hp at 2300 rpm
Net Torque: (max)	895 ft-lb at 1600 rpm
Gross Torque: (max)	980 ft-lb at 1700 rpm
Weight:	2442 pounds, dry
Fuel: 40 cetane diesel oil	300 gallons
Engine Oil:	36 quarts, 28 quarts at refill

POWER TRAIN
Transmission: X-drive, XTG-411-2A, 4 ranges forward, 2 reverse

Single stage hydraulic torque converter w/lock-up clutch

Stall Multiplication: 3.3:1

Overall Usable Ratios:	1st 4.69:1	4th	0.79:1
	2nd 3.18:1	reverse 1	5.60:1
	3rd 1.58:1	reverse 2	3.79:1

Steering Control: Mechanical, steering bar

Steering System: Clutch-brake (1st, 2nd, and 1st reverse)

Geared steer (3rd, 4th, and 2nd reverse)

Steering Ratio: 1.477:1

Brakes: Multiple disc

Final Drive: Planetary gear   Gear Ratio: 5.35:1

Drive Sprocket: At front of vehicle with 11 teeth

Pitch Diameter: 21.297 inches

RUNNING GEAR
Suspension: Flat track, torsion bar

10 individually sprung dual road wheels (5/track)

Tire Size: 32 x 4 inches

All road wheels fitted with hydraulic lock-out cylinders which serve as shock absorbers when the vehicle is moving. The rear road wheels also serve as adjustable trailing idlers.

Tracks: Center guide T132

Type: (T132) Single pin, 18 inch width, steel w/detachable rubber pad

Pitch: 6 inches

Shoes per Vehicle: 151 (76 right, 75 left)

Ground Contact Length: 148 inches

ELECTRICAL SYSTEM
Nominal Voltage: 24 volts DC

Main Generator: (1) 24 volts, 300 amperes, driven by main engine

Auxiliary Generator: None

Battery: (4) 12 volts, 2 sets of 2 in series connected in parallel

COMMUNICATIONS
Radio: None

Interphone: AN/UIC-1, 3 stations

FIRE PROTECTION
(2) 10 pound carbon dioxide, fixed

(1) 5 pound carbon dioxide, portable

PERFORMANCE

Maximum Speed: Level road		34 miles/hour
Maximum Tractive Effort: TE at stall		49,200 pounds
Per Cent of Vehicle Weight: TE/W, M107		79 per cent
	M110	84 per cent
Maximum Grade:		60 per cent
Maximum Trench:		7 feet
Maximum Vertical Wall:		40 inches
Maximum Fording Depth:		42 inches
Minimum Turning Circle: (diameter)		pivot
Cruising Range: Roads		approx. 450 miles

8 inch SELF-PROPELLED HOWITZER M110A2

GENERAL DATA

Crew:	5 men
Length: Howitzer in travel position	422.5 inches
Length: Without howitzer	254.3 inches
Howitzer Overhang:	168.2 inches
Width:	124.0 inches
Height: Howitzer in travel position	123.8 inches
Tread:	106.0 inches
Ground Clearance:	17.4 inches
Fire Height:	approx. 80 inches
Weight, Combat Loaded:	62,500 pounds
Weight, Unstowed:	57,500 pounds
Power to Weight Ratio: Net	11.0 hp/ton
Gross	13.0 hp/ton
Ground Pressure: Zero penetration	11.7 psi

ARMOR
Type: Rolled homogeneous steel; Welded assembly

Hull Thickness:	Actual	Angle w/Vertical
Front	0.5 inches (13mm)	0 degrees
Driver's Compartment		
Sides and Rear	0.5 inches	(13mm) 0 degrees
Top	0.5 inches (13mm)	90 degrees
Remainder of Hull	Unarmored	

ARMAMENT
Primary: 8 inch Howitzer M201A1 in Mount M158

Traverse: Hydraulic and manual	60 degrees (30 degrees left or right)
Traverse Rate: (max)	8 degrees/second
Elevation: Hydraulic and manual	+65 to −2 degrees
Elevation Rate: (max)	6 degrees/second
Firing Rate: (max)	1.5 rounds/minute
Loading System:	Semiautomatic
Stabilizer System:	None

Secondary:
Provision for (5) 5.56mm Rifle M16A1

AMMUNITION
2 rounds 8 inch 8 hand grenades
750 rounds 5.56mm

FIRE CONTROL AND VISION EQUIPMENT

Primary	Direct	Indirect Weapon:
	Telescope M116	Panoramic Telescope M115
		Elevation Quadrant M15
		Gunner's Quadrant M1A1

Vision Devices:	Direct	Indirect
Driver	Hatch	Periscope M17 (3)
Commander	Open vehicle	None
Gunner	Open vehicle	None
Loaders	Open vehicle	None

Total Periscopes: M17 (3)

ENGINE
Make and Model: General Motors 8V71T
Type: 8 cylinder, 2 cycle, vee, supercharged
Cooling System: Liquid   Ignition: Compression

Displacement:	567.4 cubic inches
Bore and Stroke:	4.25 x 5 inches
Compression Ratio:	17:1
Net Horsepower: (max)	345 hp at 2300 rpm
Gross Horsepower: (max)	405 hp at 2300 rpm
Net Torque: (max)	895 ft-lb at 1600 rpm
Gross Torque: (max)	980 ft-lb at 1700 rpm
Weight:	2442 pounds, dry
Fuel: 40 cetane diesel oil	260 gallons
Engine Oil:	36 quarts, 28 quarts at refill

POWER TRAIN
Transmission: X-drive, XTG-411-2A, 4 ranges forward, 2 reverse
Single stage hydraulic torque converter w/lock-up clutch
Stall Multiplication: 3.3:1

Overall Usable Ratios:	1st 4.69:1	4th	0.79:1
	2nd 3.18:1	reverse 1	5.60:1
	3rd 1.58:1	reverse 2	3.79:1

Steering Control: Mechanical, steering bar
Steering System: Clutch-brake (1st, 2nd, and 1st reverse)
Geared steer (3rd, 4th, and 2nd reverse)
Steering Ratio: 1.477:1
Brakes: Multiple disc
Final Drive: Planetary gear   Gear Ratio: 5.35:1
Drive Sprocket: At front of vehicle with 11 teeth
Pitch Diameter: 21.297 inches

RUNNING GEAR
Suspension: Flat track, torsion bar
10 individually sprung dual road wheels (5/track)
Tire Size: 32 x 4 inches
All road wheels fitted with hydraulic lock-out cylinders which serve as shock absorbers when the vehicle is moving. The rear road wheels also serve as adjustable trailing idlers.
Tracks: Center guide, T132
Type: (T132) Single pin, 18 inch width, steel w/detachable rubber pad
Pitch: 6 inches
Shoes per Vehicle: 151 (76 right, 75 left)
Ground Contact Length: 148 inches

ELECTRICAL SYSTEM
Nominal Voltage: 24 volts DC
Main Generator: (1) 24 volts, 300 amperes, driven by main engine
Auxiliary Generator: None
Battery: (4) 12 volts, 2 sets of 2 in series connected in parallel

COMMUNICATIONS
Radio: None
Interphone: AN/UIC-1, 3 stations

FIRE PROTECTION
(2) 10 pound carbon dioxide, fixed
(1) 5 pound carbon dioxide, portable

PERFORMANCE

Maximum Speed: Level road	34 miles/hour
Maximum Tractive Effort: TE at stall	49,200 pounds
Per Cent of Vehicle Weight: TE/W	79 per cent
Maximum Grade:	60 per cent
Maximum Trench:	6 feet
Maximum Vertical Wall:	40 inches
Maximum Fording Depth:	42 inches
Minimum Turning Circle: (diameter)	pivot
Cruising Range: Roads	approx. 325 miles

LIGHT RECOVERY VEHICLE M578 (T120E1)

GENERAL DATA

Crew: 3 men
Length: Crane in travel position 250.3 inches
Length: Without crane 219.8 inches
Crane Overhang: Crane in travel position 30.5 inches
Width: 124.0 inches
Height: Over MG, crane in travel position 130.5 inches
Tread: 106.0 inches
Ground Clearance: 17.4 inches
Weight, Combat Loaded: 54,000 pounds
Weight, Unstowed: For air transport 47,000 pounds
Power to Weight Ratio: Net 12.8 hp/ton
Gross 15.0 hp/ton
Ground Pressure: Zero penetration 10.1 psi

ARMOR

Type: Rolled homogeneous steel; Welded assembly

Hull Thickness:	Actual	Angle w/Vertical
Front	0.5 inches (13mm)	0 degrees
Driver's Compartment		
Sides and Rear	0.5 inches (13mm)	0 degrees
Top	0.5 inches (13mm)	90 degrees
Remainder of Hull	Unarmored	
Cab Thickness:		
Front	0.5 inches (13mm)	20 degrees
Sides	0.5 inches (13mm)	0 degrees
Rear	0.5 inches (13mm)	0 degrees
Top	0.5 inches (13mm)	90 degrees

ARMAMENT

(1) .50 Caliber MG HB M2 flexible AA mount on rigger's cupola
Provision for (3) 7.62mm Rifle M14

AMMUNITION

500 rounds .50 caliber
450 rounds 7.62mm

RECOVERY EQUIPMENT

Spade: Hydraulically operated, on rear of vehicle
Tow Winch: 60,000 pound capacity, hydraulically operated, located in cab front w/225 feet of 1 inch diameter cable
Boom: Hydraulically operated box boom pivoted on upper cab front
Boom Length: 171 inches
Boom Traverse: 360 degrees
Boom Turning Radius: (around cab center) 98.4 inches
Boom Winch: 30,000 pound capacity, hydraulically operated, located in cab front w/350 feet of ⅜ inch diameter cable

VISION EQUIPMENT

	Direct	Indirect
Driver	Hatch	Periscope M17 (3)
Crane Operator	Hatch	Periscope M17 (6)
Rigger	Hatch	Periscope M17 (6)

Total Periscopes: M17 (15)

ENGINE

Make and Model: General Motors 8V71T
Type: 8 cylinder, 2 cycle, vee, supercharged
Cooling System: Liquid   Ignition: Compression
Displacement: 567.4 cubic inches
Bore and Stroke: 4.25 x 5 inches
Compression Ratio: 17:1
Net Horsepower: (max) 345 hp at 2300 rpm
Gross Horsepower: (max) 405 hp at 2300 rpm
Net Torque: (max) 895 ft-lb at 1600 rpm
Gross Torque: (max) 980 ft-lb at 1700 rpm
Weight: 2442 pounds, dry
Fuel: 40 cetane diesel oil 300 gallons
Engine Oil: 36 quarts, 28 quarts at refill

POWER TRAIN

Transmission: X-drive, XTG-411-2A, 4 ranges forward, 2 reverse
Single stage hydraulic torque converter w/lock-up clutch
Stall Multiplication: 3.3:1

Overall Usable Ratios:	1st 4.69:1	4th	0.79:1
	2nd 3.18:1	reverse 1	5.60:1
	3rd 1.58:1	reverse 2	3.79:1

Steering Controls: Mechanical, steering bar
Steering System: Clutch-brake (1st, 2nd, and 1st reverse)
Geared steer (3rd, 4th, and 2nd reverse)
Steering Ratio: 1.477:1
Brakes: Multiple disc
Final Drive: Planetary gear   Gear Ratio: 5.35:1
Drive Sprocket: At front of vehicle with 11 teeth
Pitch Diameter: 21.297 inches

RUNNING GEAR

Suspension: Flat track, torsion bar
10 individually sprung dual road wheels (5/track)
Tire Size: 32 x 4 inches
All road wheels fitted with hydraulic lock-out cylinders which serve as shock absorbers when vehicle is moving. The rear road wheels also serve as adjustable trailing idlers
Tracks: Center guide, T132
Type: (T132) Single pin, 18 inch width, steel w/detachable rubber pad
Pitch: 6 inches
Shoes per Vehicle: 151 (76 right, 75 left)
Ground Contact Length: 148 inches

ELECTRICAL SYSTEM

Nominal Voltage: 24 volts DC
Main Generator: (1) 24 volts, 300 amperes, driven by main engine
Auxiliary Generator: None
Battery: (4) 12 volts, 2 sets of 2 in series connected in parallel

COMMUNICATIONS

Radio: AN/VRC-46 at rear of cab
Interphone: C-2298/VRC, 3 stations

FIRE PROTECTION

(2) 10 pound carbon dioxide, fixed
(2) 5 pound carbon dioxide, portable

PERFORMANCE

Maximum Speed: Level road 37 miles/hour
Maximum Tractive Effort: TE at stall 49,200 pounds
Per Cent of Vehicle Weight: TE/W 91 per cent
Maximum Grade: 60 per cent
Maximum Trench: 7 feet
Maximum Vertical Wall: 40 inches
Maximum Fording Depth: 42 inches
Minimum Turning Circle: (diameter) pivot
Cruising Range: Roads approx. 450 miles

The primary light tank weapon after World War II was the high velocity 76mm gun. During the 1960s, it was replaced by the 152mm gun-launcher either firing combustible case conventional ammunition or launching a guided missile. Both types of ammunition depended upon an explosive shaped charge warhead to destroy the target. With the development of the new XM8 Armored Gun System, the main armament is once again a high velocity gun using a kinetic energy projectile. These weapons, as well as many of those employed as self-propelled artillery on lightweight chassis, are included in these data sheets. The dimensions of the various cannon have been simplified and are defined as indicated in the sketches below. For example, the forcing cone and the muzzle counterbore have been neglected. Shot travel is defined as the distance from the projectile base in the chamber to the muzzle.

The ammunition is listed in the data sheets according to the U.S. Army nomenclature in use during its period of greatest service. Since this did change and was sometimes confusing, a standard nomenclature is added in parentheses based on the following terms.

APBC	Armor piercing with ballistic cap
APCR	Armor piercing, composite rigid
APDS	Armor piercing, discarding sabot
APFSDS	Armor piercing, fin stabilized, discarding sabot
HE	High explosive
HEAT	High explosive, antitank, shaped charge
HESH	High explosive, squash head
HERA	High explosive, rocket assisted
CLGP	Cannon launched, guided projectile
APERS	Antipersonnel
MP	Multipurpose
CP	Concrete piercing
TP	Target practice
TPBC	Target practice with ballistic cap
TPCR	Target practice, composite rigid
-T	Tracer

The penetration performance for the armor piercing rounds has been omitted from the data sheets since some of the later weapons and their ammunition are still subject to security restrictions. Other details of some types of ammunition also have been left out for the same reason.

Muzzle brakes were installed on the 76mm tank guns as well as many of the self-propelled artillery weapons to reduce the recoil force on the lightweight chassis. Except for the late version of the 152mm gun-launcher, bore evacuators were fitted on all weapons fired from enclosed turret or barbette mounts. The late gun-launchers were equipped with a closed breech, compressed air, scavenger system to remove powder gases and residue from the bore.

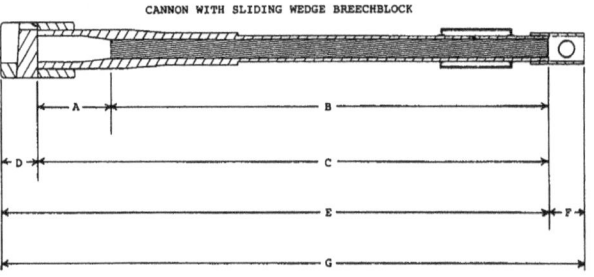

CANNON WITH SLIDING WEDGE BREECHBLOCK

A. Length of Chamber (to rifling)
B. Length of Rifling
C. Length of Bore
D. Depth of Breech Recess
E. Length, Muzzle to Rear Face of Breech
F. Additional Length, Blast Deflector, Etc.
G. Overall Length
H. Length, Breechblock and Firing Lock
I. Length of Tube
J. Length of Separable Chamber
K. Length of Tube and Chamber

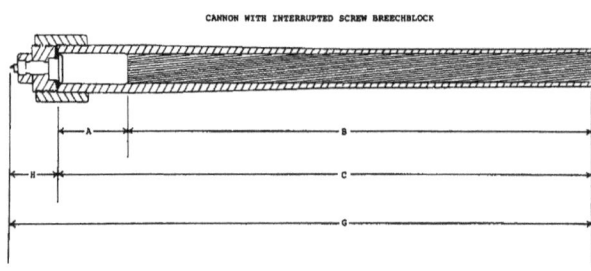

CANNON WITH INTERRUPTED SCREW BREECHBLOCK

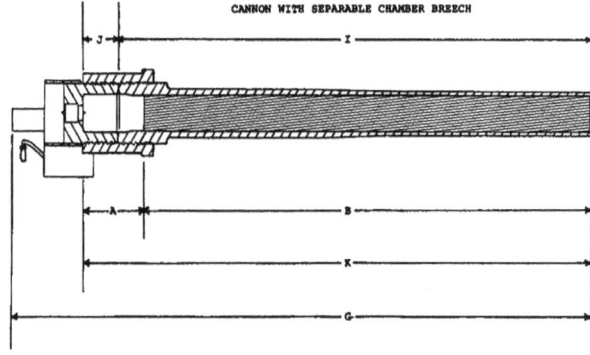

CANNON WITH SEPARABLE CHAMBER BREECH

40mm DUAL AUTOMATIC GUN M2A1

Carriage and Mount	Twin 40mm Self-Propelled Guns M42 and M42A1 in Mount M4E1
Length to Chamber (to rifling)	12.73 inches
Length of Rifling	75.85 inches
Length of Chamber	11.2 inches (square base shot AP-T M81A1)
	9.8 inches (boat-tail shell HE-T Mk II)
Length of Chamber (to projectile base)	77.4 inches (square base shot AP-T M81A1)
	78.8 inches (boat-tail shell HE-T Mk II)
Length of Bore	88.58 inches, 56.3 calibers
Dept of Breech Recess	5.9 inches, approx.
Length, Muzzle to Rear Face of Breech	95 inches, approx.
Length of Flash Suppressor	10 inches, approx.
Length of Automatic Loader Assembly	39 inches, approx.
Overall Length	144 inches, approx.
Diameter of Bore	1.573 inches (40mm)
Chamber Capacity	29.9 cubic inches
Weight of Barrel Assembly (each)	296 pounds
Total Weight	2000 pounds, approx.
Type of Breechblock	Semiautomatic, vertical sliding wedge
Rifling	16 grooves, increasing twist, one turn in 45 to 30 calibers
Automatic Loader	Each w/7 round magazine loaded from 4 round clips
Ammunition	Fixed
Primer	Percussion

Weight, Complete Round	AP-T M81A1 Shot (AP-T)	4.57 pounds (2.07 kg)
	HEI-T MkII Shell (HEI-T)	4.70 pounds (2.13 kg)
Weight, Projectile	AP-T M81A1 Shot (AP-T)	1.96 pounds (0.89 kg)
	HEI-T MkII Shell (HEI-T)	1.93 pounds (0.88 kg)
Maximum Rate of Fire	240 rounds/minute (120 rounds/gun)	
Muzzle Velocity	AP-T M81A1 Shot (AP-T)	2870 ft/sec (875 m/sec)
	HEI-T MkII Shell (HEI-T)	2870 ft/sec (875 m/sec)
Muzzle Energy of Projectile KE=½MV²	AP-T M81A1 Shot (AP-T)	112 ft-tons
Rotational energy is neglected and values are based on long tons (2240 pounds)	HEI-T MkII Shell (HEI-T)	110 ft-tons
Maximum Range (independent of mount)	AP-T M81A1 Shot (AP-T)	9475 yards (8664 m)
	HEI-T MkII Shell (HEI-T)	10,850 yards* (9921 m)

Penetration Performance   Homogeneous steel armor at 30 degrees obliquity

	Range	500 yards	1000 yards	1500 yards	2000 yards
AP-T M81A1 Shot (AP-T)		1.9 inches (48mm)	1.6 inches (41mm)	1.2 inches (30mm)	1.0 inches (25mm)

Face-hardened steel armor at 30 degrees obliquity

	Range	500 yards	1000 yards	1500 yards	2000 yards
AP-T M81A1 Shot (AP-T)		1.8 inches (46mm)	1.5 inches (38mm)	1.2 inches (30mm)	1.0 inches (25mm)

*Actual range limited by shell destroying tracer to approximately 5200 yards horizontal and 5100 yards vertical

76mm GUNS M32 (T91E3), T185, AND T185E1

Carriage and Mount	76mm Gun Tanks M41 and M41A2 in Mount M76 (T138E1) (M32 Gun)
	76mm Gun Tanks M41A1 and M41A3 in Mount M76A1 (T138E2) (M32 Gun)
	76mm Gun Tank T71 in Mount T138E2 (T185 Gun)
	76mm Gun Tank T92 in experimental mount (T185E1 Gun)
Length of Chamber (to rifling)	23.6 inches
Length of Rifling	156.4 inches
Length of Chamber (to projectile base)	19.6 inches (boat-tailed projectiles)
Travel of Projectile in Bore	160.4 inches (boat-tailed projectiles)
Length of Bore	180.0 inches, 60.0 calibers
Depth of Breech Recess	6.6 inches
Length, Muzzle to Rear Face of Breech	186.6 inches, 62.2 calibers
Additional Length	4.6 inches, w/early muzzle brake
	5.8 inches, w/late muzzle brake
Overall Length	191.2 inches, w/early muzzle brake
	192.4 inches, w/late muzzle brake
Diameter of Bore	3.000 inches
Chamber Capacity	197 cubic inches
Total Weight	1709 pounds, M32
	1425 pounds, T185 and T185E1
Type of Breechblock	Semiautomatic, vertical sliding wedge
	T185E1 inverted, breechblock moves up to open
Rifling	28 grooves, uniform right-hand twist, one turn in 25 calibers
Ammunition	Fixed
Primer	Percussion
Weight, Complete Round	AP-T M339 Shot (APBC-T)
	HVAP-T M319 (T66E3) Shot (APCR-T)
	HVAP-DS-T M331A2 Shot (APDS-T)
	HEAT-T M496 Shell (HEAT-T)
	HE M352 Shell (HE)
	WP M361 (T140) Shell, Smoke
	Canister M363 (T3E7)(909 steel balls)

Weight, Complete Round	AP-T M339 Shot (APBC-T)	*27.32 pounds (12.4 kg)
	HVAP-T M319 (T66E3) Shot (APCR-T)	*19.33 pounds (8.8 kg)
	HVAP-DS-T M331A2 Shot (APDS-T)	**20.72 pounds (9.4 kg)
	HEAT-T M496 Shell (HEAT-T)	†20.41 pounds (9.3 kg)
	HE M352 Shell (HE)	*25.83 pounds (11.7 kg)
	WP M361 (T140) Shell, Smoke	*25.82 pounds (11.7 kg)
	Canister M363 (T3E7)(909 steel balls)	*27.18 pounds (12.3 kg)
Weight, Projectile	AP-T M339 Shot (APBC-T)	14.56 pounds (6.6 kg)
	HVAP-T M319 (T66E3) Shot (APCR-T)	7.13 pounds (3.2 kg)
	HVAP-DS-T M331A2 Shot (APDS-T)	8.22 pounds (3.7 kg)
	HEAT-T M496 Shell (HEAT-T)	7.15 pounds (3.2 kg)
	HE M352 Shell (HE)	15.00 pounds (6.8 kg)
	WP M361 (T140) Shell, Smoke	15.71 pounds (7.1 kg)
	Canister M363 (T3E7)(909 steel balls)	15.00 pounds (6.8 kg)
Maximum Powder Pressure	46,000 psi	
Maximum Rate of Fire	12 rounds/minute	
Muzzle Velocity	AP-T M339 Shot (APBC-T)	3200 ft/sec (975 m/sec)
	HVAP-T M319 (T66E3) Shot (APCR-T)	4139 ft/sec (1262 m/sec)
	HVAP-DS-T M331A2 Shot (APDS-T)	4125 ft/sec (1257 m/sec)
	HEAT-T M496 Shell (HEAT-T)	3550 ft/sec (1082 m/sec)
	HE M352 Shell (HE)	2400 ft/sec (732 m/sec)
	WP M361 (T140) Shell, Smoke	2400 ft/sec (732 m/sec)
	Canister M363 (T3E7)(909 steel balls)	2900 ft/sec (884 m/sec)
Muzzle Energy of Projectile, KE=½MV²	AP-T M339 Shot (APBC-T)	1034 ft-tons
Rotational energy is neglected and	HVAP-T M319 (T66E3) Shot (APCR-T)	847 ft-tons
values are based on long tons	HVAP-DS-T M331A2 Shot (APDS-T)	970 ft-tons
(2240 pounds)	HEAT-T M496 Shell (HEAT-T)	625 ft-tons
	HE M352 Shell (HE)	574 ft-tons
	WP M361 (T140) Shell, Smoke	627 ft-tons
	Canister M363 (T3E7)(909 steel balls)	875 ft-tons
Maximum Range (independent of mount)	AP-T M339 Shot (APBC-T)	16,080 yards (14,704 m)
	HVAP-T M319 (T66E3) Shot (APCR-T)	10,810 yards (9,885 m)
	HVAP-DS-T M331A2 Shot (APDS-T)	23,630 yards (21,607 m)
	HEAT-T M496 Shell (HEAT-T)	effective 2,190 yards (2,003 m)
	HE M352 Shell (HE)	15,680 yards (14,338 m)
	WP M361 (T140) Shell, Smoke	16,070 yards (14,694 m)
	Canister M363 (T3E7)(909 steel balls)	170 yards (155 m)

*Assembled with M88 (T19E1) brass cartridge case (weight 6.66 pounds)
**Assembled with M88B1 (T19E1B1) steel cartridge case (weight 6.22 pounds)
†Assembled with M171E1 brass cartridge case

90mm GUN M54 (T125)

Carriage and Mount	90mm Self-Propelled Gun M56 (T101) in Mount M88 (T70E1)	
Length of Chamber (to rifling)	24.75 inches	
Length of Rifling	152.4 inches	
Length of Chamber (to projectile base)	20.75 inches	
Travel of Projectile in Bore	156.4 inches	
Length of Bore	177.15 inches, 50.0 calibers	
Depth of Breech Recess	9.00 inches	
Length, Muzzle to Rear Face of Breech	186.15 inches, 52.5 calibers	
Additional Length, Cylindrical Blast Deflector	6.5 inches	
Overall Length	192.7 inches	
Diameter of Bore	3.543 inches	
Chamber Capacity	300 cubic inches	
Weight, Tube	1473 pounds	
Total Weight	2440 pounds	
Type of Breechblock	Vertical sliding wedge	
Rifling	32 grooves, uniform right-hand twist, one turn in 25 calibers	
Ammunition	Fixed	
Primer	Percussion	
Weight, Complete Round	AP-T M318 (T33E7) Shot (APBC-T)	** 43.91 pounds (19.9 kg)
	HEAT-T M431 Shell (HEAT-T)	32.25 pounds (14.6 kg)
	HE-T T91E3 Shell (HE-T)	* 36.25 pounds (16.5 kg)
	HE-T M71A1 Shell (HE-T)	† 39.54 pounds (17.9 kg)
	APERS-T XM580E1 (4100 flechettes)	41.25 pounds (18.7 kg)
	Canister M336 (1280 pellets)	* 42.50 pounds (19.3 kg)
	Canister M377 (5600 flechettes)	** 39.30 pounds (17.8 kg)
	TP-T M353 (T225E1) Shot (TPBC-T)	** 43.91 pounds (19.9 kg)
Weight, Projectile	AP-T M318 (T33E7) Shot (APBC-T)	24.18 pounds (11.0 kg)
	HEAT-T M431 Shell (HEAT-T)	12.75 pounds (5.8 kg)
	HE-T T91E3 Shell (HE-T)	20.25 pounds (9.2 kg)
	HE-T M71A1 Shell (HE-T)	23.57 pounds (10.7 kg)
	APERS-T XM580E1 (4100 flechettes)	approx. 20 pounds (9 kg)
	Canister M336 (1280 pellets)	23.24 pounds (10.5 kg)
	Canister M377 (5600 flechettes)	20.44 pounds (9.3 kg)
	TP-T M353 (T225E1) Shot (TPBC-T)	24.18 pounds (11.0 kg)
Maximum Powder Pressure	47,000 psi	
Maximum Rate of Fire	10 rounds/minute	
Muzzle Velocity	AP-T M318 (T33E7) Shot (APBC-T)	3000 ft/sec (914 m/sec)
	HEAT-T M431 Shell (HEAT-T)	4000 ft/sec (1219 m/sec)
	HE-T T91E3 Shell (HE-T)	2400 ft/sec (732 m/sec)
	HE-T M71A1 Shell (HE-T)	2400 ft/sec (732 m/sec)
	APERS-T XM580E1 (4100 flechettes)	3000 ft/sec (914 m/sec)
	Canister M336 (1280 pellets)	2870 ft/sec (875 m/sec)
	Canister M377 (5600 flechettes)	2950 ft/sec (899 m/sec)
	TP-T M353 (T225E1) Shot (TPBC-T)	3000 ft/sec (914 m/sec)
Muzzle Energy of Projectile, KE=½MV²	AP-T M318 (T33E7) Shot (APBC-T)	1509 ft-tons
Rotational energy is neglected and	HEAT-T M431 Shell (HEAT-T)	1414 ft-tons
values are based on long tons	HE-T T91E3 Shell (HE-T)	809 ft-tons
(2240 pounds)	HE-T M71A1 Shell (HE-T)	941 ft-tons
	APERS-T XM580E1 (4100 flechettes)	approx. 1250 ft-tons
	Canister M336 (1280 pellets)	1327 ft-tons
	Canister M377 (5600 flechettes)	1230 ft-tons
	TP-T M353 (T225E1) Shot (TPBC-T)	1509 ft-tons
Maximum Range (independent of mount)	AP-T M318 (T33E7) Shot (APBC-T)	23,000 yards (21,031 m)
	HEAT-T M431 Shell (HEAT-T)	8,900 yards (8,138 m)
	HE-T T91E3 Shell (HE-T)	14,500 yards (13,259 m)
	HE-T M71A1 Shell (HE-T)	16,800 yards (15,362 m)
	APERS-T XM580E1 (4100 flechettes)	4,800 yards (4,389 m)
	Canister M336 (1280 pellets)	200 yards (183 m)
	Canister M377 (5600 flechettes)	440 yards (402 m)
	TP-T M353 (T225E1) Shot (TPBC-T)	23,000 yards (21,031 m)

* Assembled with the M108 (T24) brass cartridge case (weight 11.0 pounds)
** Assembled with the M108B (T24B1) steel cartridge case (weight 10.3 pounds)
† Assembled with the M19 brass cartridge case (weight 11.0 pounds)
The HEAT-T M431 and APERS-T XM580E1 rounds were assembled with the M114E1 and XM200 cartridge cases respectively. In addition to the ammunition assembled with the M108 or M108B1 cartridge cases, this weapon could fire any of the rounds for the lower pressure M1, M2, and M3 series of 90mm guns fitted in the M19 or M19B1 cartridge cases.

105mm GUNS M68, M68A1, AND XM35

Carriage and Mount	105mm Gun Tanks M1, M60, M60A1, and M60A3 (M68 and M68A1 Guns), 105mm Gun Tanks M48A1E1 and M48A5 (M68 Gun), and 105mm Armored Gun System XM8 (XM35 Gun)	
Length of Chamber (to rifling)	24.9 inches	
Length of Rifling	185.557 inches (M68), 195.607 inches (XM35)	
Length of Chamber (to projectile base)	23.42 inches (APDS shot)	
Travel of Projectile in Bore	187.08 inches (APDS shot), M68	
Length of Bore	210.50 inches, 50.92 calibers (M68), 220.55 inches, 53.35 calibers (XM35)	
Depth of Breech Recess	8.00 inches (M68), 9.05 inches (XM35)	
Length, Muzzle to Rear Face of Breech	218.50 inches, 52.85 calibers (M68)	
	229.6 inches, 55.54 calibers (XM35 including integral muzzle brake)	
Diameter of Bore	4.134 inches	
Chamber Capacity	403 cubic inches	
Weight, Tube	1660 pounds (M68)	
Total Weight	2492 pounds (M68), 2080 pounds (XM35 including mount)	
Type of Breechblock	Semiautomatic, vertical sliding wedge	
Rifling	28 grooves, uniform right-hand twist, one turn in 18 calibers	
Ammunition	Fixed	
Primer	Electric	
Weight, Complete Round	APDS-T M392A2 Shot (APDS-T)	41.0 pounds (18.6 kg)
	APFSDS-T M735 Shot (APFSDS-T)	38 pounds (17 kg)
	HEP-T M393A1 Shell (HESH-T)	46.7 pounds (21.2 kg)
	HEAT-T M456 Shell (HEAT-T)	48.0 pounds (21.8 kg)
	APERS-T XM494E3 (5000 flechettes)	55.0 pounds (25.0 kg)
	WP-T M416 Shell (Smoke)	45.5 pounds (20.7 kg)
	TP-T M393A1 Shell (TP-T)	46.7 pounds (21.2 kg)
	TP-T M490 Shell (TP-T)	48.0 pounds (21.8 kg)
Weight, Projectile	APDS-T M392A2 Shot (APDS-T)	12.75 pounds (5.8 kg)
	APFSDS-T M735 Shot (APFSDS-T)	12.78 pounds (5.8 kg)
	HEP-T M393A1 Shell (HESH-T)	24.8 pounds (11.3 kg)
	HEAT-T M456 Shell (HEAT-T)	22.4 pounds (10.2 kg)
	APERS-T XM494E3 (5000 flechettes)	approx. 31 pounds (14 kg)
	WP-T M416 Shell (Smoke)	25.17 pounds (11.4 kg)
	TP-T M393A1 Shell (TP-T)	24.8 pounds (11.3 kg)
	TP-T M490 Shell (TP-T)	22.4 pounds (10.2 kg)
Maximum Powder Pressure	60,000 psi (M68), 83,000 psi (XM35)	
Maximum Rate of Fire	7 rounds/minute (M68), 12 rounds/minute (XM35 w/automatic loader)	
Muzzle Velocity	APDS-T M392A2 Shot (APDS-T)	4850 ft/sec (1478 m/sec)
	APFSDS-T M735 Shot (APFSDS-T)	4925 ft/sec (1501 m/sec)
	HEP-T M393A1 Shell (HESH-T)	2400 ft/sec (732 m/sec)
	HEAT-T M456 Shell (HEAT-T)	3850 ft/sec (1173 m/sec)
	APERS-T XM494E3 (5000 flechettes)	2700 ft/sec (823 m/sec)
	WP-T M416 Shell (Smoke)	2400 ft/sec (732 m/sec)
	TP-T M393A1 Shell (TP-T)	2400 ft/sec (732 m/sec)
	TP-T M490 Shell (TP-T)	3850 ft/sec (1173 m/sec)
Muzzle Energy of Projectile, KE=½MV² Rotational energy is neglected and values are based on long tons (2240 pounds)	APDS-T M392A2 Shot (APDS-T)	2079 ft-tons
	APFSDS-T M735 Shot (APFSDS-T)	2149 ft-tons
	HEP-T M393A1 Shell (HESH-T)	990 ft-tons
	HEAT-T M456 Shell (HEAT-T)	2302 ft-tons
	APERS-T XM494E3 (5000 flechettes)	1567 ft-tons
	WP-T M416 Shell (Smoke)	1005 ft-tons
	TP-T M393A1 Shell (TP-T)	990 ft-tons
	TP-T M490 Shell (TP-T)	2302 ft-tons
Maximum Range (independent of mount)	APDS-T M392A2 Shot (APDS-T)	40,162 yards (36,724 m)
	HEP-T M393A1 Shell (HESH-T)	10.400 yards (9510 m)
	HEAT-T M456 Shell (HEAT-T)	8,975 yards (8207 m)
	APERS-T XM494E3 (5000 flechettes)	4,800 yards (4389 m)
	WP-T M416 Shell (Smoke)	10,400 yards (9510 m)
	TP-T M393A1 Shell (TP-T)	10,400 yards (9510 m)
	TP-T M490 Shell (TP-T)	8,975 yards (8207 m)

The M68A1 differed in only minor details from the M68 and it could be fitted with a muzzle reference system. The XM35 was a lightweight weapon designed for use with a soft recoil system and it featured an integral muzzle brake consisting of holes bored through the rifled tube near the muzzle. Ammunition for these weapons was assembled with cartridge cases M115 (brass), M150 (brass), M150B1 (steel), M148A1 (brass), and M148A1B1 (steel).

105mm HOWITZER M49 (T96E1)

Carriage and Mount	105mm Self-Propelled Howitzer M52 (T98E1) and M52A1 in Mount M85 (T67E1)
Length of Chamber (to rifling)	15.0 inches
Length of Rifling	78 inches
Length of Chamber (to projectile base)	11.4 inches
Travel of Projectile in Bore	81.6 inches
Length of Bore	93.0 inches, 22.5 calibers
Depth of Breech Recess	6.8 inches
Length, Muzzle to Rear Face of Breech	99.8 inches, 24.1 calibers
Additional Length, Counterweight etc.	None
Overall Length	99.8 inches
Diameter of Bore	4.134 inches
Chamber Capacity	154 cubic inches
Total Weight	972 pounds
Type of Breechblock	Manually operated, vertical sliding wedge
Rifling	36 grooves, uniform right-hand twist, one turn in 20 calibers
Ammunition	Semifixed, variable charge except for HEAT-T M67
Primer	Percussion

Weight, Complete Round	HE M1 Shell (HE), Charge 7	42.07 pounds (19.1 kg)
	HEAT-T M67 Shell (HEAT-T)	36.85 pounds (16.7 kg)
	HC BE M84 Shell, Smoke, Charge 7	41.94 pounds (19.0 kg)
	WP M60 Shell, Smoke, Charge 7	43.77 pounds (19.9 kg)
Weight, Projectile	HE M1 Shell (HE)	33.00 pounds (15.0 kg)
	HEAT-T M67 Shell (HEAT-T)	29.22 pounds (13.3 kg)
	HC BE M84 Shell, Smoke	32.97 pounds (15.0 kg)
	WP M60 Shell, Smoke	34.31 pounds (15.6 kg)
Maximum Powder Pressure	32,000 psi	
Maximum Rate of Fire	8 rounds/minute	
Muzzle Velocity	HE M1 Shell (HE), Charge 7	1550 ft/sec (472 m/sec)
	HEAT-T M67 Shell (HEAT-T)	1250 ft/sec (381 m/sec)
	HC BE M84 Shell, Smoke, Charge 7	1550 ft/sec (472 m/sec)
	WP M60 Shell, Smoke, Charge 7	1550 ft/sec (472 m/sec)
Muzzle Energy of Projectile, KE=½MV²	HE M1 Shell (HE), Charge 7	550 ft-tons
Rotational energy is neglected and	HEAT-T M67 Shell (HEAT-T)	317 ft-tons
values are based on long tons	HE BE M84 Shell, Smoke, Charge 7	547 ft-tons
(2240 pounds)	WP M60 Shell, Smoke, Charge 7	571 ft-tons
Maximum Range (independent of mount)	HE M1 Shell (HE), Charge 7	12,205 yards (11,160 m)
	HEAT-T M67 Shell (HEAT-T)	8,590 yards (7,855 m)
	HC BE M84 Shell, Smoke, Charge 7	12,205 yards (11,160 m)
	WP M60 Shell, Smoke, Charge 7	12,150 yards (11,110 m)

Penetration Performance
HEAT-T M67

Homogeneous steel armor at 0 degrees obliquity
    4.0 inches at any range
Concrete at 0 degrees obliquity

	Range	0 yards	500 yards	1000 yards	2000 yards
HE M1 Shell, Charge 7 w/Concrete Piercing Fuze M78A1		1.5 feet	1.4 feet	1.3 feet	1.1 feet

105mm HOWITZER M103 (XM103)

Carriage and Mount	105mm Self-Propelled Howitzer M108 (T195E1) in Mount M139 (XM139) and 105mm Light Self-Propelled Howitzer XM104	
Length of Chamber (to rifling)	15.0 inches	
Length of Rifling	108.7 inches	
Muzzle Counterbore	0.5 inches	
Length of Chamber (to projectile base)	11.4 inches (boat-tailed projectiles)	
Travel of Projectile in Bore	112.8 inches (boat-tailed projectiles)	
Length of Bore	124.2 inches, 30 calibers	
Depth of Breech Recess	7.4 inches	
Length, Muzzle to Rear Face of Breech	131.6 inches (31.8 calibers)	
Additional Length	None	
Overall Length	131.6 inches	
Diameter of Bore	4.134 inches	
Chamber Capacity	153.8 cubic inches	
Total Weight	986 pounds	
Type of Breechblock	Manually operated vertical sliding wedge	
Rifling	36 grooves, increasing twist from one turn in 35 calibers at the breech to one turn in 18 calibers at the muzzle	
Ammunition	Semifixed, variable charge except for HEAT-T M67, HEP-T M327, and APERS-T M546	
Primer	Percussion	
Weight, Complete Round	HE M1 Shell (HE), Charge 7	42.07 pounds (19.1 kg)
	HEAT-T M67 Shell (HEAT-T)	36.85 pounds (16.7 kg)
	HEP-T M327 Shell (HESH-T)	33.45 pounds (15.2 kg)
	HERA M548 Shell (HERA), Charge 7 w/RA	38.49 pounds (17.5 kg)
	APERS-T M546 (8000 flechettes)	38.25 pounds (17.3 kg)
	HE M444 Projectile (18 M39 grenades), Charge 7	42.00 pounds (19.1 kg)
	WP M60 Shell, Smoke, Charge 7	43.77 pounds (19.9 kg)
	HC M84 Shell, Smoke, Charge 7	41.94 pounds (19.0 kg)
Weight, Projectile	HE M1 Shell (HE)	33.00 pounds (15.0 kg)
	HEAT-T M67 Shell (HEAT-T)	29.22 pounds (13.3 kg)
	HEP-T M327 Shell (HESH-T)	23.28 pounds (10.6 kg)
	HERA M548 Shell (HERA)	29.34 pounds (13.3 kg)
	APERS-T M546 (8000 flechettes)	28.50 pounds (12.9 kg)
	HE M444 Projectile (18 M39 grenades)	33.00 pounds (15.0 kg)
	WP M60 Shell, Smoke	34.31 pounds (15.6 kg)
	HC M84 Shell, Smoke	32.97 pounds (15.0 kg)
Maximum Powder Pressure	45,600 psi	
Maximum Rate of Fire	10 rounds/minute	
Muzzle Velocity	HE M1 Shell (HE), Charge 7	1621 ft/sec (494 m/sec)
	HEAT-T M67 Shell (HEAT-T)	1320 ft/sec (402 m/sec)
	HEP-T M327 Shell (HESH-T)	1970 ft/sec (600 m/sec)
	HERA M548 Shell (HERA), Charge 7	1800 ft/sec (549 m/sec)
	APERS-T M546 (8000 flechettes)	1800 ft/sec (549 m/sec)
	HE M444 Projectile (18 M39 grenades), Charge 7	1621 ft/sec (494 m/sec)
	WP M60 Shell, Smoke, Charge 7	1621 ft/sec (494 m/sec)
	HC M84 Shell, Smoke, Charge 7	1621 ft/sec (494 m/sec)
Muzzle Energy of Projectile, KE=½MV² Rotational energy is neglected and values are based on long tons (2240 pounds)	HE M1 Shell (HE), Charge 7	601 ft-tons
	HEAT-T M67 Shell (HEAT-T)	353 ft-tons
	HEP-T M327 Shell (HESH-T)	626 ft-tons
	HERA M548 Shell (HERA), Charge 7	659 ft-tons
	APERS-T M546 (8000 flechettes)	640 ft-tons
	HE M444 Projectile (18 M39 grenades), Charge 7	601 ft-tons
	WP M60 Shell, Smoke, Charge 7	625 ft-tons
	HC M84 Shell, Smoke, Charge 7	601 ft-tons
Maximum Range (independent of mount)	HE M1 Shell (HE), Charge 7	12,577 yards (11,500 m)
	HEAT-T M67 Shell (HEAT-T)	8,590 yards (7,855 m)
	HEP-T M327 Shell (HESH-T)	9,500 yards (8,687 m)
	HERA M548 Shell (HERA), Charge 7 w/RA	16,404 yards (15,000 m)
	APERS-T M546 (8000 flechettes)	13,560 yards (12,400 m)
	HE M444 Projectile (18 M39 grenades), Charge 7	12,577 yards (11,500 m)
	WP M60 Shell, Smoke, Charge 7	12,577 yards (11,500 m)
	HC M84 Shell, Smoke, Charge 7	12,577 yards (11,500 m)

152mm GUN-LAUNCHERS M81 MODIFIED AND M81E1

Carriage and Mount	AR/AAV M551 and M551A1	
Length of Chamber (to rifling)	10.5 inches	
Length of Rifling	94.55 inches	
Length of Chamber (to projectile base)	9 inches	
Travel of Projectile in Bore	96 inches	
Length of Tube and Chamber	105.1 inches, 17.52 calibers	
Overall Length	116 inches	
Diameter of Bore	6.000 inches	
Chamber Capacity	285 cubic inches	
Total Weight, M81 Modified	1125 pounds (w/bore evacuator)	
M81E1	1097 pounds (w/o bore evacuator)	
Type of Breechblock	Semiautomatic, separable chamber, electrically operated	
Rifling	48 grooves, uniform right-hand twist, one turn in 41.2 calibers	
Ammunition	Fixed with combustible case or Shillelagh missile	
Primer	Electric	
Weight, Complete Round	MGM-51C Missile (as fired)	61.5 pounds (28.0 kg)
	MTM-51C Missile (as fired)	61.5 pounds (28.0 kg)
	HEAT-T-MP M409 Shell (HEAT-T-MP)	49.8 pounds (22.6 kg)
	HE-T XM657E2 Shell (HE-T)	50.0 pounds (22.7 kg)
	Canister M625 (10,000 flechettes)	48.0 pounds (21.8 kg)
	APERS XM617 (8,200 flechettes)	48.0 pounds (21.8 kg)
	TP-T M411A1 Shell (TP-T)	49.8 pounds (22.6 kg)
Weight, Projectile	HEAT-T-MP M409 Shell (HEAT-T-MP)	42.8 pounds (19.5 kg)
	HE-T XM657E2 Shell (HE-T)	43.1 pounds (19.6 kg)
	Canister M625 (10,000 flechettes)	41.8 pounds (19.0 kg)
	APERS XM617 (8,200 flechettes)	41.8 pounds (19.0 kg)
	TP-T M411A1 Shell (TP-T)	42.8 pounds (19.5 kg)
Maximum Powder Pressure	38,400 psi	
Maximum Rate of Fire	4 rounds/minute	
Muzzle Velocity	HEAT-T-MP M409 Shell (HEAT-T-MP)	2240 ft/sec (683 m/sec)
	HE-T XM657E2 Shell (HE-T)	2240 ft/sec (683 m/sec)
	Canister M625 (10,000 flechettes)	2240 ft/sec (683 m/sec)
	APERS XM617 (8,200 flechettes)	2000 ft/sec (610 m/sec)
	TP-T M411A1 shell (TP-T)	2240 ft/sec (683 m/sec)
Muzzle Energy of Projectile, $KW=\frac{1}{2}MV^2$	HEAT-T-MP M409 Shell (HEAT-T-MP)	1489 ft-tons
Rotational energy is neglected and	HE-T XM657E2 Shell (HE-T)	1499 ft-tons
values are based on long tons	Canister M625 (10,000 flechettes)	1454 ft-tons
(2240 pounds)	APERS XM617 (8,200 flechettes)	1159 ft-tons
	TP-T M411A1 Shell (TP-T)	1489 ft-tons
Maximum Range (independent of mount)	HEAT-T-MP M409 Shell (HEAT-T-MP)	9850 yards (9007 m)
	HE-T XM657E2 Shell (HE-T)	9850 yards (9007 m)
	Canister M625 (10,000 flechettes)	437 yards (400 m)
	APERS XM617 (8,200 flechettes)*	3280 yards (3000 m)
	TP-T M411A1 Shell (TP-T)	9850 yards (9007 m)
Penetration Performance	Homogeneous steel armor at 60 degrees obliquity	
HEAT-T-MP M409	7 inches at any range	

*Fuze Settings: Muzzle action and 100 meter increments starting at 200 meters

The M409, M625, and M411A1 rounds were assembled with the M157 combustible case and the M189 charge. The XM657E2 and the XM617 rounds were assembled with the XM157 combustible case and used the XM190 and M26 charges respectively.

155mm HOWITZER M45 (T186E1)

Carriage and Mount	155mm Self-Propelled Howitzers M44 (T194) and M44A1 in Mount M80 (T167)
Length of Chamber (to rifling)	28.7 inches
Length of Rifling	113.1 inches
Length of Chamber (to projectile base)	21.1 inches
Travel of Projectile in Bore	120.7 inches
Length of Bore	141.8 inches
Length, Breechblock and Firing Lock	14.8 inches
Length, Muzzle to Rear of Firing Lock	156.6 inches
Additional Length, Muzzle Brake, Etc.	None
Overall Length	156.6 inches
Diameter of Bore	6.102 inches (155mm)
Chamber Capacity	795 cubic inches
Weight, Tube	2140 pounds
Total Weight	2970 pounds
Type of Breechblock	Stepped thread, interrupted screw, horizontal swing
Rifling	48 grooves, uniform right-hand twist, one turn in 25 calibers
Ammunition	Separate loading
Primer	Percussion and electric

Weight, Complete Round	HE M107 Shell (HE), Charge M4A1	108.91 pounds (49.40 kg)
	HC BE M116 Shell, Smoke, Charge M4A1	109.01 pounds (49.45 kg)
	H M110 Shell, Chemical, Charge M4A1	109.11 pounds (49.49 kg)
Weight, Projectile	HE M107 Shell (HE)	95.00 pounds (43.09 kg)
	HC BE M116 Shell, Smoke	95.10 pounds (43.14 kg)
	H M110 Shell, Chemical	95.20 pounds (43.18 kg)
Maximum Powder Pressure	32,000 psi	
Maximum Rate of Fire	4 rounds/minute	
Muzzle Velocity	HE M107 Shell (HE), Charge M4A1	1850 ft/sec (564 m/sec)
	HC BE M116 Shell, Smoke, Charge M4A1	1850 ft/sec (564 m/sec)
	H M110 Shell, Chemical, Charge M4A1	1850 ft/sec (564 m/sec)
Muzzle Energy of Projectile, KE=½MV² Rotational energy is neglected and values are based on long tons (2240 pounds)	HE M107 Shell (HE), Charge M4A1	2254 ft-tons
	HC BE M116 Shell, Smoke, Charge M4A1	2256 ft-tons
	H M110 Shell, Chemical, Charge M4A1	2259 ft-tons
Maximum Range (independent of mount)	HE M107 Shell (HE), Charge M4A1	16,355 yards (14,955 m)
	HC BE M116 Shell, Smoke, Charge M4A1	16,355 yards (14,955 m)
	H M110 Shell, Chemical, Charge M4A1	16,374 yards (14,972 m)
Penetration Performance	Concrete at 0 degrees obliquity	

Range	0 yards	1000 yards	3000 yards	5000 yards
HE M107 Shell (HE) w/Concrete Piercing Fuze M78A1	2.9 feet	2.6 feet	2.0 feet	1.6 feet

Carriage and Mount	155mm Self-Propelled Howitzer M109 in Mount M127	
Length of Chamber (to rifling)	29.70 inches	
Length of Rifling	113.10 inches	
Length of Chamber (to base of M107 shell)	24.35 inches	
Travel of Projectile in Bore (M107 shell)	118.45 inches	
Length of Bore	142.80 inches, 23.4 calibers	
Length, Breechblock and Firing Mechanism	10.37 inches	
Length, Muzzle Brake	23.70 inches	
Overall Length	176.87 inches	
Diameter of Bore	6.100 +.002 inches	
Chamber Capacity	795 cubic inches (M107 shell)	
Weight of Tube	2006 pounds (M126), 2069 pounds (M126A1)	
Total Weight	3137 pounds (M126), 3200 pounds (M126A1)	
Type of Breechblock	Semiautomatic, Welin-step thread	
Rifling	48 grooves, uniform right-hand twist, one turn in 20 calibers	
Ammunition	Separate loading	
Primer	Percussion, M82	
Weight, Complete Round	HE M107 Shell (HE), Charge M4A2/7	109 pounds (49.4 kg)
	HE M483A1 Projectile (88 grenades), Charge M4A2/7	116 pounds (52.6 kg)
	HERA M549A1 Shell (HERA), Charge M4A2/7	114 pounds (51.7 kg)
	WP M110A1 Shell, Smoke, Charge M4A2/7	112 pounds (50.8 kg)
	ILLUM M485A2 Shell (ILLUM), Charge M4A2/7	107 pounds (48.5 kg)
Weight, Projectile	HE M107 Shell (HE)	95.0 pounds (43.1 kg)
	HE M483A1 Projectile (88 grenades)	102.6 pounds (46.5 kg)
	HERA M549A1 Shell (HERA)	96.0 pounds (43.5 kg)
	WP M110A1 Shell, Smoke	98.5 pounds (44.7 kg)
	ILLUM M485A2 Shell (ILLUM)	93.7 pounds (42.5 kg)
Maximum Powder Pressure	42,700 psi	
Maximum Rate of Fire	4 rounds/minute	
Muzzle Velocity	HE M107 Shell (HE), Charge M4A2/7	1844 ft/sec (562 m/sec)
	HE M483A1 Projectile (88 grenades), Charge M4A2/7	1761 ft/sec (537 m/sec)
	HERA M549A1 Shell (HERA), Charge M4A2/7	1840 ft/sec (561 m/sec)
	WP M110A1 Shell, Smoke, Charge M4A2/7	1844 ft/sec (562 m/sec)
	ILLUM M485A2 Shell (ILLUM), Charge M4A2/7	1891 ft/sec (576 m/sec)
Muzzle Energy of Projectile, KE=½MV²	HE M107 Shell (HE), Charge M4A2/7	2239 ft-tons
Rotational energy is neglected and	HE M483A1 Projectile (88 grenades), Charge M4A2/7	2206 ft-tons
values are based on long tons	HERA M549A1 Shell (HERA), Charge M4A2/7	2253 ft-tons
(2240 pounds)	WP M110A1 Shell, Smoke, Charge M4A2/7	2322 ft-tons
	ILLUM M485A2 Shell (ILLUM), Charge M4A2/7	2323 ft-tons
Maximum Range (independent of mount)	HE M107 Shell (HE), Charge M4A2/7	15,967 yards (14,600m)
	HE M483A1 Projectile (88 grenades), Charge M4A2/7	15,420 yards (14,100m)
	HERA M549A1 Shell (HERA), Charge M4A2/7	21,107 yards (19,300m)
	WP M110A1 Shell, Smoke, Charge M4A2/7	15,967 yards (14,600m)
	ILLUM M485A2 Shell (ILLUM), Charge M4A2/7	14,858 yards (13,586m)

The M126 and M126A1 howitzers were identical except for the tube. On the M126A1 the recoil keyway was modified and the cross section was increased in the bore evacuator area to improve the fatigue life.

155mm HOWITZERS M185 AND M284

Carriage and Mount	155mm Self-Propelled Howitzer M109A1 (M185 Howitzer in Mount M127); SP Howitzers M109A2, M109A3, and M109A4 (M185 Howitzer in Mount M178); SP Howitzers M109A5 and M109A6 (M284 Howitzer in Mount M182)	
Length of Chamber (to rifling)	39.33 inches (M185), 41.60 inches (M284)	
Length of Rifling	198.0 inches	
Length of Chamber (to base of M107 Shell)	34.4 inches (M185), 36.3 inches (M284)	
Travel of Projectile in Bore (M107 Shell)	203.65 inches (M185), 203.40 inches (M284)	
Length of Bore	238.05 inches (M185), 39.0 calibers; 240.00 inches (M284), 39.3 calibers	
Length, Breechblock and Firing Mechanism	10.3 inches	
Length, Muzzle Brake	23.7 inches	
Overall Length	272.12 inches (M185), 274.0 inches (M284)	
Diameter of Bore	6.100 +.002 inches	
Chamber Capacity	1167 cubic inches (M107 Shell)	
Weight, Tube	3166 pounds (M185)	
Total Weight	4320 pounds (M185)	
Type of Breechblock	Semiautomatic, Welin-step thread	
Rifling	48 grooves, uniform right-hand twist, one turn in 20 calibers	
Ammunition*	Separate loading	
Primer	Percussion, M82	
Weight, Complete Round	HE M107 Shell (HE), Charge M119A1/8	116 pounds (52.6 kg)
	HE M483A1 Projectile (88 grenades), Charge M119A1/8	123 pounds (55.8 kg)
	HERA M549A1 Shell (HERA) Charge M203A1/8s (M284)	117 pounds (53.1 kg)
	CLGP M712 Copperhead (CLGP, HEAT), Charge M119A1/8	160 pounds (72.6 kg)
	WP M110A2 Shell, Smoke, Charge M119A1/8	120 pounds (54.4 kg)
Weight, Projectile	HE M107 Shell (HE)	95.0 pounds (43.1 kg)
	HE M483A1 Projectile (88 grenades)	102.6 pounds (46.5 kg)
	HERA M549A1 Shell (HERA)	96.0 pounds (43.5 kg)
	CLGP M712 Copperhead (CLGP, HEAT)	138.4 pounds (62.8 kg)
	WP M110A2 Shell Smoke	98.5 pounds (44.7 kg)
Maximum Powder Pressure	39,400 psi (M185)	
Maximum Rate of Fire	4 rounds/minute	
Muzzle Velocity	HE M107 Shell (HE), Charge M119A1/8	2245 ft/sec (684 m/sec)
	HE M483A1 Projectile (88 grenades), Charge M119A1/8	2155 ft/sec (657 m/sec)
	HERA M549A1 Shell (HERA), Charge M203A1/8s (M284)	2710 ft/sec (826 m/sec)
	CLGP M712 Copperhead (CLGP, HEAT), Charge M119A1/8	1950 ft/sec (594 m/sec)
	WP M110A2 Shell, Smoke, Charge M119A1/8	2245 ft/sec (684 m/sec)
Muzzle Energy, KE=½MV²	HE M107 Shell (HE), Charge M119A1/8	3319 ft-tons
Rotational energy is neglected and	HE M483A1 Projectile (88 grenades), Charge M119A1/8	3303 ft-tons
values are based on long tons	HERA M549A1 Shell (HERA), Charge M203A1/8s (M284)	4887 ft-tons
(2240 pounds)	CLGP M712 Copperhead (CLGP, HEAT), Charge M119A1/8	3648 ft-tons
	WP M110A1 Shell, Smoke, Charge M119A1/8	3441 ft-tons
Maximum Range (independent of mount)	HE M107 Shell (HE), Charge M119A1/8	19,794 yards (18,100m)
	HE M483A1 Projectile (88 grenades), Charge M119A1/8	19,138 yards (17,500m)
	HERA M549A1 Shell (HERA), Charge M203A1/8s (M284)	32,918 yards (30,100m)
	CLGP M712 Copperhead (CLGP, HEAT), Charge M119A1/8	15,310 yards (14,000m)
	WP M110A1 Shell, Smoke, Charge M119A1/8	19,794 yards (18,100m)

*Nuclear capability is provided by the M454NUC round with a maximum range of 14,800 meters.

175mm GUN M113 (T256E3)

Carriage and Mount	175mm Self-Propelled Gun M107 (T235E1) in Mount M158	
Length of Chamber (to rifling)	64.2 inches	
Length of Rifling	349.2 inches	
Length of Chamber (to projectile base)	52.3 inches	
Travel of Projectile in Bore	361.1 inches	
Length of Bore	413.4 inches, 60 calibers	
Length, Breechblock and Firing Mechanism	14.6 inches	
Length, Muzzle to Rear of Firing Mechanism	428 inches	
Additional Length, Muzzle Brake	None	
Overall Length	428 inches	
Diameter of Bore	6.890 inches	
Chamber Capacity	2898 cubic inches	
Weight of Tube	12,050 pounds	
Total Weight	13,800 pounds	
Type of Breechblock	Manually operated, Welin-step thread	
Rifling	48 grooves, uniform right-hand twist, one turn in 20 calibers	
Ammunition	Separate loading	
Primer	Percussion	
Weight, Complete Round	HE M437A2 Shell (HE), Charge M86A1/3	202.3 pounds (91.8 kg)
Weight, Projectile	HE M437A2 Shell (HE)	147.3 pounds (66.8 kg)
Maximum Powder Pressure	50,000 psi	
Maximum Rate of Fire	1.5 rounds/minute	
Muzzle Velocity	HE M437A2 Shell (HE), Charge M86A1/3	3000 ft/sec (914 m/sec)
Muzzle Energy of Projectile, KE=½MV² Rotational energy is neglected and values are based on long tons (2240 pounds)	HE M437A2 Shell (HE), Charge M86A1/3	9190 ft-tons
Maximum Range (independent of mount)	HE M437A2 Shell (HE), Charge M86A1/3	35,760 yards (32,700 m)

106mm RIFLE M40A1C (RECOILLESS)

Carriage and Mount	106mm Multiple Self-Propelled Rifle M50 and M50A1	
Length of Rifling	105.9 inches	
Length of Tube	112.0 inches	
Overall Length	134.0 inches	
Diameter of Bore	4.134 inches	
Weight without Spotting Rifle M8C	251 pounds	
Weight with Spotting Rifle M8C	288 pounds	
Type of Breechblock	Interrupted thread	
Rifling	36 grooves, uniform right-hand twist, one turn in 20 calibers	
Ammunition	Fixed	
Primer	Percussion	
Weight, Complete Round	HEAT M344A1 Shell (HEAT)	37.23 pounds (16.88 kg)
	HEP-T M346A1 Shell (HESH-T)	37.37 pounds (16.95 kg)
	APERS-T M581 (9500 flechettes)	41.29 pounds (18.73 kg)
Weight, Projectile	HEAT M344A1 Shell (HEAT)	17.55 pounds (7.96 kg)
	HEP-T M346A1 Shell (HESH-T)	17.22 pounds (7.81 kg)
	APERS-T M581 (9500 flechettes)	21.61 pounds (9.80 kg)
Muzzle Velocity	HEAT M344A1 Shell (HEAT)	1650 ft/sec (503 m/sec)
	HEP-T M346A1 Shell (HESH-T)	1635 ft/sec (498 m/sec)
	APERS-T M581 (9500 flechettes)	1440 ft/sec (439 m/sec)
Muzzle Energy of Projectile, KE=½MV² Rotational energy is neglected and values are based on long tons (2240 pounds)	HEAT M344A1 Shell (HEAT)	331 ft-tons
	HEP-T M346A1 Shell (HESH-T)	319 ft-tons
	APERS-T M581 (9500 flechettes)	311 ft-tons
Maximum Range	HEAT M344A1 Shell (HEAT) @ 118 mils	3000 yards (2740 m)
	HEP-T M346A1 Shell (HESH-T)	7515 yards (6870 m)
	APERS-T M581 (9500 flechettes)	3600 yards (3300 m)

8 inch HOWITZER M2A2

Carriage and Mount	8 inch Self-Propelled Howitzer M110 in Mount M158
Length of Chamber (to rifling)	37.7 inches
Length of Rifling	164.8 inches
Length of Chamber (to base of M106 shell)	28.2 inches
Travel of Projectile in Bore	174.3 inches
Length of Bore	202.5 inches, 25.3 calibers
Length, Breechblock and Firing Mechanism	12.4 inches
Additional Length, Muzzle Brake	214.9 inches
Overall Length	None
Diameter of Bore	214.9 inches
Chamber Capacity	8.000 inches
Weight of Tube	1545 cubic inches
Total Weight	8490 pounds
Type of Breechblock	10,240 pounds
	Manually operated, stepped thread, interrupted screw
Rifling	64 grooves, uniform right-hand twist, one turn in 20 calibers
Ammunition	Separate loading
Primer	Percussion
Weight, Complete Round	HE M106 Shell (HE), Charge M2/7	228.8 pounds (103.8 kg)
	HE M404 Projectile (104 grenades), Charge M2/7	228.8 pounds (103.8 kg)
	VX M426 Shell (Gas), Charge M2/7	227.8 pounds (103.3 kg)
	GB M426 Shell (Gas), Charge M2/7	227.8 pounds (103.3 kg)
Weight, Projectile	HE M106 Shell (HE)	200.0 pounds (90.7 kg)
	HE M404 Projectile (104 grenades)	200.0 pounds (90.7 kg)
	VX M426 Shell (Gas)	199.0 pounds (90.3 kg)
	GB M426 Shell (Gas)	199.0 pounds (90.3 kg)
Maximum Powder Pressure	39,600 psi
Maximum Rate of Fire	1.5 rounds/minute
Muzzle Velocity	HE M106 Shell (HE), Charge M2/7	1950 ft/sec (594 m/sec)
	HE M404 Projectile (104 grenades), Charge M2/7	1903 ft/sec (580 m/sec)
	VX M426 Shell (Gas), Charge M2/7	1950 ft/sec (594 m/sec)
	GB M426 Shell (Gas), Charge M2/7	1950 ft/sec (594 m/sec)
Muzzle Energy, KE=½MV²	HE M106 Shell (HE), Charge M2/7	5272 ft-tons
Rotational energy is neglected and	HE M404 Projectile (104 grenades), Charge M2/7	5021 ft-tons
values are based on long tons	VX M426 Shell (Gas), Charge M2/7	5246 ft-tons
(2240 pounds)	GB M426 Shell (Gas), Charge M2/7	5246 ft-tons
Maximum Range (independent of mount)	HE M106 Shell (HE), Charge M2/7	18,373 yards (16,800 m)
	HE M404 Projectile (104 grenades), Charge M2/7	18,359 yards (16,788 m)
	VX M426 Shell (Gas), Charge M2/7	18,373 yards (16,788 m)
	GB M426 Shell (Gas), Charge M2/7	18,373 yards (16,788 m)

8 inch HOWITZER M201A1

Carriage and Mount	8 inch Self-Propelled Howitzer M110A2 in Mount M158	
Length of Chamber (to rifling)	42.56 inches	
Length of Rifling	273.3 inches	
Length of Chamber (to base of M106 shell)	36.16 inches	
Travel of Projectile in Bore	279.70 inches	
Length of Bore	315.86 inches, 39.5 calibers	
Length, Breechblock and Firing Mechanism	12.4 inches	
Length, Muzzle to Rear of Firing Mechanism	328.3 inches	
Additional Length, Muzzle Brake	15.4 inches	
Overall Length	343.7 inches	
Diameter of Bore	8.000 inches	
Chamber Capacity	1950 cubic inches	
Weight of Tube	12,450 pounds	
Total Weight	14,650 pounds	
Type of Breechblock	Manually operated, stepped thread, interrupted screw	
Rifling	64 grooves, uniform right-hand twist, one turn in 20 calibers	
Ammunition*	Separate loading	
Primer	Percussion, M82	
Weight, Complete Round	HE M106 Shell (HE), Charge M188A1/9	250 pounds (113.4 kg)
	HE M509A1 Projectile (180 grenades), Charge M188A1/9	258 pounds (117.0 kg)
	HERA M650 Shell (HERA), Charge M188A1/9	250 pounds (113.4 kg)
Weight, Projectile	HE M106 Shell (HE)	200.0 pounds (90.7 kg)
	HE M509A1 Projectile (180 grenades)	207.7 pounds (94.2 kg)
	HERA M650 Shell (HERA)	200 pounds (90.7 kg)
Maximum Powder Pressure	39,600 psi	
Maximum Rate of Fire	1.5 rounds/minute	
Muzzle Velocity	HE M106 Shell (HE), Charge M188A1/9	2530 ft/sec (771 m/sec)
	HE M509A1 Projectile (180 grenades), Charge M188A1/9	2510 ft/sec (765 m/sec)
	HERA M650 Shell (HERA), Charge M188A1/9	2520 ft/sec (768 m/sec)
Muzzle Energy, KE=½MV²	HE M106 Shell (HE), Charge M188A1/9	8874 ft-tons
Rotational energy is neglected and	HE M509A1 Projectile (180 grenades), Charge M188A1/9	9071 ft-tons
values are based on long tons	HERA M650 Shell (HERA), Charge M188A1/9 8804 ft-tons	
(2240 pounds)		
Maximum Range (independent of mount)	HE M106 Shell (HE), Charge M188A1/9	26,200 yards (24,000 m)
	HE M509A1 Projectile (180 grenades), Charge M188A1/9	26,250 yards (24,000 m)
	HERA M650 Shell (HERA), Charge M188A1/9	32,800 yards (30,000 m)

*Nuclear capability is provided by the M422A1NUC and the M753NUC rounds with maximum ranges of 18,100 meters and 30,000 meters respectively.

REFERENCES AND SELECTED BIBLIOGRAPHY

Books and Published Articles

Cole, Edward N., "Walker Bulldog - Its Design and Production", Cleveland Tank Plant, General Motors Corporation, Cleveland, Ohio, 22 October 1951

Corrigan, Captain Sean, "The 82nd Airborne in Saudia Arabia", Armor, September-October 1993, p 32

Hammond, Captain Kevin J. and Sherman, Captain Frank, "Sheridans in Panama", Armor, March-April 1990, p 8

Mackenzie, Captain Bob, "Airborne Armor", Armor September-October 1981, p 30

Simpson, Colonel Paul A., "Is the Sheridan the Latest of the Technological Breakthroughs ?", Armor, July-August 1965, p 42

Starry, General Donn A., "Mounted Combat in Vietnam", Department of the Army, Washington, D.C., 1978

Womack, Captain Scott, "The AGS in Low-Intensity Conflicts: Flexibility is the Key to Victory", Armor, March-April 1994, p 42

Reports and Official Documents

Anderson, Gerald, "Sheridan Weapon System M551, Memo History 1954 - 1971", 28 April 1971

"AR/AAV XM551 Armament Systems", Shillelagh Task Force, Ordnance Tank Automotive Command, Center Line, Michigan, 5 March 1962

"Armored Combat Vehicle Technology Study", Tank Automotive Concepts Laboratory, U.S. Army TARADCOM, Warren, Michigan, March 1980

"Conference on Automotive Development Materiel", Detroit Arsenal, Center Line, Michigan, March 1948

"Elevated KE Weapon Technology Study, Final Report", Pacific Car And Foundry Company, Renton, Washington, June 1979

Engelman, Rose, "History of the Tank Development Program (1945 - 1956)", Ordnance Tank Automotive Command, Center Line, Michigan, 10 January 1957

"Engineering Presentation of Recommended Changes - M41 and M41A1 Vehicles", Cleveland Ordnance Plant, Cadillac Motor Car Division, GMC, Cleveland, Ohio, 25 November 1958 and April 1959

"Evaluation of the 76mm Gun Tank T92", U.S. Army Armor School, Fort Knox, Kentucky, 19 April 1960

"Fabrication of Two Prototypes of Tank, Light Gun, 76mm, T92", Aircraft Armaments, Inc., Report ER-1937, Cockeysville, Maryland, April 1960

"Final Technical Report: Gun, Field Artillery, Self-Propelled, Full Tracked, 175mm T235; Howitzer, Field Artillery, Self-Propelled, Full Tracked, 8 inch, T236; Gun, Field Artillery, Self-Propelled, Full Tracked, 155mm T245; Recovery Vehicle, Full Tracked, Unarmored, T119; Recovery Vehicle, Full Tracked, Armored, T120; Pacific Car and Foundry Company, Renton, Washington, 31 October 1965

"HIMAG High Mobility Agility Vehicle, Volume III, System Description", Teledyne Continental Motors, General Products Division, Muskegon, Michigan, September 1982

"History of Automatic Loading Equipment", Ordnance Tank Automotive Command, Center Line, Michigan, April 1960

"History of the Shillelagh Missile System 1958 - 1982", U.S. Army Missile Command, Redstone Arsenal, Alabama, 17 August 1984

"Ideas and Concepts for M551 Future Improvements", U.S. Army Materiel Command, Rock Island Arsenal, Illinois, 9 November 1966

"New Tank Concepts", Cleveland Tank Plant, Cadillac Motor Car Division, GMC, Cleveland, Ohio, 1955

"Notes on Development Type Materiel, Cannon, 152mm, Gun-Launcher, XM81", Watervliet Arsenal, Watervliet, New York, March 1961

"Notes on Development Type Materiel, Cannon, 152mm, Gun-Launcher, XM81E3", Watervliet Arsenal, Watervliet, New York, September 1962

"Notes on Development Type Materiel, Light Tank T41", Detroit Arsenal, Center Line, Michigan, October 1950

"Notes on Development Type Materiel, Automatic Loading Equipment for 76mm Gun in T41E1 Tank", Rheem Manufacturing Company, Philadelphia, PA, 15 April, 1953

"Notes on Development Type Materiel, 76mm Gun Tank T92", Aircraft Armaments, Inc., Cockeysville, Maryland, 15 April 1957

"Notes on Development Type Materiel and Preliminary Technical Manual, Armored Reconnaissance/Airborne Assault Vehicle XM551 (General Sheridan)", Cleveland Ordnance Plant, General Motors Corporation, Cleveland, Ohio, 1 February 1963

"155mm Artillery Weapon Systems, Reference Data Book", Office of the Project Manager - Paladin, Picatinny Arsenal, New Jersey, December 1990

"Project Management of the Sheridan Weapon System (Fiscal years 1963 to 1972)", Historical Highlights, LTC John Barrowclough, August 1971

"Proposal for Armored Reconnaissance Airborne Assault Vehicle", Aircraft Armaments, Inc., Cockeysville, Maryland and Allis-Chalmers Manufacturing Company, Milwaukee, Wisconsin, Report ER-1780, October 1959

"Questionmark", presentation at the Ordnance Tank Automotive Command, Center Line, Michigan, April 1952

"Questionmark II", presentation at the Ordnance Tank Automotive Command, Center Line, Michigan, August 1952

"Questionmark III", presentation at the Ordnance Tank Automotive Command, Center Line, Michigan, June 1954

"Questionmark IV", presentation at the Ordnance Tank Automotive Command, Center Line, Michigan, August 1955

"Report of the Ad Hoc Group on Armament for Future Tanks or Similar Combat Vehicles (ARCOVE)", 20 January 1958

"Report of the Army Ground Forces Equipment Review Board", Washington, D.C., 20 June 1945

"Report of the War Department Equipment Board" (Stilwell Board), Washington, D.C., 19 January 1946

"Report on Evaluation of Concept Studies and Feasibility Analysis for the High Survivability Test Vehicle - Lightweight (HSTV-L)", U.S. Army TARADCOM, Warren, Michigan, 17 November 1977

"Research Engineering Program Review", Ordnance Tank Automotive Command, Center Line, Michigan, July 1960

"Special Text 17-181, Armored Reconnaissance/Airborne Assault Vehicle M551, General Sheridan", U.S. Army Armor School, Fort Knox, Kentucky, September 1967

"Tank Development", presentation for the Research and Development Board, Detroit Arsenal, Center Line, Michigan, 16 May 1951

"Tank, 76mm Gun T71, Preliminary Concept and Development Schedule", Detroit Arsenal, Center Line, Michigan, 1953

"Technical Manual TM9-308A 76mm Gun T91E3", Department of the Army, Washington, D.C., March 1952

"Technical Manual TM9-730 76mm Gun Tank T41E1", Department of the Army, Washington, D.C., June 1951

"Technical Manual TM9-761A Self-Propelled Twin 40mm Gun T141", Department of the Army, Washington, D.C., 8 May 1952

"Technical Manual TM9-2300-216-10 Gun, Field Artillery, Self-Propelled, Full Tracked, 175mm M107 (T235E1) and Howitzer, Heavy, Self-Propelled, Full Tracked, 8 inch M110 (T236E1)", Department of the Army, Washington, D.C., 28 June 1962

"Technical Manual TM9-2350-201-12 76mm Gun Full Tracked Combat Tank M41(T41E1) and M41A1(T41E2)", Department of the Army, Washington, D.C., 29 July 1958

"Technical Manual TM9-2350-212-12 Rifle, Self-Propelled, Full Tracked, Multiple 106mm M50", Department of the Army, Washington, D.C., 11 December 1959

"Technical Manual TM9-2350-213-10 90mm Full Tracked Self-Propelled Gun M56, Department of the Army, Washington, D.C., 22 May 1958

"Technical Manual TM9-2350-217-10 Howitzer, Light, Self-Propelled, 105mm T195E1 and Howitzer, Medium, Self-Propelled, 155mm T196E1", Department of the Army, Washington, D.C., 11 September 1962

"Technical Manual TM9-2350-217-10 Howitzer, Light, Self-Propelled, 105mm M108 and Howitzer, Medium, Self-Propelled, 155mm M109", Department of the Army, Washington, D.C., December 1969

"Technical Manual TM9-2350-217-20N Howitzer, Medium, Self-Propelled, 155mm M109 and M109A1", Department of the Army, Washington, D.C., December 1977 with Change C1 to include 155mm Howitzer M109A3, 26 June 1981

"Technical Manual TM9-2350-230-12 Armored Reconnaissance/Airborne Assault Vehicle: 152mm XM551", Department of the Army, Washington, D.C., 1 June 1966

"Technical Manual TM9-7004 Self-Propelled 155mm Howitzer M44 (T194)", Department of the Army, Washington, D.C., 4 June 1958

"Technical Manual TM9-7204 105mm Full Tracked Self-Propelled Howitzer M52 (T98E1), Department of the Army, Washington, D.C., 27 September 1957

"Technical Manual TM9-7218 Twin 40mm Full Tracked Self-Propelled Gun M42 (T141)", Department of the Army, Washington, D.C., 21 May 1957

"Technical Manual TM43 0001-28 Artillery Ammunition, Guns, Howitzers, Mortars, Recoilless Rifles, Grenade Launchers, and Artillery Fuzes", Department of the Army, Washington, D.C., April 1977

"Test of Light Tank T37, First Report on Project TT2-674" Aberdeen Proving Ground, Maryland, 28 December 1950

"The T92 Light Tank, 76mm Gun, Report No. ER-361", Aircraft Armaments, Inc., Cockeysville, Maryland, May 1954

"Vehicle Concepts Utilizing General Sheridan Components", General Motors Corporation, undated

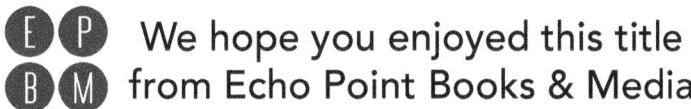 We hope you enjoyed this title
from Echo Point Books & Media

Before Closing this Book, Two Good Things to Know

1. Buy Direct & Save

Go to www.echopointbooks.com to see our complete list of titles. We publish books on a wide variety of topics—from spirituality to auto repair.

Buy direct and save 10% at www.echopointbooks.com

DISCOUNT CODE: EPBUYER

2. History Buff? Tank Lover? We've got you covered!

Echo Point Books & Media is proud to announce the release of new, top quality reprints of R.P. Hunnicutt's tank books. Finally back in print at an affordable price!

Echo Point Books & Media is the official publisher of the R.P. Hunnicutt tank series. Easily the most comprehensive series of reference books on the developmental history of American military vehicles, Hunnicutt's 10-volume compendium of tank information is an absolute must-have for every serious military and history buff's bookshelf. These books contain a wealth of detailed information including:

- o   Important facts and figures
- o   Thousands of photographs
- o   Hundreds of detailed line drawings and diagrams

View our entire catalog of military history books at:
www.echopointbooks.com

Follow us to keep up with new releases in the Hunnicutt Series!

echopointbooks   @EPBM   echopointbooks